Hematology and Coagulation

Hematology and Coagulation

A Comprehensive Review for Board Preparation, Certification and Clinical Practice

Second Edition

Amer Wahed, MD

Associate Professor of Pathology and Laboratory Medicine
University of Texas McGovern Medical School at Houston
Houston, Texas

Andres Quesada, MD

Assistant Attending Memorial Sloan Kettering Cancer Center
New York, NY

Amitava Dasgupta, PhD

Professor of Pathology and Laboratory Medicine
University of Texas McGovern Medical School at Houston
Houston, Texas

ELSEVIER

Elsevier
Radarweg 29, PO Box 211, 1000 AE Amsterdam, Netherlands
The Boulevard, Langford Lane, Kidlington, Oxford OX5 1GB, United Kingdom
50 Hampshire Street, 5th Floor, Cambridge, MA 02139, United States

Notices

Knowledge and best practice in this field are constantly changing. As new research and experience broaden our understanding, changes in research methods, professional practices, or medical treatment may become necessary.

Practitioners and researchers must always rely on their own experience and knowledge in evaluating and using any information, methods, compounds, or experiments described herein. In using such information or methods they should be mindful of their own safety and the safety of others, including parties for whom they have a professional responsibility.

To the fullest extent of the law, neither the Publisher nor the authors, contributors, or editors, assume any liability for any injury and/or damage to persons or property as a matter of products liability, negligence or otherwise, or from any use or operation of any methods, products, instructions, or ideas contained in the material herein.

Library of Congress Cataloging-in-Publication Data
A catalog record for this book is available from the Library of Congress

British Library Cataloguing-in-Publication Data
A catalogue record for this book is available from the British Library

ISBN: 978-0-12-814964-5

For information on all Elsevier publications visit our website at https://www.elsevier.com/books-and-journals

Publisher: Stacy Masucci
Acquisition Editor: Tari Broderick
Editorial Project Manager: Pat Gonzalez
Production Project Manager: Sreejith Viswanathan
Cover Designer: Mark Rogers

Typeset by TNQ Technologies

Working together
to grow libraries in
developing countries

www.elsevier.com • www.bookaid.org

This book is dedicated to my wife Tania and my sons Arub, Ayman, and Abyan.

Amer Wahed

This book is dedicated to my father Jorge Quesada, MD, and my mother Rocio Quesada for all their love and support throughout my life.

Andres Quesada

This book is dedicated to my wife Alice.

Amitava Dasgupta

Contents

Preface

The first edition of *Hematology and Coagulation: A Comprehensive Review for Board Preparation, Certification and Clinical Practice* was published by Elsevier in 2015. This book was the second book in a series of books, designed for board review for pathology residents. The first book in this series *Clinical Chemistry, Immunology and Laboratory Quality Control: A Comprehensive Review for Board Preparation, Certification and Clinical Practice* was published, also by Elsevier in January 2014. Subsequently we published the third book in the series *Microbiology and Molecular Diagnosis in Pathology: A Comprehensive Review for Board Preparation, Certification and Clinical Practice* in 2017. The last book in the series *Transfusion Medicine for Pathologists: A Comprehensive Review for Board Preparation, Certification, and Clinical Practice* was published also by Elsevier in 2018. All of these books are coauthored by faculties in the clinical pathology division of our department.

The aim of both the first and second edition of the hematology and coagulation book was to provide a strong foundation for students, residents, and fellows embarking on the journey of mastering hematology. It is expected that the book will also act as a valuable resource for residents preparing for the clinical pathology board exam. It was and still is not a text book of hematology as there are numerous excellent text books in this field. At the end of each chapter, we have included a section, denoted as "key points." We hope that this section will be a good resource for reviewing information, when time at hand is somewhat limited.

We received good feedback from residents and fellows for our first edition of hematology and coagulation book. To incorporate the 2017 WHO guidelines as well as to improve on the content, we decided to proceed publishing a second edition of the book. Elsevier has been very supportive in this endeavor. A new chapter (Chapter 11: Precursor Lymphoid Neoplasms, Blastic Plasmacytoid Dendritic Cell Neoplasm and Acute Leukemias of Ambiguous Lineage) has also been added to complete the updated second edition.

We have recruited Andres Quesada, MD, to be a coauthor with us in the second edition. He completed his pathology residency in our department and had also helped us with the first edition of the book. Currently he is an assistant attending at Memorial Sloan Kettering Cancer Center at New York.

We hope that readers will find the second edition useful and if so, our work for the last year will be duly rewarded.

Amer Wahed
Andres Quesada
Amitava Dasgupta

Complete blood count and peripheral smear examination

1

Introduction

A complete blood count (CBC) is one of the most common laboratory tests ordered by clinicians. Even for a routine health checkup of a healthy person, CBC is ordered to ensure there is no underlying disease when the individual may be asymptomatic. Tefferi et al. commented that in Mayo Clinic, Rochester, MN, approximately 10%−20% of CBC results are reported as abnormal. Common abnormalities associated with an abnormal CBC include anemia, thrombocytopenia, leukemia, polycythemia, thrombocytosis, and leukocytosis [1]. For CBC analysis, the specimen must be collected in an EDTA (ethylenediamine tetraacetic acid) tube (lavender or purple top).

CBC consists of certain numbers that are printed out from the hematology analyzer. In addition, the printout contains certain graphs and "flags." Flags are essentially messages provided by the analyzer to the interpreting person that certain abnormalities may be present. For example, an analyzer may flag that blasts are present. Therefore, reviewing blood smear slide is required to ensure presence of blasts. To make a meaningful interpretation of the peripheral smear, the CBC printout should be reviewed along with patient's electronic medical records. CBC parameters that are printed from an automated hematology analyzer are red blood cell (RBC)−related numbers, white blood cell (WBC)−related numbers, and platelet−related numbers (Table 1.1).

Analysis of various parameters by hematology analyzers

A modern hematology analyzer is capable of counting and determining size of various circulating blood cells in blood, including RBC, WBC, and platelets [2].

Different hematology analyzers may use different methods for counting. Examples of different methods include (one analyzer may employ multiple methods) the following:

- Impedance: The traditional method for counting cells is electrical impedance, which was first used by Wallace Coulter in 1956. This is also known as the Coulter principle. It is used in almost all hematology analyzer. Whole blood passes between two electrodes through an aperture. This aperture allows only one cell to pass through at a time. The impedance changes as each cell passes through. The change in impedance is proportional to the volume of the cell. The cell is counted, and the volume of the cell is measured. This method is unable to distinguish between the three granulocytes accurately.

Hematology and Coagulation. https://doi.org/10.1016/B978-0-12-814964-5.00001-2

1

Table 1.1 Various parameters printed by a hematology analyzer following complete blood count analysis.

Parameter	Individual number
RBC (red blood cell)-related numbers	• RBC count • Hemoglobin level • Hematocrit • Red cell differential width • Mean corpuscular volume • Mean corpuscular hemoglobin • Mean corpuscular hemoglobin concentration • Reticulocyte count
WBC (white blood cell)-related numbers	• Total WBC count corrected • Total WBC count uncorrected • WBC differential • Absolute count of each type of WBC
Platelet-related numbers	• Platelet count • Mean platelet volume • Platelet differential width

- Conductivity measurements with high frequency electromagnetic current (depends on the internal structure including nuclear cytoplasmic ratio, nuclear density to granularity ratio).
- Light scatter: Cells are made to pass in a single file in front of a light source. Light is scattered by the cells passing through the light beam. The amount of light scatter is detected, and electrical impulses are generated for counts.
- Flow cytometry is an excellent method to determine the five-part WBC differential.
- Fluorescence flow cytometry: It is useful for analysis of platelets, nucleated RBCs, and reticulocytes.
- Peroxidase-based cell counter.
- Immunological-based cell counters.

Modern hematology analyzers are capable of multimodal assessment of cell size and cell count, thus providing additional information regarding various categories of WBCs such as neutrophils, lymphocytes, monocytes, eosinophils, and basophils.

Hematology analyzers utilize various channels to perform certain parts of the CBC. For example, to measure the hemoglobin concentration, the red cells have to be lysed. If the red cells are lysed, then the RBC count cannot be performed. Thus the channel used for hemoglobin concentration measurement does not count RBC number.

Examples of various channels in a hematology analyzer:

- Channel for red cells (and also platelets): This channel is capable of analyzing RBCs and platelets
- Channel for WBC and hemoglobin measurement: Here lytic agents lyse red cells first before analysis
- Channel for WBC differential count

- Channel for reticulocyte count
- Other channels: NRBC (nucleated RBC channel), separate hemoglobin (Hb) channel, WBC/basophil channel, immature granulocyte channel

Histograms

Based on the different numerical data obtained during analysis by the hematology analyzer, graphical representations are produced. These are histograms. So, the analyzers produce a histogram for RBCs, another for platelets, and a third histogram for WBCs. Abnormalities of these histograms will result in flags.

The x axis of the histogram represents cell size, and the y axis represents the number of cells.

Typically one channel is used to detect RBCs and platelets. The detector is set such that any cell between 2 and 30 fL will be counted as a platelet, and any cell between 40 and 250 fL will be counted as a red cell (Fig. 1.1).

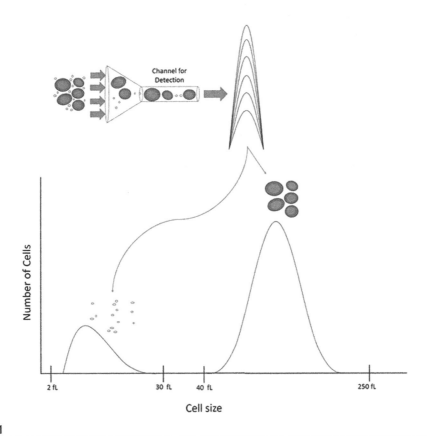

FIGURE 1.1

Cell discrimination by size: different cell types are sorted based on a typical size range.

Red blood cell histogram

- The normal RBC distribution curve is a Gaussian bell-shaped curve.
- The RBC histogram has an ascending slope, a peak, and a descending slope.
- The peak of the curve should fall within the normal mean corpuscular volume (MCV) range (approximately 80–100 fL)
- The MCV is obtained by drawing a perpendicular line form the peak to the baseline.
- Red cell distribution width (RDW) values are also obtained from the RBC histogram

The RDW denotes the extent of variation of size of RBCs, i.e., it is a measure of degree of anisocytosis. Two RDW measurements are currently in use, RDW-CV (coefficient of variation) and RDW-SD (standard deviation).

RDW-CV = One SD of mean cell size/MCV, multiplied by 100. Normal range is 11%–15%.

RDW-SD: It is an actual measurement of the width of the red cell distribution in femtoliter at the point, 20% above baseline. Normal range: 40–55 fL (Fig. 1.2).

There are two flexible discriminators in an RBC histogram:

Lower discriminator (LD) is set at 25–75 fL, and upper discriminator (UD) is set at 200–250 fL. The distribution curve should always start and end at the baseline and should be located between the two discriminators. Abnormalities of RBC histograms will result in flags (Fig. 1.3).

Red cell lower (RL) flag:

- When the LD exceeds the preset height by 10%.

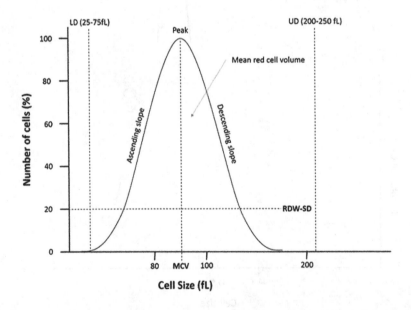

FIGURE 1.2

Red blood cell histogram. *LD*, lower discriminator; *MCV*, mean corpuscular volume is calculated by tracing a perpendicular line from the peak to the baseline; mean red cell volume is the total area of curve; *RDW*, red cell distribution width is 20% of peak height (normal range: 11–14 fL); *UD*, upper discriminator.

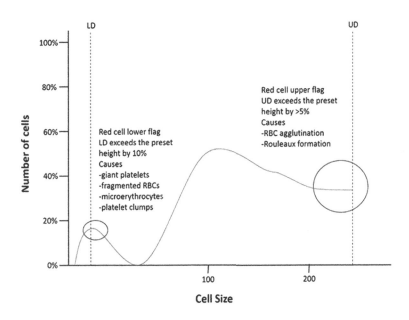

FIGURE 1.3

Red blood cell flags. *LD*, lower discriminator; *RBC*, red blood cell; *UD*, upper discriminator.

Causes of RL flag are as follow:

- Giant platelets
- Fragmented RBCs
- Small RBCs
- Platelet clumps

Red cell upper (RU) flag:

- When the UD exceeds the preset height by greater than 5%.

Causes of RU flag are as follow:

- RBC agglutination
- Rouleaux formation

Platelet histogram

- Platelets are counted if they are between 2 fL and 30 fL (Fig. 1.4).
- The platelet LD is set at 2−6 fL, and platelet UD is set at 12−30 fL.
- A third discriminator is set at 12 fL.

From the platelet histogram we obtain the following values:

- MPV (mean platelet volume)

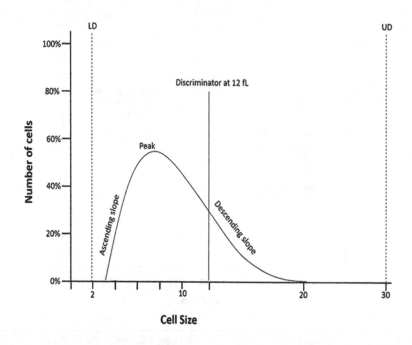

FIGURE 1.4

Platelet histogram. *LD*, lower discriminator; *RBC*, red blood cell; *UD*, upper discriminator.

- PDW (platelet differential width)
- P-LCR (ratio of large platelets): This is the percentage of platelets which are greater than 12 fL (the third discriminator)

Platelet lower (PL) flag:

- When the LD exceeds preset height by 10% (Fig. 1.5).

Causes of PL flag are as follow:

- Cell fragments
- Bacteria
- High blank value

Platelet upper (PU) flag:

- When the UD exceeds the preset height by more than 40%.

Causes of PU flag are as follow:

- Platelet clumps
- Giant platelets
- Small or fragmented RBCs

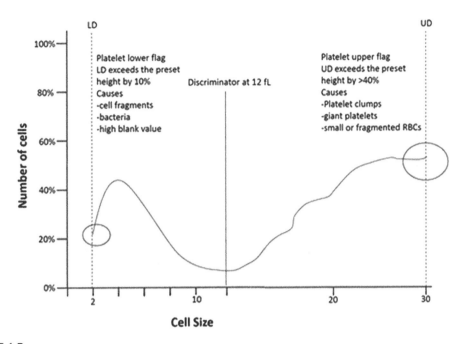

FIGURE 1.5

Platelet flags.

White blood cell histogram

The WBC histogram has three peaks. The first peak corresponds to lymphocytes and the third peak corresponds to neutrophils, whereas the second peak corresponds to the remainder types of WBCs (Fig. 1.6).

- The LD is set at 30–60 fL.
- The UD is set at 300 fL.
- The number of cells between LD and UD is the WBC count.
- The WBC histogram has two troughs discriminators, T1 (78–114 fL) and T2 (<150 fL).
- The peak between LD and T1 represents lymphocytes.
- The peak after T2 represents neutrophils.
- The peak between T1 and T2 represents all other WBCs.

    WBC lower (WL) flag:

- This is when the WBC histogram does not start at baseline (Fig. 1.7).

    Causes of WL flag are as follows:

- Lysis resistant RBCs
- Platelet clumps
- NRBCs

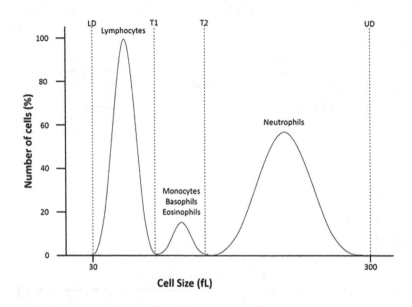

FIGURE 1.6

White blood cell histogram. *LD*, lower discriminator; *UD*, upper discriminator.

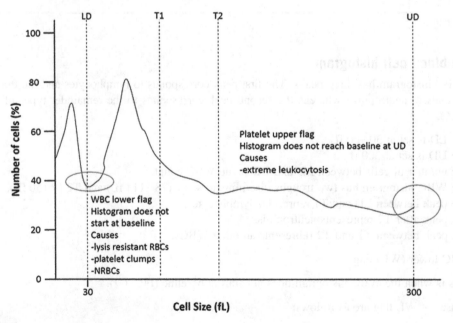

FIGURE 1.7

White blood cell flags. *LD*, lower discriminator; *RBC*, red blood cell; *UD*, upper discriminator; *WBC*, white blood cell.

Please note that for hemoglobin estimation and WBC count, red cells have to be lysed. In certain hemoglobinopathies, red cells are resistant to lysis. This is why the machine needs separate channels for hemoglobin estimation and WBC counts.

WBC upper (WU) flag:

- The curve does not reach the baseline at the UD point.

Causes of WU flag are as follow:

- Extreme leukocytosis

T1 and T2 flags:

- These flags occur when discrimination between various cell populations cannot be made due to presence of abnormal WBCs.

Red blood cell count and hemoglobin measurement

Typically one channel is used to detect RBCs and platelets. The detector is set such that any cell between 2 and 30 fL will be counted as a platelet and any cell between 40 and 250 fL will be counted as a red cell. If there are large platelets, these will be counted as red cells and will also result in a falsely low platelet count. Similarly, if there are fragmented red cells, these smaller red cells will be counted as platelets.

For hemoglobin measurement, spectrophotometric method is used after the red cells are lysed. The principle of the method is oxidation of ferrous ion of hemoglobin by potassium ferricyanide to the ferric ion forming methemoglobin, which is then converted into stable cyanomethemoglobin by potassium cyanide. However, sulfhemoglobin, if present, is not converted into cyanomethemoglobin under this reaction condition. Usually dihydrogen potassium phosphate (added to lower pH and accelerate the reaction) is used in the reaction mixture. Nonionic detergents are also used to accelerate lysis and reduce turbidity. Finally, absorbance of light at 540 nm is measured, and the intensity of signal corresponds to hemoglobin concentration. Smokers have higher than usual carboxyhemoglobin concentration. This is because carboxyhemoglobin takes longer for conversion into cyanmethemoglobin and also absorbs more light at 540 nm than cyanomethemoglobin. Thus the hemoglobin value in smokers may be falsely elevated. Nordenberg et al. commented that cigarette smoking seems to cause a generalized upward shift of the hemoglobin distribution curve, which reduces the diagnostic value of detecting anemia in smokers using the hemoglobin value. The authors suggested that minimum hemoglobin cutoff level for anemia should be adjusted for smokers [3]. Fetal hemoglobin may interfere with spectrophotometric measurement of carboxyhemoglobin, thus falsely indicating carbon monoxide poisoning in an infant [4].

Hyperlipidemia and hypergammaglobulinemia can also falsely elevate hemoglobin levels. In cold agglutinin disease, red cell agglutination usually takes place. In such situations, a clump of red cells may be counted as one red cell. Thus the RBC count may be falsely low and the MCV will be falsely high. However, when the red cells are lysed, a true hemoglobin result will be available. Thus a clue to cold agglutinin disease is a disproportionate low RBC count when compared with the hemoglobin level.

Hematocrit, RDW, MCV, MCH, and MCHC

The values for MCV and RDW are derived from the RBC histogram. This has been explained earlier in this chapter. Values for hematocrit (Hct), mean corpuscular hemoglobin (MCH), and mean corpuscular hemoglobin concentration (MCHC) are calculated. The Hct value is calculated from the MCV and the RBC count, using the following formula (normal Hct value approximately 45%):

$$Hct = MCV \times RBC \text{ count}$$

RDW is a measure of the degree of variation of size of red cells, i.e., it reflects extent of anisocytosis. The RDW is elevated in iron deficiency anemia, myelodysplastic syndrome (MDS), and macrocytic anemia secondary to B12 or folate deficiency. In contrast, the RDW is usually normal or mildly elevated in thalassemia. MCH refers to the average amount of hemoglobin found in RBCs. MCHC represents the concentration of hemoglobin in RBCs. Both MCH and MCHC are also calculated values (Table 1.2). In cold agglutinin disease, the RBC count will be low, and therefore, the Hct will also be low. The MCHC will be high. The laboratory scientist uses abnormally high MCHC as an indicator of possible cold agglutinin disease and warms the blood before repeating the CBC run on the analyzer. In hyperosmolar states, cells swell causing increased MCV.

The MCH is decreased in patients with anemia caused by impaired hemoglobin synthesis. The MCH may be falsely elevated in blood specimens with turbid plasma (usually caused by hyperlipidemia) or severe leukocytosis.

The MCHC is decreased in microcytic anemias where the decrease in hemoglobin mass exceeds the decrease in the size of the RBCs. It is increased in hereditary spherocytosis and in patients with hemoglobin variants, such as sickle cell disease and hemoglobin C disease.

Table 1.2 Various formulas used for calculating different parameters in complete blood count analysis.

Parameter	Formula
Hematocrit (Hct)	Hct = MCV × RBC count
Mean corpuscular hemoglobin (MCH)	MCH = Hb/RBC count
Mean corpuscular hemoglobin concentration (MCHC)	MCHC = Hb/Hct
Corrected reticulocyte count	Corrected reticulocyte count = reticulocyte count × patient Hct/normal Hct
Reticulocyte production index (RPI)	RPI = ([% reticulocyte × hematocrit]/45) × (1/correction factor) Correction factor: 1 (hematocrit 40−45), 1.5 (hematocrit 35−39), 2.0 (hematocrit 25−34), 2.5 (hematocrit 15−24), or 3 (hematocrit <15)

Hb, *hemoglobin;* RBC, *red blood cell.*

Reticulocyte count

Reticulocytes are immature red cells. They are named as such as they contain a reticular material that is actually RNA. These RNA can be seen with special stains such as new methylene blue. Reticulocyte count is used to assess bone marrow response to anemia. It is important to use the corrected reticulocyte count when making such assessments. The formula is provided in Table 1.2. Another way to assess reticulocytes is to use the reticulocyte production index, and the formula is also provided in Table 1.2.

White blood cell count and differential

The WBC histogram has three peaks. The first peak corresponds to lymphocytes and the third peak corresponds to neutrophils, whereas the second peak corresponds to the remainder types of WBCs. When nucleated RBCs are present, then these cells may be counted as WBCs, especially lymphocytes. The total WBC count may thus be falsely elevated. For an accurate WBC count, the analyzer must be run in the NRBC mode. This is referred to as the corrected WBC count. In some printouts, UWBC represents uncorrected WBC count, and WBC represents corrected WBC count. In some printouts, the corrected WBC count is denoted by the sign "&" before the WBC count. When significant myeloid precursors are present, the downward slope of the neutrophil peak may not touch the baseline.

Platelet count, mean platelet volume, platelet differential width

Just like RBC values, the hematology analyzers provide platelet counts along with MPV and PDW. The MPV is calculated from the platelet histogram by drawing a vertical line from the peak to the baseline. PDW denotes degree of variation of size of platelets.

Pseudothrombocytopenia is an important issue. Causes of falsely low platelet count include the following:

- Traumatic venipuncture and activation of clotting
- Significant number of large platelets (platelets being counted as RBCs)
- EDTA-induced platelet clump
- EDTA-dependent platelet satellitism (here platelets form a satellite around neutrophils)

The last two conditions are typically diagnosed when the slide is reviewed, although hematology analyzers are capable of flagging platelet clumps. Blood should be recollected in citrate or heparin. If thrombocytopenia is due to peripheral destruction or consumption of platelets, then the bone marrow responds to the thrombocytopenia by releasing immature platelets that are larger than normal. This would increase the MPV and also the PDW. If thrombocytopenia is due to reduced production by the bone marrow, large platelets are not seen and thus MPV and PDW are not increased.

Review of peripheral smear

A microscopic examination of appropriately prepared and well-stained blood smear slide by a pathologist or a knowledgeable laboratory professional is useful for clinical diagnosis. A blood smear analysis takes into account flagged automated hematology results to determine if a manual differential count should be performed. Therefore, peripheral blood smear examination along with manual differential leukocyte count (if necessary) and CBC provides the complete hematological picture of the patient [5]. Correlating findings of peripheral smear review with bone marrow biopsy report may provide additional information for proper diagnosis [6].

Review of the smear should start with ensuring that the name and accession number on the slide matches with that of the CBC printout. Sometimes naked eye examination of the slide may provide some important clues. For example, if the slide appears blue, there is a possibility of underlying paraproteinemia or myeloma. When paraproteins are present in significant amounts in blood, they are stained blue by the Wright Giemsa stain. Sometimes tumor emboli are visible as clumps on the slide. Cryoglobulinemia may appear as "blobs" on the slide.

The slide should be scanned at first in low power to assess overall cellularity (especially of white cells), to find an appropriate area where red cell morphology is best assessed (under higher power), and to check for platelet clumps. Red cell morphology is typically best assessed where the cells are evenly distributed, and this area is away from the tail, toward the body. Rouleaux formation and red cell agglutination can also be appreciated under low power. With experience, blasts can also be picked up on low power. It is then best to assess each cell line under higher power.

Red cells are assessed for size, shape, anisocytosis, central pallor, and red cell inclusions, if any. WBC should be checked for reactive (toxic) changes, left shift, presence of blasts, and degree of segmentation as well as presence of dysplasia. Platelets should be checked for clumps, size, and adequacy of granules.

Red cell variations and inclusions

Normal red cells are normocytic normochromic. By this we mean that the average size of a red cell in an adult is the size of the nucleus of a mature lymphocyte. Only one-third of the central portion of the red cell has central pallor. Increased pallor means that the red cell is hypochromic. Thus red cells may be

- Normocytic normochromic
- Microcytic hypochromic
- Macrocytic

Variation is shape refers to poikilocytosis. Examples of poikilocytosis are sickle cells, target cells, ovalocyte, elliptocyte, stomatocyte, echinocyte (burr cells), acanthocyte, schistocyte, spherocyte, dacryocyte (tear drop red cell), and bite cells. Each of these poikilocytes is associated with one or more underlying clinical conditions and is discussed in the chapter dealing with RBC disorders. There are also various red cell inclusions that can be observed during peripheral blood smear examination. These are listed in Table 1.3. Please see chapter on RBC disorders (Chapter 3).

White blood cell morphology

Reactive changes are predominantly appreciated in neutrophils and lymphocytes. Reactive neutrophils have prominent azurophilic granules, cytoplasmic vacuoles, and Dohle bodies (blue cytoplasmic bodies). Dohle bodies are named after a German pathologist, Karl Gottfried Paul Dohle, and these bodies represent rough endoplasmic reticulum. Dohle bodies seen in reactive neutrophils are typically

Table 1.3 Various red cell inclusions.
Howell−Jolly bodies
Pappenheimer bodies
Cabot rings
Basophilic stippling (punctate basophilia)
Heinz bodies
Hemoglobin C crystals
Malarial Parasite
Nucleated red blood cell

seen at the periphery of the cell. Another condition where we may see "Dohle-like" bodies are May−Hegglin anomaly. In May−Hegglin anomaly, "Dohle-like" bodies are seen randomly distributed throughout the cell. They are devoid of organelles. Rather they are thought to consist of a mutant form of the nonmuscle myosin heavy chain protein.

Reactive lymphocytes (also known as Downey cells) may be of three types:

• A larger than usual lymphocyte with abundant cytoplasm that appears to surround (or hug) the red cells (Downey type II cell) is the most common type of reactive lymphocyte
• A small lymphocyte with nuclear membrane irregularity (Downey type I cell)
• A larger lymphocyte with blue cytoplasm and nucleoli (Downey type III cell)

When there is neutrophilic leukocytosis, a pathologist may observe presence of immature myeloid precursors in the peripheral blood. This is referred to as left shift. Rare blasts may also be present. On occasion, we may witness patients with very high WBC count and left shift mimicking leukemia, although the process is reactive. This is referred to as leukemoid reaction. Neutrophilic leukocytosis without reactive changes and with basophilia with or without eosinophilia is suspicious for chronic myeloid leukemia (CML). Numerous blasts may indicate an acute leukemic process.

When the WBC count is high, smudge cells may be observed, which are distorted white cells produced as an artifact during the process of making the slide. Presence of smudge cells with a high lymphocyte count should make us suspicious for chronic lymphocytic leukemia (CLL). Pretreatment (before making the slide) with albumin ensures that smudge cells are reduced.

Documenting dysplasia is probably the hardest feature to establish from the peripheral blood. It is most often assessed in the neutrophils. Normal mature neutrophils have two to five nuclear segments and have fine granules in the cytoplasm. Hypogranulation is a feature of dysplasia. Hyposegmentation and hypersegmentation, if present, may represent dysplasia. A bilobed polymorphonuclear leukocyte (PMN) with hypogranulation is referred to as "Pseudo−Pelger−Huët" cell. This cell is considered to be dysplastic. Hypersegmented PMN is a PMN with more than five segments. This can be seen in megaloblastic anemia and MDS. It may also be inherited, without any clinical significance. Here majority (>75%) of the neutrophils are hypersegmented.

Benign disorders of WBC such as May−Hegglin anomaly, Alder−Reilly, and Chediak−Higashi diseases may all be diagnosed from the peripheral smear. Rarely cells such as hairy cells representing hairy cell leukemia may be seen. Patients with lymphoma may have lymphoma cells circulating in the peripheral blood. These will appear as atypical lymphocytes, i.e., lymphoid cells that are neither mature nor reactive in appearance. Presence of a Barr body, which is a nuclear appendage, denotes the inactivated X chromosome and implies female sex of the patient.

Platelets

The normal platelet count is 150,000−450,000 per microliter. Thrombocytopenia is defined as platelet count below 2.5th lower percentile. Results of the third US National Health and Nutrition Examination Survey support the traditional value of platelet count below 150,000 per microliter, as the definition of thrombocytopenia, but adoption of cutoff value below 100,000 may be more practical. Thrombocytopenia is a common hematological finding with variable clinical expression or may reflect a life-threatening disorder such as thrombotic microangiopathy [7]. Thrombocytopenia is also a common hematological abnormality found in newborns [8]. Thrombocytopenias can be broadly categorized into

- Thrombocytopenias due to decreased production: congenital and acquired (any cause of bone marrow failure)
- Thrombocytopenia due to increased destruction: e.g., ITP (idiopathic thrombocytopenia purpura)
- Thrombocytopenias due to increased consumption: e.g., DIC (disseminated intravascular coagulation), TTP (Thrombotic thrombocytopenia purpura), and HUS (hemolytic uremic syndrome)
- Thrombocytopenias due to sequestration: sequestration in hemangiomas (Kasabach−Merritt syndrome)

Congenital thrombocytopenias can be broadly divided into three groups including thrombocytopenia with small platelets, thrombocytopenia with normal-sized platelets, and thrombocytopenia with large platelets.

- Thrombocytopenia with small platelets can be associated with Wiskott−Aldrich syndrome, X-linked thrombocytopenia, or inherited macrothrombocytes.
- Thrombocytopenia with normal-sized platelets can be due to Fanconi's anemia, thrombocytopenia with absent radii (TAR syndrome), amegakaryocytic thrombocytopenia, or Quebec platelet disorder.
- Thrombocytopenia with large platelets could be associated with Bernard−Soulier syndrome, May−Hegglin anomaly, Sebastian syndrome, Epstein syndrome, Fechtner syndrome, or gray platelet syndrome. Normally the alpha granules of the platelets are stained by the Wright Giemsa stain. The delta granules are not stained. Absence of the alpha granules will result in gray-appearing platelets, i.e., gray platelet syndrome. The platelets are dysfunctional.

Thrombocytosis is defined as platelet counts over 450,000 per microliter. Thrombocytosis may be due to reactive thrombocytosis (associated with infection, inflammation, neoplasms, or iron deficiency), rebound thrombocytosis following thrombocytopenia, redistributional diorder (e.g., post-splenectomy), and myeloproliferative disorders such as essential thrombocythemia or familial thrombocytosis. Platelet disorders are discussed in greater detail in the chapter dealing with platelet disorders in Chapter 5.

Special situations with complete blood count and peripheral smear examination

There are also special situations involving CBC and peripheral blood smear review. Pancytopenia is an important hematological finding where all three major cells present in blood (RBCs, WBCs, and platelets) are decreased in number. Pancytopenia itself may not be a disease entity but a triad of findings that may result from a number of diseases primarily or secondarily involving bone marrow. The severity of pancytopenia determines the course of therapy [9]. Important causes of pancytopenia include the following:

- Bone marrow failure: any cause of bone marrow failure may result in pancytopenia. Examples include aplastic anemia, bone marrow fibrosis, leukemias, metastatic diseases, and granulomas
- Vitamin B12 or folate deficiency
- MDS
- Autoimmune destruction of cells
- Hypersplenism

  However, case studies of pancytopenia of unknown causes have also been reported [10].

Splenic atrophy or postsplenectomy

Absence of spleen is characterized by presence of Howell–Jolly bodies (in RBCs), acanthocytes, and target cells. There may be transient thrombocytosis and leukocytosis as well.

Microangiopathic hemolysis

There are three important causes of microangiopathic hemolysis:

- TTP
- HUS
- DIC

  All are characterized by low platelets and presence of schistocytes in the peripheral smear. The number of schistocytes in TTP and HUS are typically numerous, in contrast to DIC where they may present in fewer numbers. In DIC, the coagulation profile (such as PT, PTT) is abnormal. In TTP and HUS, the coagulation profile is typically normal. TTP is a medical emergency and requires urgent therapeutic plasma exchange (TPE).

Leukoerythroblastic blood picture

This term refers to presence of red cell precursors (i.e., nucleated red cells) in the peripheral blood and WBC precursors (i.e., left shift with blasts). This may be seen in patients with significant hemolysis or hemorrhage. In such a situation, we should also search for tear drop red cells. Leukoerythroblastic blood picture with tear drop red cells may be due to a bone marrow infiltrative process. Anemias due to such bone marrow infiltrative processes are known as myelophthisic anemia. This infiltration may be due to many causes such as fibrosis, infiltration by tumor, or even leukemia.

Parasites, microorganisms and nonhematopoietic cells in the peripheral blood

Several parasites and microorganisms may be seen in the peripheral blood. These include malaria parasites, various other parasites, and microorganisms (Table 1.4).

On occasions, nonhematopoietic cells may be seen in the peripheral blood. These include the following:

- Epithelial cells
- Fat cells
- Endothelial cells
- Malignant cells: e.g., neuroblastoma cells, rhabdomyosarcoma cells, and medulloblastoma cells. These cells may resemble lymphoblasts. Presence of carcinoma cells is referred to as carcinocythemia. This is most often observed with carcinoma of the lung and breast. Melanoma cells and Reed–Sternberg cells have been described in the peripheral blood.

Buffy coat preparation

Buffy coat films are sometimes made to concentrate nucleated cells (i.e., white cells). This is done to look for low frequency abnormal cells or bacteria or other microorganisms.

Table 1.4 Parasites and microorganisms that may be present in the peripheral blood smear.

Parasites	• Malarial parasites (discussed in chapter 3).
	• Toxoplasmosis: Trophozoite forms are present in monocytes and may also be seen free from ruptured monocytes.
	• Babesiosis: Trophozoites are seen within red cells. The trophozoites are ring form and similar to that of *Plasmodium falciparum*. They have one, two, or three chromatin dots. Sometimes they are pear shaped. The pointed ends of four parasites may come into contact and assume the shape of a Maltese cross, helping in their identification.
	• Trypanosomiasis: Flagellated parasites of *Trypanosoma brucei* or *Trypanosoma cruzi* may be seen in the peripheral smear.
Parasitic microfilaria	• *Bancroftian filariasis*: Caused by human parasitic roundworm.
	• *Wuchereria Bancroft*: The microfilaria form may be seen in the peripheral blood.
	• Loiasis: Caused by parasitic worm, *Loa loa*. Here again the microfilaria may be seen in the peripheral blood.
Bacteria	• Bartonellosis: Rod-shaped coccobacilli may be seen within the red cells due to infection with bacteria of genus Bartonellosis.
	• Borreliosis: Borrelia spirochete may be seen in the peripheral blood due to infection with Lyme borreliosis (Lyme disease).
Fungi	Individuals with impaired immunity and also individuals with indwelling catheters may show fungi in their peripheral blood smear. *Candida, Histoplasma capsulatum*, and *Cryptococcus neoformans* may be seen within neutrophils.

Key points

- For hemoglobin measurement, the red cells should be lysed first. Then hemoglobin (and methemoglobin and carboxyhemoglobin) is converted to cyanmethemoglobin (sulfhemoglobin is not converted). The absorbance of light at 540 nm is then measured.
- Smokers have higher than usual carboxyhemoglobin. Carboxyhemoglobin takes longer to be converted to cyanmethemoglobin. Carboxyhemoglobin absorbs more light at 540 nm than cyanmethemoglobin, giving rise to a falsely higher hemoglobin level.
- Hyperlipidemia, hypergammaglobulinemia, cryoglobulinemia, fat droplets from hyperalimentation, and leukocytosis ($> 50,000$) can falsely elevate hemoglobin levels.
- In cold agglutinin disease, red cell agglutination is observed. In such situations, a clump of red cells may be counted as one red cell. Thus the RBC count will be falsely low, and the MCV will be falsely elevated. However, when the red cells are lysed, a true hemoglobin result should be observed. Thus a clue to cold agglutinin disease is disproportionate to low RBC count compared with hemoglobin level. In cold agglutinin disease, MCHC should also be high. The lab uses abnormally high MCHC as an indicator of possible cold agglutinin disease and warms the blood before repeating the CBC run on the analyzer.
- In patients with severe hyperglycemia (glucose > 600 mg/dL), osmotic swelling of the RBCs may spuriously elevate the MCV.
- The MCH is decreased in patients with anemia caused by impaired hemoglobin synthesis. The MCH may be falsely elevated in blood specimens with turbid plasma (usually caused by hyperlipidemia) or severe leukocytosis.
- The RDW is elevated in iron deficiency anemia, MDSs, and macrocytic anemia secondary to vitamin B12 or folate deficiency. In contrast, the RDW is usually normal or only mildly elevated in thalassemia.
- The MCHC is decreased in microcytic anemias where the decrease in hemoglobin mass exceeds the decrease in the size of the RBC. It is increased in hereditary spherocytosis and in patients with hemoglobin variants, such as sickle cell disease and hemoglobin C disease
- Corrected reticulocyte count is reticulocyte count x patient Hct/normal Hct
- Pseudothrombocytopenia may be due to traumatic venipuncture and activation of clotting, significant number of large platelets (platelets being counted as RBCs), EDTA-induced platelet clump, or EDTA-dependent platelet satellitism (here platelets form a satellite around neutrophils)
- Features of reactive neutrophils are as follow: prominent granules, cytoplasmic vacuoles, and Dohle bodies (blue cytoplasmic bodies). Dohle bodies represent rough endoplasmic reticulum. Dohle bodies seen in reactive neutrophils are typically seen at the periphery of the cell. Another condition where we may see "Dohle-like" bodies are May–Hegglin anomaly. In May–Hegglin anomaly, "Dohle-like" bodies are seen randomly distributed throughout the cell. They are devoid of organelles. Rather they are thought to consist of a mutant form of the nonmuscle myosin heavy chain protein.
- Reactive lymphocytes (also known as Downey cells) may be of three types: a larger than usual lymphocyte with abundant cytoplasm appearing to be "hugging" the red cells (Downey type II cell), which is the most common type of reactive lymphocyte; a small lymphocyte with nuclear

membrane irregularity (Downey type I cell); and a larger lymphocyte with blue cytoplasm and nucleoli (Downey type III cell).

- Important causes of pancytopenia are bone marrow failure, B12 or folate deficiency, MDS, autoimmune destruction, and hypersplenism.
- Absence of the spleen is characterized by presence of Howell−Jolly bodies (in RBCs), acanthocytes, and target cells. There may be transient thrombocytoses and leukocytosis as well.
- There are three important causes of microangiopathic hemolysis: TTP, HUS, and DIC. All are characterized by low platelets and presence of schistocytes in the peripheral smear. In DIC, the coagulation profile (e.g., PT, PTT) are abnormal. In TTP and HUS, the coagulation profile is typically normal. TTP is a medical emergency and requires urgent TPE.
- Leukoerythroblastic blood picture: This term refers to presence of red cell precursors (i.e., nucleated red cells) in the peripheral blood and WBC precursors (i.e., left shift with blasts). This may be seen in patients with significant hemolysis or hemorrhage. In such situation, we should also search for tear drop red cells. Leukoerythroblastic blood picture with tear drop red cells may be due to a bone marrow infiltrative process. This infiltration may be due to many causes such as fibrosis, infiltration by tumor, or even leukemia.

References

[1] Tefferi A, Hanson CA, Inwards DJ. How to interpret and pursue and abnormal complete blood count in adults. Mayo Clin Proc 2005;80:923−36.

[2] Gulati GL, Hyun BH. The automated CBC: a current perspective. Hematol Oncol Clin N Am 1994;8: 593−603.

[3] Nordenberg D, Yip R, Binkin NJ. The effect of cigarette smoking on hemoglobin levels and anemia screening. J Am Med Assoc 1990;264:1556−9.

[4] Mehrotra S, Edmonds M, Lim RK. False elevation of carboxyhemoglobin: a case report. Pediatr Emerg Care 2011;27:138−40.

[5] Gulati G, Song J, Florea AD, Ging J. Purpose and criteria for blood smear scan, blood smear examination and blood smear review. Ann Lab Med 2013;33:1−7.

[6] Bersabe AR, Aden JK, Shumway NM, Osswald MB. Peripheral smear review and bone marrow biopsy correlation. J Clin Diagn Res 2017;11:JC01−3.

[7] Stasi R. How to approach thrombocytopenia. Hematology Am Soc Hematol Educ Program 2012;2012: 191−7.

[8] Gunnink SF, Vlug R, Fijnvandraat K, van der Bom JG, et al. Neonatal thrombocytopenia: etiology, management and outcome. Expert Rev Hematol 2014;7:387−95.

[9] Gayathri BN, Rao KS. Pancytopenia: a clinico-hematological study. J Lab Physicians 2011;3:15−20.

[10] Gudina EK, Amare H, Benti K, Ibrahim S, et al. Pancytopenia of unknown cause in adult patients admitted to a tertiary hospital in Ethiopia: case series. Ethiop J Health Sci 2018;28:375−82.

Bone marrow examination and interpretation

Introduction

Complete blood count (CBC), examination of peripheral blood smear, and other routine laboratory tests may not provide enough information for unambiguous diagnosis of hematological or non-hematological disease in certain patients. For these patients, direct microscopic examination of the bone marrow is required for a proper diagnosis. The bone marrow that is disseminated within the intertrabecular and medullary spaces of bone is a complex organ with dynamic hematopoietic and immunological function. The role of bone marrow in hematopoiesis was first described by Neumann in 1868, and since then methods for bone marrow procedures have undergone many improvements. Following development of newer techniques and equipment, bone marrow aspirate and bone marrow biopsy have become an important medical procedure for diagnosis of hematological malignancies and other diseases and also for follow-up evaluation of patients undergoing chemotherapy, bone marrow transplantation, and other forms of therapy [1]. Bone marrow trephine biopsy should be carried out by a trained health-care professional, and bone marrow aspirate should be collected during the same procedure. Because diagnostic specimen is a small representation of the total marrow, it is important that material should be adequate and representative of the entire marrow. The specimen must also be of high technical quality. Cytochemical analysis and various other diagnostic procedures can be performed on the liquid bone marrow aspirate, while bone marrow biopsy can be stained using immunoperoxidase and other stains. The recent development of bone marrow biopsy needles with specially sharpened cutting edges and core-securing devices has reduced the discomfort of procedure and improved the quality of the specimen obtained [2]. Today, bone marrow examination is considered as an important and effective way of diagnosing and evaluating primary hematological and metastatic neoplasm as well as nonhematological disorders [3]. Common indications for performing bone marrow examination are listed in Table 2.1.

Fundamentals of bone marrow examination

Before a bone marrow examination, relevant history of the patient, CBC, and report of peripheral blood smear examination must be reviewed [4]. During a routine bone marrow examination, slides obtained from the aspirate, slides obtained from the clot sections, slides from the trephine biopsy, touch preparation slides obtained from the trephine biopsy, and iron strains must be carefully examined for proper interpretation of results. At times, examination of a well-prepared aspirate slide, core biopsy

---

Table 2.1 Common indications for bone marrow examination.

Diagnosis of Diseases
- Acute or chronic unexplained anemia including hypoplastic or aplastic anemia
- Differentiating megaloblastic anemia from normoblastic maturation
- Unexplained leukopenia
- Unexplained thrombocytopenia, pancytopenia
- Myelodysplastic syndrome
- Myeloproliferative disease
- Plasma cell dyscrasia
- Hodgkin and non-Hodgkin lymphoma (for staging)
- Suspected leukemia
- Disseminated granulomatous disease
- Primary amyloidosis
- Metabolic bone disease
- Suspected multiple myeloma
- Suspected storage diseases (e.g., Gaucher disease, etc.)
- Fever of unknown origin
- Confirmation of normal marrow in a potential allogenic donor

Follow-up of medical treatment
- Chemotherapy/bone marrow transplant follow-up
- Treatment of isolated cytopenia

Table 2.2 Additional steps that may be performed as part of a bone marrow examination.

- Immunophenotyping by flow cytometry (performed on the aspirate specimen)
- Immunophenotyping by immunohistochemistry (performed on the biopsy or clot section slides)
- Special stains: e.g., AFB (acid fast bacilli), GMS (Grocott's methenamine silver), reticulin, trichrome, Wright–Giemsa stain, Prussian blue stain
- Cytogenetic studies
- Molecular studies (e.g., polymerase chain reaction, fluorescence in situ hybridization, next generation sequencing)
- Electron microscopy

---

specimen, and iron train by a well-trained professional may be adequate for arriving at a diagnosis [1]. However, additional tests such as flow cytometry and cytogenetics studies may be needed for other cases. Additional steps that may be performed during bone marrow examination are listed in Table 2.2.

The aspirate slides are typically used to assess morphology, performing a differential count and thus obtaining the myeloid:erythroid (M: E) ratio. If the aspirate lacks particulates or is unsatisfactory, morphology may be assessed from the touch prep slides.

The architecture of the bone marrow is best assessed from the trephine biopsy slides. Infiltrates (e.g., granulomas, lymphomatous infiltrates, metastatic tumors), if any, and their distribution can also be assessed from the biopsy slides. The cellularity of the bone marrow is usually assessed from the biopsy slides. In addition, reticulin or collagen fibrosis is also assessed from the biopsy slides. Bone marrow stroma and the bone itself are assessed form the biopsy slides. In the absence of a good trephine biopsy specimen, the slides from the clot section may be used as an alternate means of assessment.

Dry tap

Causes of dry tap while performing a bone marrow procedure include the following:

- Faulty technique
- Packed marrow (e.g., with leukemia)
- Fibrotic marrow (e.g., myelofibrosis)
- Hairy cell leukemia

In cases of dry tap, we would need to improvise to obtain the most possible information. One way of achieving this is to obtain two trephine biopsies and to submit one part for flow cytometry and another part for cytogenetic studies. Good touch preps from the second one and the second biopsy itself should provide adequate morphological and architectural information.

Granulopoiesis

Granulopoiesis involves maturation of myeloblasts into mature polymorphonuclear neutrophils, basophils, and eosinophils. The steps include maturation of myeloblasts to promyelocytes to myelocytes to metamyelocytes to bands to mature granulocytes. Granulopoiesis in a normal marrow is seen adjacent to the bony trabecular surface (as a layer two to three cell thick) and to arterioles.

Myeloblasts are large cells with high nuclear to cytoplasmic ratio, moderately blue cytoplasm (less blue than the cytoplasm of an erythroblast), and prominent nucleoli. Promyelocytes are larger cells compared with myeloblasts with prominent nucleoli, a Golgi hof, and granules. These granules are primary granules and appear reddish purple. The promyelocytes of the three granulocytic lineages cannot be distinguished by routine light microscopy. Myelocytes no longer have nucleoli but continue to have granules. However, these granules are secondary and are also known as specific granules. Thus, cells of the three granulocytic lineages can now be distinguished. Myeloblasts, promyelocytes, and myelocytes are all capable of cell division. Metamyelocytes have indented nuclei and cannot undergo cell division. The nuclei of bands are U-shaped.

Myeloblasts are found in the paratrabecular areas and close to arterioles. Metamyelocytes are found close to the sinusoids. Mature neutrophils can cross the wall of sinusoids and enter into the circulation.

One approach to accurate identification of various stages of granulopoietic cells include the following:

- If the cell has nucleoli then it is either a blast or a promyelocyte. If there is a Golgi hof and many granules, it is a promyelocyte; otherwise it is a blast (high nuclear-cytoplasmic ratio). Erythroblasts may resemble myeloblasts, but the cytoplasm of a myeloblast is moderately blue and may have granules. The cytoplasm of an erythroblast is deep blue and does not have granules.
- If the nucleus is indented but not U-shaped, it is a metamyelocyte. If there are no nucleoli and no nuclear indentation, then it is a myelocyte. Cells with a U-shaped nucleus without proper nuclear segmentation represent a band.

Erythropoiesis

The stages of erythropoiesis are proerythroblast to basophilic normoblast (early normoblast) to polychromatic normoblast (intermediate normoblast) to orthochromatic normoblast (aka late normoblast) to reticulocyte to mature red cell. Proerythroblast is the largest cell and has deep blue cytoplasm, high N: C (nuclear to cytoplasm) ratio, and a prominent nucleolus. Subsequent to proerythroblasts, nucleoli are no longer seen. In the three normoblasts, the chromatin in the nuclei becomes progressively clumped. In the late normoblast stage, the chromatin is dark, dense and clumped, and ready to be extruded. Once extruded, the cell is now known as a reticulocyte that is converted to mature red cell in approximately 3 days. Within the three normoblasts, the cytoplasm will change color from blue (basophilic normoblast) to gray to gray orange (late normoblast). These changes denote an increased amount of hemoglobin the cytoplasm. One normoblast typically gives rise to eight reticulocytes. Reticulocytes circulate for about 1—3 days before converting in to mature red cells.

Erythropoiesis typically occurs close to the sinusoids in the bone marrow. Erythropoiesis can be seen as erythropoietic islands, where cells of the erythropoietic series are seen surrounding a macrophage. The macrophage is referred to as the nurse cell, providing nourishment to the maturing cells. The red cells closest to the nurse cells are less mature than those which are more distant. The macrophage also engulfs defective erythroblasts and the extruded nuclei from late normoblasts.

Monopoiesis, megakaryopoiesis, thrombopoiesis, and other cells in bone marrow

The steps involved in monopoiesis are monoblasts to promonocytes to mature monocytes. These are present in very small numbers in a normal bone marrow. Megakaryopoiesis is the process by which mature megakaryocytes develop from hematopoietic stem cell [5]. Thrombopoiesis is the generation of platelets from megakaryocytes.

Megakaryocytes comprise approximately 0.05%—0.1% of hematopoietic cells in a normal bone marrow and are highly specialized large nuclear cells (50—100 μm in diameter) that differentiate to produce platelets. Each megakaryocyte gives rise to 1000—3000 platelets. Megakaryocytes mature through endomitosis (endoreduplication) where the nuclear DNA content increases in multiples of two without nuclear or cytoplasmic division. Thus, in the bone marrow megakaryocytes which are 2N or 4N or 8N or 16N and 32N (32N DNA or 16 copies of normal complement of DNA), the dominant ploidy category is 16 N. The nucleus is large and multilobated. Megakaryocytes development and formation are regulated by a multitude of cytokines, principally Tpo (thrombopoietin), which is produced in the liver and marrow stroma [6]. Megakaryocytes can be classified into three stages of maturation:

- Group I: These have strong basophilic cytoplasm and high N:C ratio
- Group II: Less basophilic cytoplasm with some azurophilic granules and lesser N:C ratio
- Group III: Plentiful cytoplasm; cytoplasm is weakly basophilic; the cytoplasm has abundant azurophilic granules. The cytoplasm of the cell margin is agranular. This is the mature megakaryocyte.

Megakaryocytes may apparently engulf other hematopoietic cells and this is known as emperipolesis. Megakaryocytes are typically found close to the sinusoids; a paratrabecular location is considered abnormal and may be a feature of dysplasia. On average, a high-power field should demonstrate two to four megakaryocytes. Megakaryocytes do not cluster. Clustering of more than three megakaryocytes may be observed in a regenerating marrow, following chemotherapy and bone marrow transplantation or in pathological states, e.g., myelodysplasia. Various other cells are also present in the bone marrow, such as mast cells, and lymphocytes and are listed in Table 2.3.

Table 2.3 Various other cells found in the bone marrow.
• Mast cells: These are seen occasionally and have deep purple granules in their cytoplasm. The granules do not obscure the nucleus, and the nucleus is not lobulated. These features help to distinguish mast cells from basophils. • Lymphocytes: These are a small population of bone marrow cells (approximately 10%) where T cells outnumber B cells. Bone marrow of children and recovering marrow exhibit cells that resemble lymphoblasts. These are actually B lymphocyte precursors and are known as hematogones. • Plasma cells: Representing typically 1% or less in a normal bone marrow. • Fat cells • Osteoblasts: These may resemble plasma cells. However, osteoblasts are larger and the Golgi hof is separate from the nucleus. • Osteoclasts: These are multinucleated giant cells. They are derived from hematopoietic stem cells and are formed from the fusion of monocytic cells. They may resemble megakaryocytes. The nuclei of osteoclasts are separate and the azurophilic granules are coarser than that of megakaryocytes.

Various bone marrow examination findings and bone marrow failure

Bone marrow findings can be discussed under the various categories:

- Leukemias
- Lymphomas
- Bone marrow failure
- Disorders of erythropoiesis, granulopoiesis, or thrombopoiesis
- Infections
- Granulomatous changes
- Storage disorders
- Metabolic bone diseases
- Metastatic tumors
- Miscellaneous: Hemophagocytic syndrome, necrosis/infarction, serious atrophy, bone marrow fibrosis, reactive lymphoid surrogate, and amyloidosis

Bone marrow findings in various leukemia and lymphomas are discussed in the sections dealing with various leukemias and lymphomas.

There are also various scenarios for bone marrow failure including the following:

- In aplastic anemia, trephine biopsy is crucial to document bone marrow hypocellularity. Hematopoietic cells are markedly reduced, and fat cells are increased. Lymphocytes are increased and reactive lymphoid aggregates may be seen. Other cells that may be apparently increased are plasma cells, mast cells, and macrophages. Erythroid cells may show features of dysplasia.
- Anticancer and immunosuppressive agents may also cause bone marrow hypoplasia. Dyserythropoiesis may be seen along with megaloblastoid changes. In the early stages of chemotherapy administration, there may be interstitial edema. Subsequently, serous atrophy (also known as gelatinous transformation) may become evident.
- With radiation damage, there may be necrosis of the bone marrow with bone necrosis. Cells next to bony trabeculae are particularly vulnerable to radiation. Hematopoietic cells may be permanently replaced by fat or fibrous tissue.

Mirzal et al. commented that bone marrow examination had the highest diagnostic value with respect to suspicion of leukemia, multiple myeloma, myeloproliferative disorders, and lymphoma and was least helpful in diagnosis of infection and storage disorders [7]. Juneja and Westerman commented that for diagnosis of hematological malignancies including myelodysplastic syndromes (MDS), peripheral blood smear examination, and bone marrow morphological correlation should be considered as the gold standard [8].

Disorders of erythropoiesis, granulopoiesis, and thrombopoiesis

Most of the anemias such as iron deficiency anemia, megaloblastic anemia, and hemolytic anemias will show features of erythroid hyperplasia and dyserythropoiesis in the bone marrow. In iron deficiency, the erythroblasts are smaller than normal, and in megaloblastic anemia, they are larger than normal. In megaloblastic anemia, granulopoiesis is also increased with presence of large forms. Giant myelocytes and metamyelocytes are usually present. With increased intramedullary destruction, increased number of macrophages may be seen. The increase in cell turn over may be so marked in cases with congenital dyserythropoietic anemia type II that pseudo-Gaucher cells may be apparent. Patients with sickle cell anemia may show evidence of bone marrow necrosis.

Agranulocytosis occurs typically as an idiosyncratic reaction to drugs or chemicals. Marked reductions in mature neutrophils are seen. If myeloid precursors such as promyelocytes and myelocytes are present, then bone marrow recovery occurs earlier than if they are absent.

In Kostmann's syndrome and agranulocytosis associated with thymoma, myeloid differentiation shows apparent arrest at the promyelocyte stage. In agranulocytosis with superimposed sepsis, promyelocytes may be predominate, mimicking APL (acute promyelocytic leukemia). However, these promyelocytes possess Golgi hof, and Auer rods are absent.

When thrombocytopenia is due to increased destruction or consumption and the process is sustained, megakaryocytes are increased in the bone marrow. There is also reduction in the average size of megakaryocytes. In MDS, features of megakaryocytic dysplasia are evident. In reactive thrombocytosis, megakaryocytes are increased in number, with increase in average size and increased variation in size.

Infections

Bone marrow examination is indicated for diagnosis of fever of unknown origin. In general, mycobacterial bone marrow infection is the most common diagnosis established by bone marrow examination for diagnosis of fever of unknown origin. Lin et al., in a study, investigated 24 patients and reported that 11 patients (46%) had infection due to *Mycobacterium tuberculosis* and 10 patients showed nontuberculosis mycobacteria such as *Mycobacteria avium* or *Mycobacteria kansasii*. In addition, nine patients were also positive for HIV (human immunodeficiency virus) [9]. Bone marrow abnormalities are frequently observed in all stages of patients infected with HIV. The most common abnormality is dysplasia affecting one or more cell lines, while erythroid dysplasia is the most common type of dysplasia, observed in over 50% of patients. Moreover, abnormal granulocytic and megakaryocytic development is encountered in one-third of all HIV infected patients. Plasma cells are also strikingly increased in the bone marrow [10]. Bone marrow abnormalities are also observed in children infected with HIV [11]. Various characteristics of bone marrow findings observed in an infected patient include the following:

- Bacterial infections lead to bone marrow hypercellularity due to granulocytic hyperplasia. Granulopoiesis may be left shifted, i.e., immature cells predominate. With chronic infection, increase in plasma cells is noted. Plasma cell satellitosis (macrophage surrounded by plasma cells) may be evident, as well as secondary hemophagocytic syndrome.
- Viral infections cause an increase in bone marrow lymphocytes, plasma cells, and macrophages. Hemophagocytosis may also be evident. In infections with CMV (cytomegalovirus) or HHV-6 (human herpesvirus-6), intranuclear inclusions may be seen. In posttransplant patients, CMV or HHV-6 infection can cause bone marrow hypoplasia. Chronic hepatitis B and C infection can result in the presence of reactive lymphoid aggregates.
- Fungal infections in bone marrow are typically seen in immunocompromised individuals. Usually the organisms are within macrophages or within necrotic tissue. Rarely, these may be found within megakaryocytes. Granulomas may also be present.
- Presence of parasites is an unusual finding in the bone marrow. Bone marrow examination may be done when leishmaniasis is being considered. In visceral leishmaniasis, the organisms are seen within the macrophages. There may be granuloma formation. They may be confused with *Histoplasma capsulatum*. However, *H. capsulatum* is Grocott's methenamine silver positive, whereas *Leishmania* is not. Also with the Giemsa stain, the kinetoplast of the *Leishmania* is stained giving a characteristic double dot appearance.

Granulomatous changes

Causes of bone marrow granulomas include the following:

- Infections: e.g., tuberculosis, fungal infections, brucellosis, etc.
- Sarcoidosis
- Malignancy: Hodgkin and non-Hodgkin lymphoma, metastatic disease
- Drugs
- Lipogranuloma: These are characterized by focal aggregate of macrophages with lipid vacuoles where plasma cells, lymphocytes, and eosinophils are associated with them. In rare occasions, giant cells may also be seen. Lipogranuloma is a benign disorder.

Storage disorders

Lysosomal storage disorders (caused by enzyme deficiency and transmitted in an autosomal recessive fashion) may be evident from bone marrow examinations. Clinically, these disorders may result in pancytopenia. Partially degraded lipids accumulate in macrophages of the liver, spleen, bone marrow, etc. This results in organomegaly and cytopenias. Storage diseases that show characteristic bone marrow abnormalities include the following:

- Gaucher's disease is due to lack of the enzyme glucocerebrosidase, in which "Gaucher" cells are seen in the bone marrow. These are macrophages with "wrinkled cigarette paper" appearance of its cytoplasm. The cells are PAS (periodic acid—Schiff) positive.

- Indistinguishable from Gaucher cells are pseudo-Gaucher cells that are seen in conditions with high cell membrane turnover. Examples of such states are CML (chronic myelogenous leukemia), hemoglobinopathies, and myeloma. A mimicker of Gaucher cell is when due to MAI infection (Mycobacterium avium-intracellulare infection), macrophages are packed with organisms and negatively staining organisms that mimic striations of Gaucher cells.
- In Niemann–Pick disease, which is due to lack of the enzyme sphingomyelinase, foamy macrophages with bubbly cytoplasm is seen. These macrophages are weakly PAS positive and Oil Red O and Sudan black positive. Sea blue histiocytes may also be seen in Niemann–Pick disease.
- Sea blue histiocytes are histiocytes with sea blue cytoplasm. They are macrophages containing ceroid. With hematoxylin and eosin (H&E), they appear yellow-brown, and with Giemsa, they appear bright blue green and are PAS positive. They are seen in high turnover states, CML, ITP (idiopathic thrombopoietic purpura), and sickle cell disease (SCD). They may also be seen in lipidoses and hyperlipidemia.

Metabolic bone diseases

Examples of metabolic bone diseases that may be evident in trephine biopsies include the following:

- Osteopetrosis (Albers-Schonberg or marble bone disease): Here there is functional defect in osteoclasts with resultant abnormal accumulation of dense bone (with persistence of cartilaginous cores). It may be transmitted as autosomal recessive (severe form) or autosomal dominant (may be asymptomatic). In the autosomal recessive form, there may be reduced hematopoiesis (myelophthisic anemia) with extramedullary hematopoiesis.
- Osteomalacia: This disorder is due to defective mineralization of bone as a result of vitamin D deficiency. There is hyperosteoidosis, which is increased unmineralized osteoid.
- Renal osteodystrophy: This condition is seen in chronic renal insufficiency. There is deficiency of active vitamin D which results in hyperosteoidosis. There is also 2° hyperparathyroidism with resultant increased osteoclast activity. This leads to irregular scalloping of bony trabeculae and peritrabecular fibrosis. Ultimately, there may be osteitis fibrosa cystica.
- Paget's disease (osteitis deformans): This disease is due to disordered bone remodeling, characterized by thickened trabeculae and irregular cement lines giving rise to "mosaic" pattern. Both osteoblasts and osteoclasts are prominent.

Metastatic tumors

Common metastatic tumors to the bone marrow include the following:

- Adults: Breast, prostate, lung, and gastrointestinal (GI) tract carcinoma
- Pediatric: Neuroblastoma

Hemophagocytic syndrome

Hemophagocytic syndrome, also referred to as hemophagocytic lymphohistiocytosis (HLH), occurs when there is increased hemophagocytosis (phagocytosis of nucleated cells) by macrophages with resultant cytopenias. HLH is a syndrome of excessive inflammation and tissue destruction due to a highly activated but ineffective immune response. There is thought be impairment in the ability of NK cells and cytotoxic T lymphocytes to destroy their cellular targets, resulting in a failure to produce inhibitory signals. This ultimately promotes further immune activation resulting in markedly elevated proinflammatory cytokine levels ("cytokine storm").

Primary, or genetic/familial, hemophagocytosis usually occurs before 1 year of age and is transmitted as autosomal recessive. Genes known to be mutated in familial HLH are *PRF1, UNC13D,* and *STX11.* Immune deficiencies associated with HLH include Chediak—Higashi syndrome, Griscelli syndrome 2, and X-linked proliferative syndrome. Secondary HLH may be triggered by bacterial or viral infections, especially herpes viruses including Epstein—Barr virus and CMV. Lymphomas, especially T-cell lymphomas, are also associated with hemophagocytic syndrome.

The diagnostic criteria for HLH include five or more of the following signs or symptoms:

- Ferritin > 500 μg/L
- Fever
- Splenomegaly
- Cytopenias (in at least two cell lines)
- Hypertriglyceridemia ($\geq$3 mmol/L) and/or hypofibrinogenemia (<1.5 g/L)
- Hemophagocytosis in the bone marrow, spleen, or lymph nodes
- Low or absent NK cell activity
- Soluble CD25 (IL2 receptor) $\geq$ 2400 U/mL

Bone marrow necrosis/infarction

Bone marrow necrosis is noted more often in trephine biopsies rather than aspirate slides. The aspirate slides may show amorphous debris with presence of karyorrhectic nuclear material. Bone marrow necrosis may also be accompanied by necrosis of the adjacent bone. Bone marrow necrosis may be seen in leukemias, lymphomas, metastatic disease, SCD, and infections.

Serous atrophy

This is also known as gelatinous transformation of the bone marrow. Here there occurs loss of fat cells and hematopoietic cells with replacement by an increased amount of ground substance. H&E stain show amorphous pink material. Causes of serous atrophy include cachexia due to chronic debilitating illnesses such as malignancy, renal failure, high dose radiation, and overwhelming infections.

Bone marrow fibrosis

This may be due to increased reticulin or reticulin and collagen in the bone marrow. With marked collagen fibrosis, there may also be osteosclerosis. Collagen deposition is more uncommon than reticulin deposition and is greater significance of abnormality. Increased reticulin and subsequently collagen deposition may be seen in the chronic myeloproliferative neoplasm. Increased reticulin without collagen deposition is seen in HIV infection and hairy cell leukemia. The peripheral blood changes seen in bone marrow fibrosis are presence of tear drop cells with leukoerythroblastic blood picture. Attempts at bone marrow aspirate may yield a dry tap.

Reactive lymphoid aggregate

Reactive or benign lymphoid aggregates are seen in the bone marrow especially with increasing age and the causes include infection, inflammation, and autoimmune diseases. Increased frequency of benign lymphoid aggregate has also been documented with Castleman disease. Benign aggregates are usually few in number and should not be located in the paratrabecular area. They should be well circumscribed and have a heterogenous population of cells. These cells should be small mature lymphocytes with some plasma cells, macrophages, and occasional eosinophils and mast cells. With immunohistochemistry, the lymphocytes are typically predominantly T cells or may have a central core of T cells surrounded by a rim of B cells, or a mixed distribution of B and T cells.

Bone marrow infiltration in lymphoproliferative disorders

Bone marrow infiltration is a common finding in lymphoproliferative disorders. The particular pattern of infiltration is best assessed from the trephine biopsy. The various patterns include nodular (which in turn may be paratrabecular and nonparatrabecular), interstitial (individual neoplastic cells interspersed between hematopoietic cells), intrasinusoidal (this pattern alone is quite rare; usually seen in combination with a second pattern), random focal (irregular, randomly distributed foci of neoplastic cells), and diffuse (bone marrow elements are replaced by neoplastic cells). With B cell lymphomas, bone marrow infiltration is more common in low-grade lymphomas. Bone marrow infiltration is more often seen in B-cell lymphomas than T-cell lymphomas. Paratrabecular aggregates are a classical feature of follicular lymphoma. Diffuse involvement is more often seen in T-cell lymphomas. Patterns of lymphoid infiltration of the bone marrow in lymphoproliferative disorders are listed in Table 2.4.

Amyloidosis

In amyloidosis, there is extracellular deposition of amorphous eosinophilic material. Amyloidosis can be seen in a number of conditions. In amyloidosis seen with immunocyte dyscrasias, immunoglobulin light chain is the main protein content. When amyloidosis occurs due to an underlying reactive process, then serum amyloid protein A is the major component. Presence of amyloid may be confirmed with Congo Red staining. This demonstrates apple green birefringence under polarized light.

Key points

- Causes of dry tap while performing a bone marrow procedure include faulty technique, packed marrow (e.g., with leukemia), fibrotic marrow (e.g., myelofibrosis), and hairy cell leukemia.
- Promyelocytes are large cells with prominent nucleoli, a Golgi hof, and primary granules. The promyelocytes of the three granulocytic lineages cannot be distinguished by routine light

---

Table 2.4 Patterns of lymphoid infiltration of bone marrow in lymphoproliferative disorders.

- Nodular lymphoid aggregates: Paratrabecular or non-paratrabecular
- Interstitial (individual neoplastic cells interspersed between hematopoietic cells)
- Intrasinusoidal (this pattern alone is quite rare; usually seen in combination with a second pattern)
- Random focal (irregular, randomly distributed foci of neoplastic cells)
- Diffuse (bone marrow elements are replaced by neoplastic cells)

microscopy. Myelocytes no longer have nucleoli and now contain secondary, or specific, granules that do now allow the three granulocytic lineages to be distinguished. The myelocytes are the last cell capable of cell division.

- The stages of erythropoiesis are proerythroblast → basophilic normoblast → polychromatic normoblast → orthochromatic normoblast → reticulocyte → mature red cell.
- Megakaryocytes mature through endomitosis, and the nuclear DNA content increases in multiples of two; i.e. 2N, 4N, 8N, 16N or 32N. The dominant ploidy category is 16N.
- Megakaryocytes normally do not cluster. Clustering of more than three megakaryocytes is seen in regenerating marrow, following chemotherapy and bone marrow transplantation, pathological states, e.g., myelodysplasia.
- In aplastic anemia, hematopoietic cells are markedly reduced and fat cells are increased. Lymphocytes may appear increased and reactive lymphoid aggregates may be seen. Other cells that may be apparently increased are plasma cells, mast cells and macrophages.
- Anticancer and immunosuppressive agents may also cause bone marrow hypoplasia. Dyserythropoiesis may be seen along with megaloblastoid changes. In the early stages of chemotherapy administration, there may be interstitial edema. Subsequently, serous atrophy (gelatinous transformation) may become evident.
- With radiation damage, there may be necrosis of the bone marrow with bone necrosis. Cells next to bony trabeculae are particularly vulnerable to radiation. Hematopoietic cells may be permanently replaced by fat or fibrous tissue.
- In megaloblastic anemia, erythroids are larger than normal. Granulopoiesis is also increased with large forms including giant myelocytes and metamyelocytes. With increased intramedullary destruction, increased number of macrophages may be seen.
- Causes of bone marrow granulomas include infections (e.g., tuberculosis, fungal infections, brucellosis, etc.), sarcoidosis, malignancy (Hodgkin and non-Hodgkin lymphoma, metastatic disease), drugs, and lipogranuloma (these are characterized by focal aggregate of macrophages with lipid vacuoles).
- Gaucher's disease is due to lack of the enzyme glucocerebrosidase. "Gaucher" cells are macrophages with a "wrinkled cigarette paper" appearance of its cytoplasm. The cells are PAS positive and are seen in the bone marrow.
- Indistinguishable from Gaucher cells are pseudo-Gaucher cells which are seen in conditions with high cell membrane turnover, such as CML, hemoglobinopathies, and multiple myeloma. MAI infection can result in macrophages packed with negatively staining organisms, thus mimicking the striations of Gaucher cells.
- Common metastatic tumors to the bone marrow include breast, prostate, lung, and GI carcinoma (for adults) and neuroblastoma (for children).
- Bone marrow necrosis may be seen in leukemias, lymphomas, metastatic disease, SCD, and infections.
- Serous atrophy (gelatinous transformation of the bone marrow) occurs when there is loss of fat cells and hematopoietic cells with replacement by an increased amount of ground substance. H&E show amorphous pink material, and its causes include cachexia due to chronic debilitating illnesses such as malignancy, renal failure, high dose radiation, and overwhelming infections.
- Reactive or benign lymphoid aggregates are seen in the bone marrow especially with increasing age. Causes include infection (most often viral, e.g., HIV, Epstein—Barr virus), inflammation, and

autoimmune diseases. Increased frequency of benign lymphoid aggregate has also been documented with Castleman disease. Benign aggregates are usually few in number. They should not be located in the paratrabecular area. They should be well circumscribed and have a heterogenous population of cells. These cells should be small mature lymphocytes with some plasma cells, macrophages, and occasional eosinophils and mast cells. With immunohistochemistry, the lymphocytes are typically predominantly T cells or may have a central core of T cells surrounded by a rim of B cells, or a mixed distribution of B and T cells.

Bone marrow infiltration is more often seen in B-cell lymphomas than T-cell lymphomas. Paratrabecular aggregates are a classical feature of follicular lymphoma. Diffuse involvement is more often seen in T-cell lymphomas.

References

[1] Oudat RI. Bone marrow: indications and procedure set-up. Jordon Med J 2010;44:88—94.
[2] Riley RS, Hogan TF, Pavot DR, Forysthe R, et al. A pathologist's perspective on bone marrow aspiration and biopsy; I: performing a bone marrow examination. J Clin Lab Anal 2004;18:70—90.
[3] Hyun BH, Stevenson AJ, Hanau CA. Fundamentals of bone marrow examination. Hematol Oncol Clin N Am 1994;8:651—63.
[4] Hyun BH, Gulati GL, Ashton JK. Bone marrow examination: techniques and interpretation. Hematol L Oncol Clin North Am 1988;2:513—23.
[5] Yu M, Cantor AB. Megakaryopoiesis and thrombopoiesis: an update on cytokines and linkage surface markers. Methods Mol Biol 2012;788:291—303.
[6] Italiano JE, Shivdasani RA. Megakaryocytes and beyond: the birth of platelets. J Thromb Haemost 2003;1: 1174—82.
[7] Mirzai AZ, Hosseini N, Sadeghipour A. Indications and diagnostic utility of bone marrow examination in different bone marrow disorders in Iran. Lab Hematol 2009;15:38—44.
[8] Juneja S, Westerman D. Changing role of bone marrow examination in the diagnosis of hematological malignancies. Leuk Lymphoma 2018;59:2018—20.
[9] Lin SH, Lai CC, Hunag SH, Hung CC, et al. Mycobacterial bone marrow infections at medical center in Taiwan 2001—2009. Epidemiol Infect 2014;142:1524—32.
[10] Tripathi AK, Misra R, Kalra P, Gupta N, et al. Bone marrow abnormalities in HIV disease. J Assoc Phys India 2005;53:705—10.
[11] Shah I, Murthy A. Bone marrow abnormalities in HIV infected children, report of three cases and review of the literature. J Res Med Sci 2014;19:181—3.

Red blood cell disorders

Introduction

Anemia is defined as reduction in the concentration of hemoglobin, taking into account the age and sex of the individual. Anemia is the most common blood disorder affecting millions of people worldwide and is relatively more common among the elderly population. Over the past decade, anemia has emerged as a risk factor with variety of adverse outcomes in aging adults including hospitalization, disability, and mortality. The WHO (world Health Organization) criteria for anemia are hemoglobin level below 13 gm/dL in men or 12 gm/dL in women. Endres et al. reported that among 6880 individuals (2905 men and 3975 women) aged 65−95, the prevalence of mild anemia (hemoglobin 10 mg/dL or higher) was 6.1% among woman and 8.1% among men. The authors further observed that in the typical elderly population without severe comorbidity, mild anemia was associated with greater mortality in men but not women [1]. Anemia is very common among older adults 85 years and older (26.1% men, 20.1% female), but in adults aged 75−84, the prevalence was 15.7% in male and 10.3% in female. In contrast, in adults aged 50−64, prevalence of anemia was 4.4% in male and 6.8% in female [2]. Ania reported that before age 55 years, the prevalence of anemia was lower among men than among women but after that age, anemia became more frequent in men reaching a prevalence of 44.4% among community men, 85 years of age or older [3]. Anemia and iron deficiency are common findings in hospitalized patients admitted to internal medicine service. In one study, 67% of patients had anemia when admitted. However, 80% of these hospitalized patients were 65 years of age or older [4].

Anemia: morphological and etiological classification

At birth, the hemoglobin level is highest and ranges from 16 to 20 gm/dL but then declines with lowest hemoglobin observed at 3−6 months, when values of 9−11 gm/dL are considered normal. Men have slightly higher levels of hemoglobin and this is thought to be due to stimulatory effect of androgens on the bone marrow. WHO definition of hemoglobin below 12 gm/dL in nonpregnant women 15 years of age or older, below 11 gm/dL in pregnant woman, and below 13 mg/dL in men 15 years of age or older is widely used for diagnosis of anemia worldwide. Anemia can be classified on the basis of morphology and on the basis of etiology. Morphological classification of anemia includes the following:

- Normocytic normochromic
- Microcytic hypochromic
- Macrocytic

Hematology and Coagulation. https://doi.org/10.1016/B978-0-12-814964-5.00003-6

Etiologic classifications of anemia include the following:

- Anemia due to blood loss
- Anemia due to deficiency of hematopoietic factors
- Anemia due to bone marrow failure
- Anemia due to increased red cell breakdown

Examples of etiological classification of anemia are listed in Table 3.1.

Common causes of anemia

Causes of normocytic normochromic anemia include the following:

- Anemia due to acute blood loss
- Bone marrow failure and
- Anemia of chronic disease.

Causes of microcytic hypochromic anemia include the following:

- Iron deficiency
- Anemia of chronic disease
- Hemoglobinopathies (with the exception of hemoglobin S [HbS]) and thalassemias
- Sideroblastic anemia and
- Lead poisoning.

Causes of macrocytic anemia include the following:

- Megaloblastic macrocytic anemia due to vitamin B_{12}, folate deficiency. Here the red cells are macrocytic (typically oval macrocytes and with significantly high mean corpuscular volume [MCV], typically above 120 fL), and bone marrow demonstrates megaloblastoid erythroid precursors
- Normoblastic macrocytic anemia related to hypothyroidism, chronic liver disease, alcohol, or pregnancy. Here the red cells are macrocytic and the MCV is typically less than 120 fL. Bone marrow does not demonstrate megaloblastoid erythroid precursors.

Table 3.1 Etiological examples of anemia.

Etiological cause	Specific examples
Anemia due to blood loss	Gastrointestinal blood loss, menorrhagia
Anemia due to deficiency of hematopoietic factors	Iron deficiency, folate deficiency, vitamin B_{12} deficiency, erythropoietin deficiency
Anemia due to bone marrow failure	Aplastic anemia, bone marrow infiltration (e.g., metastatic cancer, bone marrow fibrosis), myelodysplasia, bone marrow toxicity (alcohol abuse, chemotherapy)
Anemia due to increased red cell breakdown	Inherited or acquired defects (please see Table 3.3)

Dimorphic red cells

On occasions, we may see dimorphic red cells. This means that we can identify two distinct red cell populations. Causes of dimorphic red cell population are as follow:

- Red cell transfusion
- Combined deficiency
- Patient who had anemia due to deficiency of iron, folate and B_{12}, and status post replacement therapy
- Pregnancy

Anemia due to blood loss

Blood loss may be acute or chronic. In acute blood loss, replacement of plasma occurs within 24 h, and thus there is hemodilution. Anemia is normocytic normochromic. The maximum fall in hematocrit is usually observed in 3 days, while maximum reticulocytosis is seen in 10 days. There is also transient leukocytosis and thrombocytosis. With chronic blood loss, iron deficiency may occur, which results in microcytic hypochromic anemia.

Iron deficiency anemia

Iron deficiency is a common cause of anemia which may occur due to reduced dietary intake, malabsorption, chronic blood loss, or parasitic infestations. Iron deficiency anemia occurs in 2%—5% of adult men and postmenopausal women in developed countries. Although menstrual blood loss is the most common cause of iron deficiency anemia in premenopausal woman, blood loss from gastrointestinal tract is the most common cause in adult men and postmenopausal woman. Iron deficiency is the most common micronutrient deficiency worldwide with >20% of women experiencing it during their reproductive lives [5]. Asymptomatic colonic and gastric carcinoma may also present with iron deficiency anemia. Celiac disease resulting in malabsorption, use of nonsteroidal antiinflammatory drugs with resultant chronic blood loss, may also cause iron deficiency anemia [6]. Common causes of iron deficiency anemia are listed in Table 3.2.

Table 3.2 Common causes of iron deficiency anemia.

Cause	Commonly encountered conditions	Uncommonly encountered conditions
Gastrointestinal tract (GI) blood loss	Gastric ulceration	Esophagitis
	Gastric carcinoma	Esophageal carcinoma
	Colon cancer	Small bowel tumor
	GI bleeding due to chronic use of aspirin/NSAIDs	Gastric antral vascular ectasia
Non-GI bleeding	Menstruation	Hematuria
	Frequent blood donation	Epistaxis
Malabsorption	Celiac disease	Gut resection
	Gastrectomy	Gastric bypass surgery
	Helicobacter pylori colony formation	Bacterial overgrowth

Iron deficiency results in microcytic hypochromic anemia, and MCV is usually below 80 fL. In iron deficiency, increased anisocytosis is observed, and as a result, red blood cell distribution width (RDW) values are high. Anisocytosis precedes microcytosis and hypochromasia and is significant. This is in contrast to thalassemia where microcytic hypochromic anemia is observed with normal or near normal RDW. Examination of the peripheral blood smear of iron-deficient patients may also show presence of pencil cells and occasional or rare target cells. Patients with iron deficiency anemia may also have thrombocytosis. Iron studies show low serum iron levels, low serum ferritin (<12 ng/mL), and low serum transferrin saturation. Serum transferrin iron-binding capacity is increased. Free erythrocyte protoporphyrin or zinc protoporphyrin and soluble transferrin receptor levels are increased. Castel et al. in their study observed that serum ferritin is the best single-laboratory test for diagnosis of iron deficiency anemia, and a serum ferritin value below 20 ng/mL (20 μg/L) is highly indicative of iron deficiency anemia, while a serum ferritin value over 100 ng/mL (100 μg/L) usually excludes possibility of iron deficiency anemia. However, ferritin values between 20 and 100 ng/mL may be inconclusive and for these patients transferrin/log(ferritin) ratio may be useful for diagnosis of iron deficiency anemia [7]. Ferritin is also an acute-phase reactant, and thus in states of inflammation, values of serum ferritin may be misleading. It is important to note that iron deficiency anemia may mask the presence of beta thalassemia trait. This is because HbA_2 levels are falsely low in the setting of iron deficiency.

Lead poisoning

Lead is a metal associated with human civilization for over 6000 years but is also a toxic metal with its most deleterious effects on the hemopoietic, nervous, and reproductive system as well as the urinary tract. Hyperactivity, anorexia, decreased play activity; low intelligent quotient; and poor performance in school have been observed in children with high lead levels. Lead is capable of crossing placenta during pregnancy and has been associated with intrauterine death, prematurity, and low birth weight. The CDC (Center for Disease Control and Prevention) defined elevated blood level of lead as any value exceeding 10 μg/dL [8]. In lead poisoning, two important enzymes involved of heme synthesis are inhibited. These are

- Delta-aminolevulinic (5-aminolevulinic) acid dehydratase and
- Ferrochelatase.

Inhibition of 5-aminolevulinic acid dehydratase prevents the formation of porphobilinogen, and as a result, accumulation of 5-aminolevulinic acid is observed, which is also observed in acute intermittent porphyria. By also inhibiting mitochondrial ferrochelatase, lead prevents the incorporation of iron into protoporphyrin IX, which then forms a metal chelate with zinc, and as a result, elevated level of zinc protoporphyrin in serum may be seen, indicating lead poisoning. Lead poisoning may result in hypochromic anemia and punctuate basophilia in the peripheral blood with presence of ringed sideroblasts in the bone marrow. The punctate basophilia (basophilic stippling of erythrocytes) occurs due to inhibition of the enzyme 5 pyrimidine nucleotidase by lead. This enzyme is responsible for degradation of the RNA material in reticulocytes.

Anemia of chronic disease

Anemia of chronic disease is seen in the setting of chronic infection, inflammation, or malignancy. This is characterized by

- Low serum iron, reduced transferrin saturation, reduced iron binding capacity, normal or raised serum ferritin.
- It has been hypothesized that cytokines from inflammatory cells cause reduced iron release from reticuloendothelial (RE) cells (this is RE blockade). There is also reduced red cell survival and inadequate erythropoietin response.

Sideroblastic anemia

Sideroblastic anemia is anemia due to defective utilization of iron. It is characterized by the presence of ringed sideroblasts in the bone marrow. The iron is present in the mitochondria. Sideroblastic anemia can be congenital or acquired. There are several forms of congenital sideroblastic anemia:

- X-linked sideroblastic anemia due to mutation of 5-aminolevulinic acid synthase gene (ALAS2) or mutation in the ABC7 gene that encodes a half-type ATP-binding cassette transporter.
- Autosomal recessive sideroblastic anemia.
- Mitochondrial DNA defect (e.g., Pearson syndrome where there occurs sideroblastic anemia, neutropenia, thrombocytopenia, exocrine pancreatic, and hepatic dysfunction).
- Wolfram syndrome also known as DIDMOAD (diabetes insipidus, diabetes mellitus, optic atrophy, and deafness) is also accompanied by sideroblastic or megaloblastic anemia. Patients usually present with diabetes mellitus followed by optic atrophy in the first decade of life followed by cranial diabetes insipidus and sensorineural deafness in the second decade, which later leads to further complications including multiple neurological abnormalities followed by premature death [9]. This syndrome is due to mutation in the WFS1 (Wolfram syndrome 1) gene.

Congenital sideroblastic anemia usually exhibits dimorphic red cells (normocytic normochromic and microcytic hypochromic) or microcytic hypochromic red cells. This is in contrast to acquired sideroblastic anemia, which shows normocytic or macrocytic red cells. Acquired sideroblastic anemia may be primary or secondary. Primary acquired sideroblastic anemia is actually myelodysplasia. Secondary acquired sideroblastic anemia is most often due to drugs (e.g., antituberculous drugs), alcohol, and lead poisoning.

In patients with excessive alcohol intake, early red cell precursors may exhibit vacuoles. This is a characteristic morphologic finding.

Megaloblastic anemia

Megaloblastic anemias are a group or disorder characterized by peripheral blood cytopenia due ineffective hematopoiesis in the bone marrow. Most common cause is folate or vitamin B_{12} deficiency or both. Vitamin B_{12} or folate deficiency may result due to poor nutrition, malabsorption, or drugs (e.g., methotrexate, hydroxyurea). Pernicious anemia is an important cause of vitamin B_{12} deficiency.

Pernicious anemia is due to autoimmune destruction of the parietal cells. The parietal cells normally secrete intrinsic factor, required for the absorption of vitamin B_{12}. Antiparietal cell antibody and antiintrinsic factor antibodies are found in patients with pernicious anemia. In megaloblastic anemia, macrocytic red cells are observed in the peripheral blood, and these are classically oval macrocytes. Hypersegmented polymorphonuclear leukocytes may be seen. Megaloblastic anemia is a cause of pancytopenia. The bone marrow shows erythroid hyperplasia with large erythroid precursors. This is known as megaloblastoid changes. The myeloid precursors may also be large. Giant myelocytes and giant metamyelocytes may also be observed. Nuclear cytoplasmic dyssynchrony may also be seen. Overt features of dysplasia are seen in megaloblastic anemia. As such, this condition remains a differential diagnosis of myelodysplasia. Both conditions are associated with peripheral blood cytopenias, bone marrow hyperplasia, and dysplasia.

Bone marrow failure

Bone marrow failure may be due to bone marrow aplasia or related to bone marrow infiltration (myelophthisic anemia). Aplastic anemia is an example of potentially fatal bone marrow failure—related anemia. Aplastic anemia may be constitutional (inherited) or acquired (Table 3.3). Patients with aplastic anemia present with pancytopenia and reticulocytopenia. The bone marrow cellularity is markedly decreased. In the peripheral blood and bone marrow, the lymphocytes and plasma cells are intact.

Fanconi's anemia is an autosomal recessive disease characterized by bone marrow failure and represents 2/3 of all constitutional aplastic anemia. This disorder is due to DNA repair defect, and cells derived from patients with this disorder display hypersensitivity to DNA cross-linking agents resulting in increased numbers of chromosomal abnormalities including translocations and radial chromosomes. Aplasia develops by midchildhood and may progress to acute myelogenous leukemia (AML). These patients are also susceptible to developmental abnormalities (short stature, skeletal defects, short stature, etc.), renal defects, mental retardation, and hypogonadism. These patients are also have markedly increased risk of squamous cell carcinoma of head and neck and genitourinary tract [10].

Dyskeratosis congenita (also known as Zinsser-Engman-Cole syndrome) is a rare progressive bone marrow failure syndrome characterized by skin hyperpigmentation, nail dystrophy, oral

Table 3.3 Causes of aplastic anemia.

Type of aplastic anemia	Causes
Constitutional (inherited)	• Fanconi's anemia • Dyskeratosis congenita • Shwachman—Diamond syndrome • Diamond Blackfan anemia (constitutional red cell disorder)
Acquired	• Idiopathic, drugs, toxins, radiation, infections, paroxysmal nocturnal hemoglobinuria • Transient erythroblastopenia of childhood (acquired red cell disorder), human parvovirus infection that cause acquired red cell aplasia.

leukoplakia, and aplastic anemia. Chromosomal pattern is normal in this disorder, but there are evidences for telomerase dysfunction, ribosome deficiency, and protein synthesis disorder in patients with Dyskeratosis congenita. Early mortality in these patients may be related to bone marrow failure, infections, malignancy, or fatal pulmonary complications.

Shwachman—Diamond syndrome is a rare genetic disorder typically characterized by exocrine pancreatic dysfunction with malabsorption, malnutrition, growth failure, hematological abnormalities with cytopenia, susceptibility to myelodysplasia syndrome, and AML. In almost all affected children, persistent or intermittent neutropenia is seen.

Diamond Blackfan anemia is a disease with pure congenital red cell aplasia (inherited erythroblastopenia). Usually white blood cell and platelet counts are normal in these patients (different from Shwachman—Diamond syndrome where neutropenia is common). Skeletal defects are often seen in these patients, but renal or chromosomal abnormalities are not observed.

Acquired bone marrow failure causing aplastic anemia could be drug induced. Drugs such as chemotherapeutic agents and benzene cause dose-dependant bone marrow failure. Drugs such as chloramphenicol cause dose-independent, idiosyncratic bone marrow failure. However, newer drugs may also cause aplastic anemia. One female patient with glioblastoma multiforme, who was treated with the programmed cell death-1 receptor inhibitor nivolumab, developed aplastic anemia [11]. Ionizing radiation can cause bone marrow failure as an early complication. Late complications of such radiation exposure include myelofibrosis and leukemia.

Viral hepatitis is a rare but recognized cause of aplastic anemia. It typically occurs 6 weeks after onset of symptoms. This complication is not related to severity of hepatitis. Parvovirus B19—induced aplastic crisis tends to occur in individuals who have a background of chronic hemolysis. The red cells are selectively involved. Giant erythroblasts are seen. Nuclear inclusions may be seen. Immunostains are available for confirmation. The condition is self-limiting.

Aplastic anemia is characterized by pancytopenia with low corrected reticulocyte count. Anemia results in weakness and fatigue. Leukopenia predisposes to infections, and thrombocytopenia results in bleeding.

Other causes of pancytopenia include acute leukemia, myelophthisic anemia, megaloblastic anemia, myelodysplastic syndrome, and hypersplenism.

Severe aplastic anemia is characterized by any three of the following four for 2 weeks:

- <40,000 reticulocytes/μL
- <10,000 platelets/μL
- <500 granulocytes/μL
- >80% of bone marrow comprises of nonmyeloid elements

Congenital dyserythropoietic anemia

This disorder is characterized by anemia at an early age with multinucleation of erythroid precursors and internuclear bridging between erythroid precursors in the bone marrow. There are three types (I, II, and III) of which type II, transmitted as autosomal recessive, is most common. Ham's acidified serum test (checks fragility of erythrocytes when placed in mild acid), which is usually used for diagnosis of paroxysmal nocturnal hemoglobinuria (PNH), is also positive in patients with congenital dyserythropoietic anemia (CDA), type II. However, Ham's acidified serum test used for CDA does not use the patient's serum.

Hemolytic anemia

Hemolysis is the destruction or removal of red blood cell (RBC) from circulation before their normal life span of 120 days. Although hemolysis can be an asymptomatic condition, commonly patients presents with anemia when erythrocytosis cannot overcome the pace of erythrocyte destruction. Patients may complain of dyspnea or fatigue due to anemia and may also present with dark urine and jaundice [12]. Hemolytic anemia is clinically characterized anemia, jaundice, increased reticulocyte count, splenomegaly, leg ulcers, increased serum concentration of lactate dehydrogenase, and decreased serum haptoglobin if intravascular hemolysis is present. Hemolytic anemia may be due to corpuscular defect or extracorpuscular defect (Table 3.4).

Hemolytic anemia due to corpuscular defects

This category includes hemolytic anemias due to membrane defects, enzyme defects, hemoglobin-opathies, thalassemias, and PNH. All hemolytic anemias in this category are inherited with the exception of PNH. Hemoglobinopathies and thalassemias are discussed in Chapter 4.

Hemolytic anemia due to membrane defects

Various hemolytic anemias are related to erythrocyte membrane defects. These anemias are summarized in this section. Da Costa et al. reviewed anemia due to red cell membrane disorders in a review article [13].

Hereditary spherocytosis (aka Minkowski-Chauffard disease)

Spectrin is the major component of normal RBC cytoskeleton and consists of two intercoiled nonidentical filamentous subunits that form heterodimers. The chain heads of each dimer pair bind with opposite subunit heads of another dimer pair to form tetramers. The tail of spectrin tetramers bind with a protein cluster of short actin protofilaments, and this interaction is markedly enhanced by protein 4.1. Spectrin—actin—protein 4.1 assembly is secured to overlying lipid bilayer by ankyrin.

Table 3.4 Underlying causes of hemolytic anemia.

Underlying cause	Specific defect
Corpuscular defects (all inherited, except paroxysmal nocturnal hemoglobinuria [PNH])	• Membrane defects (e.g., hereditary spherocytosis) • Enzyme defects (e.g., glucose-6-phosphate dehydrogenase deficiency, pyruvate kinase deficiency) • Hemoglobinopathies and thalassemias • PNH
Extracorpuscular defects (acquired)	• Immune mediated (hemolytic disease of newborn, mismatch transfusion, autoimmune hemolytic anemia) • Nonimmune mediated (infections, march, karate, and microangiopathic hemolysis)

Hereditary spherocytosis is the most common of the hereditary hemolytic anemias among people of Northern European descent. In the United States, the incidence is approximately 1 per 5000 people. It is transmitted as autosomal dominant, but in 25% cases, it is due to a spontaneous mutation. Four abnormalities in RBC membrane proteins have been identified in this disorder including spectrin deficiency alone, combined spectrin and ankyrin deficiency, band 3 deficiency, and protein 4.1 defects. However, spectrin deficiency is the most common cause of this disorder.

Spectrin deficiency leads to loss of erythrocyte surface area, which produces spherical RBCs. Spherocytic RBCs are rapidly removed from the circulation by the spleen. Patients with hereditary spherocytosis develop splenomegaly. Spectrin deficiency is due to impaired synthesis. In some cases, there are quantitative or qualitative deficiencies of other proteins that integrate spectrin into the cell membrane. In the absence of these binding proteins, free spectrin is degraded, leading to spectrin deficiency. The degree of spectrin deficiency is reported to correlate with the extent of spherocytosis, the degree of abnormality on osmotic fragility test results, and the severity of hemolysis. Most cases of hereditary spherocytosis are heterozygous as homozygous states are lethal.

Peripheral smear from these patients will demonstrate spherocytes with polychromasia (increased reticulocytes). Spherocytes are also seen in autoimmune hemolytic anemia. It is logical to perform a DAT (direct antiglobulin test) to rule out autoimmune hemolytic anemia. The confirmatory test for hereditary spherocytosis is osmotic fragility test. The principle of this test is that spherocytes will easily lyse when placed in hypotonic solution. If the standard osmotic fragility test is negative, we may also perform incubated osmotic fragility tests to pick up more subtle cases. A normal osmotic fragility test does not necessarily exclude the diagnosis of hereditary spherocytosis. Such normal tests may be seen in 10%−20% of cases. The test may be normal in the presence of iron deficiency, obstructive jaundice, and increased reticulocytes. Patients with hereditary elliptocytosis and hemolysis may yield a false positive result.

Other initial tests that may be ordered are glycerol lysis and Pink test. The sensitivity of these tests including osmotic fragility test is low. A relatively newer test is eosin-5-maleimide (EMA) binding by flow cytometry. This test has a superior sensitivity of 92.7% and specificity of 99.1%. This test measures mean red cell fluorescence after the red cells are labeled with the dye EMA, to document the loss of surface area. EMA binds covalently to the Lys430 on the first extracellular loop of the band 3 protein. In addition it also binds with the sulfhydryl groups expressed by Rh, RhAg, and CD47. The test is able to detect defects in band 3, spectrin, and 4.2, but it is less effective in picking up defects with ankyrin. This test is also positive in hereditary pyropoikilocytosis.

Hereditary elliptocytosis

This disorder is characterized by >20% elliptocytes in the peripheral blood. The condition is transmitted as autosomal dominant. Clinical manifestations range from an asymptomatic carrier state to severe hemolytic anemia. Members of the same family may exhibit different clinical courses, and an individual's frequency and severity of hemolysis may change with time. Most patients with hereditary elliptocytosis or its variants lead healthy lives. There are three major categories:

- Common form: There are six different subtypes of the common form. Hereditary pyropoikilocytosis is one of the subtypes. In this subtype, the patient has features of pyropoikilocytosis for the first 2 years of life and then evolves in to hereditary elliptocytosis.

- Spherocytic form: This form is thought to be a hybrid form of hereditary elliptocytosis and hereditary spherocytosis.
- Stomatocytic form: This form is usually seen in Melanesians. These individuals have red cells described as shoe buckle red cells.

Hereditary elliptocytosis and its related disorders are caused by mutations that disrupt the RBC cytoskeleton; most frequent is defective spectrin dimer—dimer interaction due to an abnormal chain of spectrin. Quantitative deficiencies of protein 4.1 are associated with spherocytic form of hereditary elliptocytosis.

Hereditary pyropoikilocytosis

This is a severe form of congenital hemolytic anemia, and it is clinically similar to and now considered a subtype of homozygous hereditary elliptocytosis. RBCs show increased susceptibility to thermal injury. Mode of inheritance and exact defects are not fully understood, and similar to hereditary elliptocytosis, a defective spectrin dimer—dimer interaction due to an abnormal alpha chain of spectrin is speculated for this disorder.

The peripheral smear is striking and pathognomonic and shows bizarre forms, anisocytosis, fragments, micropoikilocytosis, microspherocytes, and budding red cells. The MCV may be as low as 25—55 fL.

Hereditary stomatocytosis

This disease is due to abnormal RBC membrane cation permeability and consequent defective cellular hydration. It is transmitted as autosomal dominant. Hereditary stomatocytosis represents a spectrum of disorders with variable phenotypic expression. At one end of the spectrum is hydrocytosis, which results when erythrocytes are swollen with water, and at the other end of the spectrum is xerocytosis, which results when cells are dehydrated. Active Na, K, and ATPase pumps are incapable of counterbalancing flow of cations across the cell membrane. The net influx of sodium brings with it water, resulting in swollen cell, that is susceptible to osmotic and mechanical lysis. The hallmark is a stomatocyte, a red cell with slit like or "fish mouth" area of central pallor. Stomatocytes, >35% and evidence of hemolysis are required to establish the diagnosis.

Rh null disease

This is when an individual lacks all Rh determinants. Stomatocytes and spherocytes are found in the peripheral smear. It is clinically characterized by a mild compensated hemolytic anemia, which requires no therapy.

Hemolytic anemia due to enzyme defects

Hemolytic anemia due to enzyme defects may be related to deficiencies of enzymes in glycolytic pathway or defective enzyme in HMP shunt (hexose monophosphate shunt or pentose phosphate pathway). The mature RBC is totally dependent on glucose as a source of energy. Glucose usually (90%) is catabolized to pyruvate and lactate in the major anaerobic glycolytic pathway

(Embden–Myerhof pathway). As mitochondria are absent in RBC, it depends entirely on anaerobic glycolysis to provide for energy. The glycolytic pathway generates ATP (adenosine triphosphate) required for the functioning of Na,/ATPase pump. The proper function of Na/K-ATPase pump ensures that sodium and water are constantly pumped out of the red cell, allowing it to maintain its biconcave shape.

The pentose phosphate pathway produces NADPH (nicotinamide adenine diphosphate) which maintains the reduced state of glutathione. If glutathione is not its reduced state, disulfide bond (S-S) formation takes place with subsequent denaturation and precipitation of globin as Heinz bodies. Heinz bodies are removed by the RE cells as the red cells pass through the spleen. The damaged red cells have the appearance of bite or blister cells. Heinz bodies are not seen in the peripheral smear with routine Wright–Giemsa stain. However, the bite cells are apparent in the peripheral smear. Presence of bite cells should be a clue to order special stains for Heinz bodies and evaluate for enzymes of the pentose phosphate pathway. Hereditary deficiency of some of the glycolytic enzymes has been documented, and several cause a hereditary nonspherocytic hemolytic anemia (HNSHA), whereas others cause multisystem disease. Most are rare, and pyruvate kinase deficiency is the most common and comprises 90% of affected patients.

Pyruvate kinase deficiency

Pyruvate kinase deficiency is transmitted as autosomal recessive. The condition is detected in infancy or childhood due to anemia, jaundice, splenomegaly, and gall stones. The severity of the condition is widely variable, even among patients with the same level of deficiency. Fluorescent screening tests are available for pyruvate kinase deficiency. Diagnosis is confirmed by enzyme assay and, more recently, DNA analysis by polymerase chain reaction or single-strand conformation polymorphism is available to confirm the diagnosis and to identify the carrier state if the need arises.

Glucose-6-phosphate dehydrogenase deficiency

Glucose 6-phosphate dehydrogenase (G6PD) deficiency is the most common enzyme deficiency in the pentose phosphate pathway affecting over 400 million people worldwide. The gene for G6PD enzyme is located on X chromosome, and 200 different mutations have been reported, which results in a variety of clinical conditions. Clinical manifestation of G6PD deficiency include neonatal jaundice and acute hemolytic anemia, arising from oxidative stress on RBC induced by some medications, an infection, or ingestion of fava beans. However, many individuals with this disorder remain asymptomatic throughout their life and may not be even aware of it. Drug-induced hemolytic anemia in patients with G6PD deficiency was first described with antimalaria drug primaquine, but many other drugs are capable of causing hemolytic anemia in these patients [14]. The highest prevalence of this disorder is in sub-Saharan Africa, followed by North–Africa and Middle East. Interestingly, this disorder protects an individual from infection by malaria parasite.

Normally, two isotypes of G6PD A and B can be differentiated based on electrophoretic mobility, and B isoform is the most common type of enzyme found in all population groups. However, A isoform, found in 20% black men in United States, migrates more rapidly on electrophoretic gels than B and has similar enzyme activity as B. About 11% of US black men have G6PD variant (G6PD A⁻). It has same electrophoretic mobility as A but is unstable, resulting in enzyme loss and ultimate enzyme

Table 3.5 Classification of glucose-6-phosphate dehydrogenase (G6PD) deficiency.

Class	Residual enzyme activity	Clinical manifestation	Mutation type
I	1% or less	Chronic hemolytic anemia, very rare condition. Total absence of G6PD deficiency which is very unusual is incompatible with life.	G6PD Buenos Aires G6PD Durham
II	<10%	Severe enzyme deficiency causing acute hemolytic anemia induced by certain drugs or ingestion of kava beans.	G6PD Mediterranean G6PD Santamaria
III	10%−60%	Moderate to mild enzyme deficiency that may cause occasional acute hemolytic anemia	G6PD A⁻ G6PD Canton
IV	60%−90%	Very mild to moderate enzyme deficiency. These individuals are asymptomatic. This occurs rarely.	G6PD Orissa G6PD Montalbano
V	>110%	Increased enzyme activity, individuals are asymptomatic, Also rarely observed.	Not known

deficiency. Older RBC have only 5%−15% enzyme, and G6PD A⁻ is the most clinically significant type of abnormal G6PD among US blacks. Various G6PD variants predominate in other racial groups: G6PD⁻Mediterranean in Sicilians, Greeks, Sephardi Jews, and Arabs, G6PD Canton Asian population, etc. Five classes of G6PD deficiency exist based on enzyme activity levels as recommended by the WHO (Table 3.5).

The two most common mutations involved with G6PD deficiency are G6PD A⁻ and G6PD-Mediterrannean [15]. In G6PD A⁻ deficiency, the hemolytic anemia is self-limited as the young RBC produced in response to hemolysis have nearly normal enzyme and are less affected by favism. In contrast, favism is associated with G6PD-Mediterrannean. Patients with G6PD develop acute hemolysis with formation of Heinz bodies (denatured globin) and bite cells (eccentrocytes). The Heinz bodies may be demonstrated by special stains such crystal violet. Further testing includes fluorescent screening tests for G6PD deficiency and G6PD enzyme activity assay [16]. Koralkova et al. reviewed rare hereditary RBC enzymopathies associated with hemolytic anemia [17].

Paroxysmal nocturnal hemoglobinuria

This disorder is the only acquired disorder amongst the intrinsic red cell disorders causing hemolytic anemia due to spontaneous mutation in a pluripotent stem cell of the PIG-A (phosphatidylinositol glycan anchor biosynthesis, class A) gene. The PIG-A gene encodes for glycophosphatidylinositol-anchored protein (GPI). GPI functions as cell membrane "anchor" and serves as attachment for about 20 cell surface proteins. Amongst the proteins that are thus lacking on the cell surface are cell surface proteins that regulate complement, and these include

- CD55/decay-accelerating factor (DAF): CD55 inhibits association of C4b and C2 and promotes dissociation of C4bC2a (C3 convertase) complex.
- Homologous restriction factor (HRF)
- CD59/membrane inhibitor of reactive lysis (MIRL): CD59 (MIRL) and HRF prevent formation of membrane attack complex and cell lysis.

PNH is thus a clonal disorder and affects all hematopoietic cells where there is increased susceptibility to complement mediated lysis. Nucleated hematopoietic cells are able to endocytose membrane attack complex that causes cell lysis and this allows them to resist complement mediated lysis. Nonnucleated cells such as red cells are unable to resist lysis. PNH is clinically characterized by

- Hemolytic anemia: This may take the form of chronic intravascular hemolysis. There may be paroxysmal episodes of acute hemolysis. It is postulated that sleep causes relative acidosis, and under acidotic conditions, there occurs red cell lysis. This result in hemoglobinuria, and when the patient wakes up in the morning and passes urine, the urine is dark due to hemoglobinuria (thus the name). The urine will clear during the day.
- Thrombosis: Complement activity creates a prothrombotic state, and thrombosis is a major cause of morbidity/mortality. Clinically significant thrombosis is seen in one-third of pateints. The thrombosis is most often venous and includes Budd−Chiari syndrome (hepatic vein thrombosis).
- Cytopenias/Bone marrow failure: Patient may develop neutropenia, thrombocytopenia, and aplastic anemia. There is increased incidence of transformation to AML/myelodysplastic syndrome (MDS).

Screening tests for PNH is sucrose lysis test, where patient's red cells are added to a solution of isotonic sucrose and serum. Sucrose aggregates globin onto red cells, and serum is a source of complement. The test is positive if >5% of red cells are lysed. Lysis makes the supernate red. Confirmatory tests for PNH include Ham's acidified serum test and flow cytometry. Characteristics of Ham's acidified serum test include the following:

- The basis of this test is that under acidified condition, patient's own complement destroys patient's red cells. Patient red cells are taken in three test tubes to which is added patient's serum. In two tubes, acid is added. One of these two tubes is heated to 56° C to destroy complement. Thus there is one tube, where there is no acid but has patient's red cells and serum. In another we have patient's red cells and acidified serum, but no complement. The third tube has all three component required for hemolysis. All are incubated for an hour. Hemolysis present only in the third, but not in the first two constitutes a positive test.
- Positive test >1% lysis in acidified serum (acid and complement both present)
- This test may also be positive in CDA, type II

Ham's acidified serum test is very sensitive but may not be specific. In flow cytometry, expression of GPI-anchored proteins CD55 and CD59 are analyzed on hematopoietic cells using monoclonal antibodies. This test is highly specific because there is no other condition where RBCs are a mosaic of normal and GPI-linked protein deficient cells.

Hemolytic anemias due to extracorpuscular defects

Extracorpuscular causes of hemolytic anemia can be divided into immune and nonimmune causes. Examples of nonimmune causes are malaria and red cell destruction due to trauma (e.g., march or karate hemoglobinuria) or very strenuous exercise. Immune hemolytic anemia is due to increased destruction of RBCs by antibody against antigens on RBCs. Examples of immune hemolytic anemias are

- Hemolytic disease of newborn
- Hemolytic transfusion reactions

- Autoimmune hemolytic anemias—patient makes antibodies to antigens on own RBCs. Autoimmune hemolytic anemia are divided into warm autoimmune hemolytic anemia (WAHA), cold hemagglutinin disease (CHAD), and paroxysmal cold hemoglobinuria (PCH)
- Drug-induced immune hemolytic anemia caused by antibody directed against a red cell membrane—drug complex (e.g., penicillin), immune complex deposition on red cell surface (e.g., quinine, rifampin), or true autoimmune hemolytic anemia (e.g., due to methyldopa)

In WAHA, IgG antibodies are formed against patient's own red cell antigens. These antibodies are often formed against the broad Rh antigens. Hemolysis is classically extravascular. The IgG coated red cells are destroyed by splenic macrophages. Splenomegaly is thus a feature. In CHAD, IgM antibodies are formed against red cell antigens. Agglutination of red cells can occur at low temperatures, and complement activation may result in intravascular hemolysis. Antibodies are often directed against I antigens. There are several subtypes of this condition:

- Acute postinfectious: acute, self-limited, younger patients
- Chronic idiopathic: insidious, older patients
- Cold agglutinin disease: Insidious, elderly women, associated with lymphoproliferative malignancy

In PCH, antibodies are typically formed against P antigens. The antibody binds to red cells at low temperature and, when warmed, activates complement causing hemolysis. These antibodies are also known as biphasic antibodies or Donath-Landsteiner antibodies. PCH is a rare condition and sometimes seen in children following viral infections with sudden onset of hemolysis.

Causes of WAHA and CHAD may be idiopathic, drug induced (more often with WAHA), related to infection (more common with CHAD), or lymphoproliferative disorders. Laboratory findings in WAHA include positive DAT to IgG or IgG and C3 and presence of spherocytes in peripheral smear. Osmotic fragility test may be positive. In CHAD, DAT test is only positive for C3. Red cell agglutination on peripheral smear may also be observed.

Red cell poikilocytosis

Variation in shape of red cells refers to poikilocytosis. Examples of poikilocytosis are

- Sickle cells (drepanocytes): Under low oxygen tension six to eight HbS molecules condense to form a tubular structure. This is called a tactoid. This distorts the red cell to form boat-shaped cells or sickle-shaped cells. Sickle cells cause occlusion of small vessels and can cause infractions. Circulating red cells are effectively decreased resulting in impaired oxygen delivery to tissue. When oxygen tension is improved, sickle cells resume the shape of normal red cells. Repeated sickling and unsickling makes the red cell membrane fragile and results in hemolysis.
- Target cells (codocytes): These are formed as a consequence of the presence of redundant membrane in relation to volume of cytoplasm. Excess cell membrane lipid occurs in cholestatic liver disease and LCAT (lecithin–cholesterol acyltransferase) deficiency resulting in target cells. Target cell formation due to reduction of cytoplasm is seen in thalassemia and hemoglobinopathies.
- Ovalocyte: This is when a red cell's long axis <twice compared with its short axis. Ovalocytes especially macroovalocytes are seen in folate and vitamin B_{12} deficiency.

- Elliptocyte: This is when a red cell's long axis >twice its short axis. Significant elliptocytosis (>20%) can be seen in hereditary elliptocytosis. Other causes include thalassemia, HbS trait, and hemoglobin C (HbC) trait. Causes of lesser degree of elliptocytosis include cirrhosis, iron deficiency anemia, megaloblastic anemia, and myelophthisic anemia.
- Stomatocyte: This is when red cells have a slit-like area of pallor. These cells may be seen in hereditary stomatocytosis. Small numbers of stomatocytes may be seen in acute alcohol intake, cirrhosis, obstructive liver disease, and Rh null disease.
- Echinocyte or burr cell: These are red cells with 10−30 short blunt spicules, and these cells may be observed in storage artifact, liver and kidney disease, or pyruvate kinase deficiency.
- Acanthocytes: The term acanthocyte is derived from the Greek *acantha,* meaning thorn. Acanthocytes are dense contracted RBC with multiple thorny projections that vary in width, length, and surface distribution. This is presumably due to increase in relative sphingomyelin content of RBC membrane, resulting in a rigid wall. Acanthocytes must be distinguished from echinocytes (Greek *echinos,* urchin). Echinocytes, also known as burr cells, have multiple small projections that are distributed uniformly on the red cell surface. Acanthocyte are red cells with 2−20 unequal, irregular spicules. Occasional acanthocytes may be seen in postsplenectomy, hemolytic anemia due to pyruvate kinase deficiency, microangiopathic hemolytic anemia, autoimmune hemolytic anemia, renal disease, thalassemias, and with McLeod phenotype. If the majority of RBCs are acanthocytes, then this may be due to abetalipoproteinemia.

Abetalipoproteinemia is where there is absence of apolipoprotein B resulting in an inability to transport triglycerides in the blood. Clinically, this presents in infancy with steatorrhea, progressive development of acanthocytosis, ataxic neuropathy, and an atypical form of retinitis pigmentosa.

Acanthocytes are also found with McLeod blood group, an X-linked disorder in which red cell Kx antigen and precursor substance for Kell blood group system are absent. The McLeod phenotype may be associated with chronic granulomatous disease because of the proximity of the genetic loci for these two disorders.

- Keratocytes: Red cells with pair/s of spicules. Seen in microangiopathic hemolytic anemia and renal disease.
- Schistocytes: These are fragmented red cells. Increased numbers suggest microangiopathic hemolytic anemia. True schistocytes do not have central pallor.
- Spherocyte: These are red cells without central pallor. Such cells are observed in autoimmune hemolytic anemia and hereditary spherocytosis.
- Dacryocyte (tear drop red cell): If present in significant numbers may imply conditions with bone marrow infiltration. Other causes include megaloblastic anemia, hemolytic anemias, and hypersplenism.
- Bite cells: These are seen in conditions with Heinz body formation, for example, G6PD deficiency. They are usually accompanied by blister cells. They are associated with oxidative stress to red cells. Bite cells can also be seen in normal individuals receiving large amount of aromatic drugs (or their metabolites).
- Blister cells: These are RBCs with vacuoles or with very thin areas at the periphery of the cell. Causes are similar to bite cells.

Red cell inclusions

Several red cell inclusions have been described. These include the following:

- Howell—Jolly bodies: These are usually single peripheral bodies within red cells representing DNA material. These bodies may be seen in postsplenectomy, megaloblastic anemia, severe hemolysis, and myelophthisic anemia.
- Pappenheimer bodies: These are smaller than Howell—Jolly bodies and are multiple often stacked like cannon balls. They represent iron material. These are found within the mitochondria.
- Cabot rings: These are mitotic spindle remnants seen as ring-shaped or as figure of eight inclusions.
- Basophilic stippling: These represent RNA material that could be fine or physiological or coarse, but all are pathological in nature. Causes include lead poisoning, hemolytic anemia, and pyrimidine 5 nucleotidase deficiency. Pyrimidine 5 nucleotidase is the enzyme responsible for degradation of the RNA material.
- Heinz bodies: These are denatured globin. They require supravital staining for their demonstration. They can be seen in G6PD deficiency. Heinz bodies are cleared by splenic macrophages. The damaged cells are known as bite cells.
- Hb C crystals: These are an in vitro phenomenon. These crystals have been described as similar to the Washington Monument. As in the monument, the crystals have a conical tip. These are seen in individuals with HbC hemoglobinopathy.
- Malarial parasite
- Nucleated RBC

Malaria parasites

Malaria is an infectious disorder of RBCs, caused by *Plasmodium* infection, and as expected, anemia is a common manifestation of this disease that is responsible for substantial morbidity and mortality of infected individuals. There are four species of malarial parasite: *Plasmodium falciparum, Plasmodium vivax, Plasmodium ovale,* and *Plasmodium malariae.* Acute falciparum malaria results in increased removal from circulation of parasitized and to a greater extent nonparasitized RBC through a combination of splenic filtration, schizont rupture, macrophage phagocytosis, complement-mediated hemolysis, and increased free radical damage. Outside of Africa, *P. falciparum* invariably coexist with other plasmodium species, the most important being *P. vivax* [18]. It is important to differentiate *P. falciparum* from others as *P. falciparum* infection is very serious and it may cause death within a short period of time. Although prevalence of malaria is very low in the United States and other developed countries, malaria deaths in industrialized countries are preponderantly due to the fact that physicians fail to obtain from their patients an appropriate travel history and, even when it is elicited, they fail to consider that malaria remains the most frequent cause of fever among travelers returning from the tropics and that *P. falciparum* malaria is a medical emergency [19]. Red cell exchange is one the therapeutic modalities for *P. falciparum* infection. The clinicians may also want to know the parasite load (this is percentage of red cells affected). Diagnostic points for identification of various malaria parasites are summarized in Table 3.6.

Table 3.6 Diagnostic points for malaria parasite identification.

Plasmodium falciparum	Plasmodium vivax	Plasmodium malariae	Plasmodium ovale
• Red cells not enlarged • Rings appear fine/delicate, several per cell • Some rings with two chromatin dots • Presence of marginal/applique forms • Unusual to see developing forms • Crescent-shaped gametocytes • Maurer's dots may be present	• Red cells containing parasites usually enlarged • Schüffner's dots frequently present • Mature ring forms large and coarse • Developing forms frequently present	• Ring forms have squarish appearance • Band forms characteristic • Mature schizonts may have typical "daisy head" appearance with <10 merozoites • Red cells not enlarged • Chromatin dot may be on inner surface of the ring	• Red cells enlarged • Comet forms common • Rings large and coarse • Schüffner's dots prominent when present • Mature schizonts similar to *P. malariae*, but larger and coarser

Key points

- Causes of normocytic normochromic anemia: Anemia due to blood loss (acute), bone marrow failure, and anemia of chronic disease
- Causes of microcytic hypochromic anemia: Iron deficiency, anemia of chronic disease, hemoglobinopathies, thalassemia, sideroblastic anemia, and lead poisoning
- Causes of macrocytic anemia: Megaloblastic macrocytic, vitamin B_{12}, and/or folate deficiency; normoblastic macrocytic: hypothyroidism, chronic liver disease, alcohol, and pregnancy
- Iron deficiency results in microcytic hypochromic anemia with anisocytosis. Anisocytosis precedes microcytosis and hypochromasia. This is in contrast to thalassemia where microcytic hypochromic anemia with normal or near normal RDW is observed. Examination of the peripheral smear may also show presence of pencil cells and occasional or rare target cells. Patients with iron deficiency anemia may also have thrombocytosis. Iron studies show low serum iron levels, low serum ferritin, and low serum transferrin saturation. Serum transferrin iron-binding capacity is increased. Free erythrocyte protoporphyrin or zinc protoporphyrin and soluble transferrin receptor levels are increased. It is important to note that iron deficiency anemia may mask the presence of beta thalassemia trait. This is because HbA2 levels are falsely low in the setting of iron deficiency.
- Sideroblastic anemia is characterized by the presence of ringed sideroblasts in the bone marrow. The iron is present in the mitochondria. Sideroblastic anemia can be congenital or acquired.
- In lead poisoning, two important enzymes of heme synthesis are inhibited. These are delta aminolevulinic acid dehydratase and ferrochelatase. The punctate basophilia occurs due to inhibition of the enzyme 5 pyrimidine nucleotidase by lead.
- Megaloblastic anemia is due to folate or vitamin B_{12} deficiency. In the peripheral blood, macrocytic red cells are seen, and these are classically oval macrocytes. Hypersegmented polymorphonuclear neutrophils (PMNs) may be seen. Megaloblastic anemia is a cause of pancytopenia. The bone marrow shows erythroid hyperplasia with large erythroid precursors.

This is known as megaloblastoid changes. The myeloid precursors may also be large. Giant myelocytes and giant metamyelocytes may be seen. Nuclear cytoplasmic dyssynchrony may also be seen.

- Aplastic anemia: The causes of aplastic anemia include constitutional (e.g., Fanconi's anemia, dyskeratosis congenita, Shwachman–Diamond syndrome, Diamond Blackfan anemia) and acquired: idiopathic, drugs, toxins, radiation, infections, PNH, transient erythroblastopenia of childhood (acquired red cell disorder), and human parvovirus infection, which cause acquired red cell aplasia.
- CDA is characterized by anemia at an early age with multinucleation of erythroid precursors and internuclear bridging between erythroid precursors in the bone marrow.
- Classification of hemolytic anemia: Corpuscular defects: membrane defects, enzyme defects, hemoglobinopathies and thalassemia, and PNH and extracorpuscular defects: this could be nonimmune or immune in nature.
- Hereditary spherocytosis is the most common of the hereditary hemolytic anemias among people of Northern European descent. It is transmitted as autosomal dominant; however, in 25% cases, it is due to a spontaneous mutation. Spectrin deficiency is the most common defect. One of the confirmatory tests for hereditary spherocytosis is osmotic fragility test.
- Hereditary stomatocytosis is due abnormal RBC membrane cation permeability and consequent defective cellular hydration. It is transmitted as autosomal dominant.
- Hereditary elliptocytosis is characterized by >20% elliptocytes in the peripheral blood. The condition is transmitted as autosomal dominant. Clinical manifestations range from an asymptomatic carrier state to severe hemolytic anemia.
- Hereditary pyropoikilocytosis is a severe form of congenital hemolytic anemia, and it is clinically similar to and now considered a subtype of homozygous hereditary elliptocytosis. RBCs show increased susceptibility to thermal injury.
- Hemolytic anemia due to enzyme defects may be due to deficiencies of glycolytic pathway or disorders of HMP shunt (pentose phosphate pathway)
- Pyruvate kinase deficiency is transmitted as autosomal recessive and is the commonest enzyme deficiency of the glycolytic pathway.
- G6PD deficiency is the most common enzyme deficiency in the pentose phosphate pathway. Normally two isotypes of G6PD A and B can be differentiated based on electrophoretic mobility. B isoform is the most common type of enzyme found in all population groups. A isoform, found in 20% black men in United States, migrates more rapidly on electrophoretic gels than B. It has similar enzyme activity as B. About 11% of US black men have G6PD variant (G6PD A⁻). It has same electrophoretic mobility as A but is unstable, resulting in enzyme loss and ultimate enzyme deficiency. Older RBC have only 5%–15% enzyme. G6PD A⁻ is the most clinically significant type of abnormal G6PD among US blacks.
- PNH is the only acquired disorder amongst the intrinsic red cell disorders causing hemolytic anemia. There occurs spontaneous mutation in a pluripotent stem cell of the PIGA gene. The PIGA gene encodes for GPI protein. GPI functions as cell membrane "anchor" and serves as attachment for about 20 cell surface proteins. Amongst the proteins that are thus lacking on the cell surface are cell surface proteins that regulate complement, and these include CD55/DAF, HRF, and CD59. PNH is clinically characterized by hemolytic anemia, thrombosis, and cytopenias/bone marrow failure. There is increased incidence of transformation to AML/MDS.

Screening tests for PNH is sucrose lysis test, while confirmatory tests are Ham's acidified serum test and flow cytometry.

- WAHA: Here IgG antibodies are formed against patient's own red cell antigens. These antibodies are often formed against broad Rh antigens. Hemolysis is classically extravascular. The IgG coated red cells are destroyed by splenic macrophages. Splenomegaly is thus a feature.
- CHAD: Here IgM antibodies are formed against red cell antigens. Agglutination of red cells can occur at low temperatures and complement activation may result in intravascular hemolysis. Antibodies are often directed against I antigens.
- PCH: Here antibodies are typically formed against P antigens. The antibody binds to red cells at low temperature and when warmed activates complement causing hemolysis. These antibodies are also known as biphasic antibodies or Donath-Landsteiner antibodies. PCH is a rare condition and sometimes seen in children following viral infections with sudden onset of hemolysis.
- Examples of poikilocytosis are sickle cells (drepanocytes), target cells (codocytes), ovalocyte (macroovalocytes are seen in folate and Vitamin B_{12} deficiency), elliptocyte (this is when a red cell's long axis >twice its short axis; can be seen in hereditary elliptocytosis), stomatocyte, echinocyte or burr cell (these are red cells with 10−30 short blunt spicules; seen in storage artifact, liver and kidney disease, pyruvate kinase deficiency), acanthocytes (these are red cells with 2−20 unequal, irregular spicules), schistocytes, spherocytes, dacryocyte (tear drop red cell), bite cells, and blister cells.
- Red cell inclusions include Howell−Jolly bodies (represent DNA material), Pappenheimer bodies (iron), Cabot rings (mitotic spindle remnants), basophilic stippling (represents RNA material; seen in lead poisoning, hemolytic anemia, and pyrimidine 5 nucleotidase deficiency), Heinz bodies (denatured globin seen in G6PD deficiency; Heinz bodies are cleared by splenic macrophages; the damaged cells are known as bite cells), Hb C crystals, nucleated RBC, and malarial parasite.
- Diagnostic points of *P. falciparum*: Red cells not enlarged; rings appear fine/delicate, several per cell; some rings are with two chromatin dots; presence of marginal/applique forms; unusual to see developing forms; crescent-shaped gametocytes and Maurer's dot's may be present

References

[1] Endres HG, Wedding U, Pittrow D, Thiem U, et al. Prevalence of anemia in elderly patients in primary care: impact on 5-year mortality risk and difference between men and women. Curr Med Res Opin 2009;25: 1143−58.
[2] Patel KV. Epidemiology of anemia in older adults. Semin Hematol 2008;45:210−7.
[3] Ania BJ, Suman VJ, Fairbanks VF, Melton 3rd LJ. Prevalence of anemia in medical practice: community versus referral patients. Mayo Clin Proc 1994;69:730−5.
[4] Fonseca C, Araújo M, Moniz P, Marques F, et al. Prevalence and prognostic impact of anemia and iron deficiency in patients hospitalized in an internal medicine ward: the PRO-IRON study. Eur J Haematol 2017; 99:505−13.
[5] Percy L, Mansour D, Fraser I. Iron deficiency and iron deficiency anaemia in women. Best Pract Res Clin Obstet Gynaecol 2017;40:55−67.
[6] Goddard AF, James MV, McIntyre AS, Scott BB, et al. Guidelines for the management of iron deficiency anemia. Gut 2011;60:1309−16.

[7] Castel R, Tax MG, Droogendijk J, Leers MP, et al. The transferrin/log(ferritin) ratio: a new tool for the diagnosis of iron deficiency anemia. Clin Chem Lab Med 2012;50:1343–9.

[8] Papanikolaou NC, Hatzidaki EG, Belivanis S, Tzanakakis GN, et al. Lead toxicity update: a brief review. Med Sci Monit 2005;11:RA329–336.

[9] Barrett TG, Bundey SE. Wolfram (DIDMOAD) syndrome. J Med Genet 1997;34:838–41.

[10] Kupfer GM. Fanconi anemia: a signal transduction and DNA repair pathway. Yale J Biol Med 2013;86:491–7.

[11] Comito RR, Badu LA, Forcello N. Nivolumab-induced aplastic anemia: a case report and literature review. J Oncol Pharm Pract 2019;25:221–5.

[12] Dhaliwai G, Cornett PA, Tierney LM. Hemolytic anemia. Am Fam Physician 2004;69:2599–606.

[13] Da Costa L, Galimand J, Fenneteau O, Mohandas N. Hereditary spherocytosis, elliptocytosis, and other red cell membrane disorders. Blood Rev 2013;27(4):167–78.

[14] Beutler E. G6PD deficiency. Blood 1994;84:3613–36.

[15] Frank JE. Diagnosis and management of G6PD deficiency. Am Fam Physician 2005;72:1277–82.

[16] Minucci A, Giardina B, Zuppi C, Capoluongo E. Glucose 6-phosphate dehydrogenase: laboratory assay: how, when and why? IUBMB Life 2009;61:27–34.

[17] Koralkova P, van Solinge WW, van Wijk R. Rare hereditary red blood cell enzymopathies associated with hemolytic anemia - pathophysiology, clinical aspects, and laboratory diagnosis. Int J Lab Hematol 2014;36:388–97.

[18] Fouglas NM, Lampah DA, Kenangalem E, Simpson JA, et al. Major burden of severe anemia from non-falciparum malaria species in Southern Papua: a hospital based surveillance study. PLoS Med 2013;10:e1001575.

[19] Antinori S, Galimberti L, Milazzo L, Corbellino M. Malaria deaths in USA and Europe: "It's the same old story". It is time for a change. Trav Med Infect Dis 2016;14:513–4.

Hemoglobinopathies and thalassemias

Introduction

Hemoglobinopathies are inherited structural disorders of hemoglobin. Thalassemia is reduced rate of globin chain synthesis. There are over a 1000 different types of hemoglobinopathies. Most are clinically insignificant. The wide variation of clinical manifestation of these disorders is attributable to both genetic and environmental factors. Interestingly α-thalassemia is very prevalent in endemic region of malaria, and it protects against severe form of *Plasmodium falciparum* infection [1]. Some hemoglobinopathies such as sickle cell disease, once a fatal disorder of childhood, are now treatable with more than 95% of patients born today with sickle cell disease in developed countries are expected to survive into adulthood, largely because of improvements in supportive and preventive care. Hydroxyurea therapy, the only oral medication currently available to prevent complications, has become more widespread over the past 20 years [2]. However, hematopoietic stem cell transplantation is the only established cure which is becoming increasingly safe and cost-effective [3].

Hemoglobin structure and synthesis

Hemoglobin, the oxygen-carrying pigment of erythrocytes, consists of a heme portion (iron-containing chelate) and four globin chains. Six distinct species of normal hemoglobin are found in human, three in normal adults, and three in fetal life. The globulins associated with hemoglobin molecule (both embryonic stage and after birth) include alpha chain (α-chain), beta chain (β-chain), gamma chain (γ-chain), delta chain (δ-chain), epsilon chain, and zeta chain. In embryonic stage, hemoglobin Grower and hemoglobin Portland are found, but these are replaced by hemoglobin F (Hb F: two α-chains and two γ-chains) in fetal life. Interestingly, Hb F has higher oxygen affinity than adult hemoglobin and is capable of transporting oxygen in peripheral tissues in hypoxic fetal environment. In the third trimester, genes responsible for β- and γ-globulin synthesis are activated and as a result adult hemoglobin such as hemoglobin A (Hb A: two α-chains and two β-chains) and hemoglobin A2 (Hb A2: two α-chains and two δ-chains) may also be found in neonates, but Hb F is still the major component. Newborn babies and infants up to 6 months old do not depend on Hb A synthesis, although switch from Hb F to Hb A occurs around 3 months of age. Therefore, disorders due to β-chain defect such as sickle cell disease tend to manifest clinically after 6 months of age although diseases due to α-chain defect are manifested in utero or following birth. Embryonic, fetal, and adult hemoglobins are

Hematology and Coagulation. https://doi.org/10.1016/B978-0-12-814964-5.00004-8

Table 4.1 Embryonic, fetal, and adult hemoglobins.

Period of life	Hemoglobin species	Globulin chains	% Present in adult
Embryonic	Gower-1	Two ζ, two ε	
	Gower-2	Two α, two ε	
	Portland-1	Two ζ, two γ	
	Portland-2	Two ζ, two β	
Fetal	Hemoglobin F	Two α, two γ	
Adult	Hemoglobin A	Two α, two β	95%–97%
	Hemoglobin A2	Two α, two δ	<3.5%
	Hemoglobin F	Two α, two γ	<1%

summarized in Table 4.1. The different types of naturally occurring embryonic, fetal, and adult hemoglobin vary in their tetramer–dimer subunit interface strength (stability) in the liganded (carboxyhemoglobin or oxyhemoglobin) state [4].

The normal hemoglobin (HbA) in adults contains two α-chains and two β-chains. Each α-chain contains 141 amino acids in length, and each β-chain contains 146 amino acids. Hemoglobin A_2 (HbA_2) contains two α-chains and two δ-chains. The gene for the α-chain is located in chromosome 16 (two genes in each chromosome, a total of four genes), whereas genes for β- (one gene in each chromosome, a total of two genes), γ-, and δ-chains are located on chromosome 11. Adults have mostly HbA and small amounts of HbA_2 (less than 3.5%) and Hb F (less than 1%). A small amount of fetal hemoglobin persists in adults due to a small clone of cells called F cells. When hemoglobin is circulating with erythrocytes, glycosylation of the globin chains may takes place. These are referred to as X1c (X being any hemoglobin, e.g., HbA1c). When hemoglobin molecule is aging, glutathione is bound to cysteine at the 93rd position of the β-chain. This is HbAIII or HbA1d. Just like Hb A1c and HbA1d, in individuals with HbS, Hb S1c and HbS1d may also exist in circulation.

To summarize

- Embryonic hemoglobins are Gower and Portland
- Fetal hemoglobin consists of two α-chains and two γ-chains
- Adult hemoglobins: HbA2, <3.5%, Hb F <1.0%, remainder is Hb A. Hb A consists of two α-chains and two β-chains. Hb A2 consists of two α-chains and two δ-chains.
- α-Chain has 141 amino acids
- β-Chain has 146 amino acids
- Genes for α-chains are located in chromosome 16
- Genes for β-, δ-, and δ-chains are located in chromosome 11
- Hemoglobin adducts glycosylated hemoglobins (e.g., HbA1c, HbS1c, HbC1c) and aged hemoglobin adducts (e.g., HbA1d, HbS1d, HbC1d)

Heme is synthesized in a complex way involving enzymes in both mitochondrion and cytosol. In the first step, glycine and succinyl CoA combine in mitochondria to form δ aminolevulinic acid which is transported into cytoplasm and is converted into porphobilinogen by the action of enzyme aminolevulinic acid dehydrogenase. Then porphobilinogen is converted finally into coproporphyrinogen

III through several steps involving multiple enzymes. Then coproporphyrinogen III is transported into mitochondria and is converted into protoporphyrinogen III by coproporphyrinogen III oxidase enzyme. Then protoporphyrinogen III is converted into protoporphyrin IX by protoporphyrinogen III oxidase enzyme, and finally protoporphyrin IX is converted into heme by ferrochelatase enzyme. Finally heme is transported into cytosol and combines with globulin to form the hemoglobin molecule.

Introduction to hemoglobinopathies and thalassemias

Out of over 1000 hemoglobinopathies reported, most of such disorders are asymptomatic. However, in other cases, significant clinical disorders can be noticed including

- Thalassemias (both α and β)
- Sickling disorders (Hb SS, Hb SC, Hb SD, Hb SO)
- Cyanosis (such as Hb Kansas)
- Hemolytic anemias (such as Hb H)
- Erythrocytosis (such as Hb Malmo)

Hemoglobinopathies are transmitted in autosomal recessive fashion. Therefore, carriers who have one affected chromosome and one normal chromosome are usually healthy or slightly anemic. When both parents are carriers, then children have 25% chance of being normal, 25% chance of being severely affected by the disease, and 50% chance of being normal. Hemoglobinopathies are caused by inherent mutation of genes coded for globin synthesis. Point mutation of the gene in coding region (exons) which cause production of defective globin that results in formation of abnormal hemoglobin (hemoglobin variants) [5].

It has been estimated that approximately 5% of the world population are carrier of hemoglobin disorders. Hemoglobinopathies affect approximately 370,000 newborn babies each year in the world. The hemoglobin variants of most clinical significance are hemoglobin S, C, and E. In West Africa, approximately 25% of individuals are heterozygous for hemoglobin S (Hb S) gene, which is related to sickle cell diseases. In addition, high frequencies of Hb S gene alleles are also found in people of Caribbean, South and Central Africa, Mediterranean, Arabian Peninsula, and East India. Hemoglobin C (Hb C) is found mostly in people living or originating from West Africa. Hemoglobin E (Hb E) is widely distributed between East India and Southeast Asia with highest prevalence in Thailand, Laos, and Cambodia but may be sporadically observed in parts of China and Indonesia.

Hemoglobinopathies can be due to α-chain defect, β-chain defect, γ-chain defect, or δ-chain defect. Commonest β-chain defect hemoglobinopathies are

- Hb S (seen most often in African Americans)
- Hb C (seen most often in African Americans)
- Hb E (seen most often in SE Asians)

Commonest α-chain defect is

- Hb G (seen most often in African Americans)

Commonest δ-chain defect is

- Hb A2' (seen most often in African Americans)

Thalassemia syndrome is not due to structural defect in globin chain but due to lack of sufficient synthesis of globin chain and is also a genetically inherited disease. This is due to mutations that disrupt gene expression. When there is reduced amount of α-chain synthesis, the condition is termed as α-thalassemia. When β-chain synthesis is reduced, the condition is termed as β-thalassemia. Similarly, we have δ- and δβ-thalassemia. Of these, α- and β-thalassemias are seen most often.

In general, β-thalassemia is observed in Mediterranean, Arabian Peninsula, Turkey, Iran, West and Central Africa, India, and other Southeast Asian countries, whereas α-thalassemia is commonly observed in parts of Africa, Mediterranean, Middle East, and throughout Southeast Asia [6].

α-thalassemia

There are two genetic loci for α gene resulting in four genes (alleles) for α hemoglobin (α/α, α/α) on chromosome 16. Two alleles are inherited from each parent. α-Thalassemia occurs when there is a defect or deletion in one or more of four genes responsible for α-globin production. When there is underproduction of the α-globin chain, then the body tries to compensate by increased production of β-, γ-, and δ-globin chains. Four β-globin chains may form novel hemoglobin called Hb H. Four γ-globin chains may form another novel hemoglobin called Hb Barts.

α-Thalassemia can be divided into four categories:

- The silent carriers: Characterized by only one defective or deleted gene but three functional genes. These individuals have no health problem. An unusual case of silent carrier is individuals carrying one defective Constant Spring mutation but three functional genes. These individuals also have no health problem.
- α-Thalassemia trait: Characterized by two deleted or defective genes and two functional genes. These individuals may have mild anemia.
- α-Thalassemia major (hemoglobin H disease): Characterized by three deleted or defective genes and only one functional gene. These patients have persistent anemia and significant health problems. In a newborn with α-thalassemia, major γ-globin chain production is still active and thus Hb Barts' presence is expected. However, when the child becomes older, γ-globin chain production ceases and is overtaken by β-globin chain production. Now, we can expect Hb H.
- Hydrops fetalis: Characterized by no functional α gene, and these individuals have hemoglobin Barts. This condition is not compatible with life unless intrauterine transfusion is initiated.

When a α gene is functional, it is denoted as "α" and if not functional or deleted as "−." There is not much difference in impaired α-globin synthesis between a deleted gene and a nonfunctioning defective gene. With deletion or defect of one gene (−/α, α/α), little clinical effect is observed because three α genes are enough to allow normal hemoglobin production. These patients are sometimes referred to as "silent carriers" as there are no clinical symptoms but mean corpuscular volume (MCV) and man corpuscular hemoglobin (MCH) may be slightly decreased. These individuals are diagnosed by deduction only when they have children with thalassemia trait or hemoglobin H disorder. An unusual case of silent carrier state is an individual carrying one hemoglobin Constant Spring mutation but three functional genes. Hemoglobin Constant Spring (hemoglobin variant isolated from a family of ethnic Chinese background from Constant Spring district of Jamaica) is hemoglobin variant where mutation of α-globin gene produces an abnormally long α-chain (172 amino acids instead of normal 141 amino acids). Hemoglobin Constant Spring is due to nondeletion mutation of α gene which results in production of unstable α-globin. Moreover, this α-globin is produced in very low quantity (approximately 1% of normal expression level) and is found in people living or originating in Southeast Asia.

When two genes are defective or deleted, α-thalassemia trait is present. There are two forms of α-thalassemia trait. α-Thalassemia 1 (−/−, α/α) results from the *cis*-deletion of both α genes on the same chromosome. This mutation is found in Southeast Asian populations. α-Thalassemia 2 (−/α, −/α) results from the trans-deletion of α genes on two different chromosomes. This mutation is found in the African and African American populations (prevalence of disease in 28% in African Americans). Only in the case of cis-deletion, ζ-globin is expressed in carriers. In α-thalassemia trait, two functioning α genes are present, and as a result, erythropoiesis is almost normal in these individuals, but a mild microcytic hypochromic anemia (low MCV and MCH) may be observed. This form of the disease can mimic iron deficiency anemia. Therefore, distinguishing α-thalassemia from iron deficiency anemia is essential.

If three genes are affected (−/−, −/α), the disease is called hemoglobin H disease, which is a severe form of α-thalassemia and patients causing severe anemia requiring blood transfusion. Because only one α gene is responsible for production of α-globin in Hb H disease, high β-globin to α-globin ratio (two- to fivefold increase in β-globin production) may result in formation of a tetramer containing only β, and this form of hemoglobin is called Hb H (four β chains). This form of hemoglobin cannot deliver oxygen in peripheral tissues because hemoglobin H has a very high affinity for oxygen. A microcytic hypochromic anemia with target cells and Heinz bodies (which represents precipitated Hb H) is present in the peripheral blood smear of these patients. Moreover, red cells that contain hemoglobin H are sensitive to oxidative stress and may be more susceptible to hemolysis, especially when oxidants such as sulfonamides are administered. More mature erythrocytes also contain increasing amounts of precipitated hemoglobin H (Heinz bodies). These are removed from the circulation prematurely, which may also cause hemolysis. Therefore clinically, these patients experience a varying severity of chronic hemolytic anemia. Because of the subsequent increase in erythropoiesis, erythroid hyperplasia may result and cause bone structure abnormalities with marrow hyperplasia, bone thinning, maxillary hyperplasia, and pathologic fractures. When hemoglobin Constant Spring is associated with Hb H disease, a more severe form of anemia is observed requiring frequent transfusion [7]. However, when a child inherits one hemoglobin Constant Spring gene from father and one from mother, then hemoglobin Constant Spring disease is present, which is less severe than hemoglobin H−hemoglobin Constant Spring disease, but severity is comparable to hemoglobin H disease. Patients with hemoglobin H and related diseases require transfusion and chelation therapy to remove excess iron.

When four genes are defective or deleted, (−/−, −/−), the result is hemoglobin Barts disease where α-globin is absent because no gene is present to promote α-globin synthesis and as a result four γ-chains form a tetramer. As in Hb H, the hemoglobin in Hb Barts is unstable, which impairs the ability of the red cells to release oxygen to the surrounding tissues. The fetus usually cannot survive gestation causing stillborn with hydrops fetalis. However, more recently, with support through intrauterine transfusion and neonatal intensive care unit, survival may be possible but survivors have severe transfusion-dependent anemia.

Diagnosis of α-thalassemias

There are two forms of α-thalassemia:

α-thalassemia trait

- Normal hemoglobin electrophoresis (as α-chain is a component of HbA, HbF, and HbA2, the relative distribution of these hemoglobins remain the same)
- Complete blood count (CBC) reveals low MCV and low MCH with or without mild anemia; red blood cell (RBC) count is high especially when compared with hemoglobin levels
- Other common causes of low MCV and low MCH are ruled out. This is done when the iron profile is normal. In iron deficiency and anemia of chronic disease, iron studies demonstrate abnormalities
- Confirmation requires genetic testing

α-thalassemia disease (major)

- Hemoglobin electrophoresis will demonstrate Hb Barts if the patient is an infant and Hb H if the patient is older
- Marked microcytic hypochromic anemia

β-thalassemia

β-Thalassemia is due to deficit or absence of of β-globin production resulting in excess production of α-, γ-, and δ-globin chains. HbA2 has two α-chains and two δ-chains. Thus, in β-thalassemia, HbA2 levels are high, typically greater than 3.5%. HbF has two α-chains and two γ-chains. Thus, in β-thalassemia, HbF levels are also high. Synthesis of β-globin may vary from near complete presence to complete absence causing various severities. β-Thalassemia is due to mutation of genes (one gene each on chromosome 11), and more than 200 point mutations have been reported. However, deletion of both genes is rare. β-Thalassemia can be broadly divided into three categories:

- β-Thalassemia trait: Characterized by one defective gene and one normal gene, and individuals may experience mild anemia but not transfusion dependent. Here, the key feature is presence of Hb A2, greater than 3.5%. Analysis of the CBC values should demonstrate microcytic hypochromic red cells with mild anemia. The RDW is not significantly increased. The RBC count is disproportionately increased when compared to the hemoglobin level.
- β-Thalassemia intermedia: Characterized by two defective genes, but some β-globin production is still observed in these individuals. However, some individuals may have significant health problems requiring intermittent transfusion.
- β-Thalassemia major (Cooley's anemia): Characterized by two defective genes, but almost no function of either gene leading to no synthesis of β-globin. These individuals have severe form of disease requiring lifelong transfusion and may have shortened life span. These individuals have significantly elevated levels of HbF. In some cases of β-thalassemia major, HbA2 may not be elevated.

If a defective gene is incapable of producing any β-globin, it is characterized as "β^0" causing more severe form of β-thalassemia. However, if the mutated gene can retain some function, it is characterized as "β^+". In the case of one gene defect, β-thalassemia minor (trait; patients are β^0/β or β^+/β) is observed and individuals are either normal or mildly anemic. These patients have increased HbA_2.

In addition, HbF may also be elevated. MCV and MCH are low, but these patients are not transfusion dependent. If both genes are affected resulting in severely impaired production of β-globin (β^0/β^0 or β^+/β^0), the disease is severe and is called β-thalassemia major (also known as Cooley's anemia). However, because of the presence of fetal hemoglobin, symptoms of β-thalassemia major are not observed before 6 months of age. Patients with β-thalassemia major have elevated HbA_2 and HbF (although in some individual HbA2 may be normal). If production of β-globins are moderately hampered, then the disease is called β-thalassemia intermediate (β^0/β or β^+/β^+). These individuals have less severe disease than β-thalassemia major. In patients with β-thalassemia major, excess α-globulin chain precipitates, leading to hemolytic anemia. These patients require lifelong transfusion and chelation therapy. Interestingly, having β^0 or β^+ does not predict the severity of disease because patients with both types have been diagnosed with β-thalassemia major or intermedia. Major features of α- and β-thalassemia are summarized in Table 4.2.

Diagnosis of β-thalassemia
β-Thalassemia trait

- HbA2 >3.5%, Hb F may or may not be raised, remainder of the hemoglobin is HbA
- CBC reveals low MCV and low MCH with or without mild anemia; RBC count is high especially when compared with hemoglobin levels. RDW is not significantly raised
- Normal iron studies

β-Thalassemia major

- Significantly elevated HbF, Hb A may or may not be present. HbA2 is greater than 3.5%. Rarely Hb A2 may even be normal
- Marked microcytic hypochromic anemia

Table 4.2 Major features of α- and β-thalassemia.

Disease	Number of deleted gene	Comments
α-Thalassemia silent carrier	One of four gene deletion	Asymptomatic May have low mean corpuscular volume, man corpuscular hemoglobin
α-Thalassemia trait	Two of four gene deletion	Asymptomatic or mild microcytic hypochromic anemia
Hemoglobin H disease	Three of four gene deletion	Microcytic hypochromic anemia Hb H is found in adults and Hb Barts in neonates.
Hydrops fetalis	Four of four gene deletion	Hemoglobin Barts disease Most severe form may cause stillbirth/hydrops fetalis
β-Thalassemia trait	One gene defect	Asymptomatic
β-Thalassemia intermedia	Both genes defective	Variable degree of severity as some β-globins are still produced
β-Thalassemia major	Both genes defective	Severe impairment or no β-globin synthesis. Severe disease with anemia or splenomegaly, requiring lifelong transfusion.

δ-thalassemia

δ-Thalassemia is due to mutation of genes responsible for synthesis of δ-chain. A mutation that prevents formation of δ-chain is called δ^0, and if a δ-chain is formed, the mutation is termed as δ+. If an individual inherits two δ^0 mutations, no δ-chain is produced and no HbA_2 can be detected in blood (normal level <3.5%). However, if an individual inherits two δ + mutations, decreased HbA_2 is observed. All patients with δ-thalassemia have normal hematological consequence. Presence of δ mutation may obscure diagnosis of β-thalassemia trait. This is because diagnosis of β-thalassemia trait requires elevated levels of HbA2 (>3.5%), and if the patient has concomitant δ-thalassemia, production of HbA2 will be low.

δβ-Thalassemia

δβ-Thalassemia is a rare disorder characterized by decreased or total absence of production of δ- and β-globin. As a compensatory mechanism, γ-chain synthesis is increased resulting in significant amount of fetal hemoglobin (Hb F) in blood, which is homogenously distributed in RBCs. This condition is found in many ethnic groups but especially observed in individuals with ancestry from Greece or Italy. Heterozygous individuals are asymptomatic with normal Hb A_2 but rarely reported homozygous individuals experience mild symptoms.

Individuals with δβ-thalassemia trait will have HbF levels typically between 5% and 15%; normal levels of HbA2 and the remainder hemoglobin will be HbA. The individual will have microcytic hypochromic red cells.

Sickle cell disease

The terminology "sickle cell disease" includes all manifestation of abnormal hemoglobin S (HbS), which includes sickle cell trait (Hb AS), homozygous sickle cell disease (Hb SS), and a range of mixed heterozygous hemoglobinopathies such as hemoglobin SC disease, hemoglobin SD disease, hemoglobin SO Arab disease, and hemoglobin S combined with β-thalassemia. Individuals with sickle cell trait are typically asymptomatic and do not require medical intervention. Rarely HbS variants such as Hb S Antilles have lower solubility and may produce sickling disorder even in the heterozygous state.

Sickle cell disease affects millions of people throughout the world and is particularly common in people or people migrating from sub-Saharan Africa, South America Caribbean, Central America, Saudi Arabia, India, and Mediterranean countries such as Turkey, Greece, and Italy. Sickle cell disease is most commonly observed hemoglobinopathy in the United States affecting 1 in every 500 birth of African-American birth and 1 in every 36,000 Hispanic American births. Sickle cell disease is a dangerous hemoglobinopathy, and symptoms of sickle disease start before age 1 with chronic hemolytic anemia, developmental disorder, crisis including extreme pain (sickle cell crisis), high susceptibility to various infections, spleen crisis, acute thoracic syndrome, and increased risk of stroke. Optimally treated individuals may have a life span of 50—60 years [8].

In sickle cell disease, normal round shape of RBC is changed into a crescent shape and hence the name "sickle cell." In the heterozygous form (Hb AS), sickle cell disease protects from infection of *P. falciparum* malaria but not in the more severe form of homozygous sickle cell disease (Hb SS). The genetic defect producing sickle hemoglobin is a single nucleotide substitution at codon 6 of the β-globin gene on chromosome 11 that results in a point mutation in β-globin chain of hemoglobin (substitution of valine for glutamic acid at sixth position). Hemoglobin S is formed when two normal α-globins combine with two mutant β-globins. Because of this hydrophobic amino acid substitution, Hb S polymerizes on deoxygenation and multiple polymers bundle to rod-like structure resulting in deformed RBC. Various possible diagnoses of patients with Hb S hemoglobinopathy include sickle cell trait (Hb AS), sickle cell disease (Hb SS), and sickle cell disease status post RBC transfusion/exchange. Patients with sickle cell trait may also have concomitant α-thalassemia, and diagnosis of HbS/β-thalassemia (0/+/++) is also occasionally made. Double heterozygous states of Hb SC, Hb SD, and Hb SO Arab are important sickling states that should not be missed.

Hemoglobin C is formed because of substitution of glutamic acid residue with a lysine residue at the sixth position of β-globin. Individuals who are heterozygous with Hb C disease are asymptomatic with no apparent disease, but homozygous individuals have almost all hemoglobin (>95%) as Hb C and experience chronic hemolytic anemia and pain crisis. However, individuals who are heterozygous with both hemoglobin C and hemoglobin S (Hb SC disease) have weaker symptoms than sickle cell disease because Hb C does not polymerize as readily as Hb S.

Patients with Hb SS disease may have increased Hb F. The distribution of Hb F amongst the haplotypes of Hb SS is Hb F 5%−7% in Bantu, Benin, or Cameroon, Hb F 7%−10% in Senegal, and Hb F 10%−25% in Arab-Indian. Hydroxyurea also causes increase in Hb F. This is usually accompanied by macrocytosis. Hb F can also be increased in Hb S/HPFH (HPFH: hereditary persistence of fetal hemoglobin). Hb A2 values are typically increased in sickle cell disease and more so on high-performance liquid chromatographic (HPLC) analysis. This is because the posttranslational modification form of Hb S, Hb S1d, produces a peak in the A2 window. This elevated value of Hb A2 may produce diagnostic confusion with Hb SS disease and Hb S/β-thalassemia. It is important to remember that microcytosis is not a feature of Hb SS disease and patients with Hb S/β-thalassemia typically exhibit microcytosis.

Hb SS patients and Hb S/β 0-thalassemia patients do not have any Hb A, unless the patient has been transfused or has undergone red cell exchange. Glycated Hb S has the same retention time (approximately 2.5 min) as Hb A in HPLC. This will produce a small peak in the A window and raise the possibility of Hb S/β +-thalassemia. Hb S/α-thalassemia is considered when the percentage of Hb S is lower than expected. Classical cases are 60% of Hb A and approximately 35%−40% of Hb S. Cases of Hb S/α-thalassemia should have lower values of Hb S, typically below 30% with microcytosis. Similar picture will also be present in patients with sickle cell trait and iron deficiency. Various features of sickle cell disease are summarized in Table 4.3.

Diagnostic approach to an individual who demonstrates HbS on electrophoresis
The various possibilities include the following:

- Sickle cell trait (HbAS)
- Sickle cell trait with concomitant α-thalassemia trait
- Sickle cell disease
- Sickle cell disease patient who has been transfused or has undergone RBC exchange
- Sickle cell/β-thalassemia (this in turn can be S/β 0- or S/β + -or S/β++ thalassemia)
- Sickle cell disease patient on hydroxyurea
- Hb SC disease
- Hb S/HPFH
- Normal individual who has been transfused RBCs from a sickle cell trait donor

Table 4.3 Major features of HbS hemoglobinopathies.

Disease	Hemoglobin variants	Features
Sickle cell trait	Hb AS	Hb A > HbS; HbA: 50%−60%; Hb S: 30%−40% No apparent illness
Sickle cell disease	Hb SS	HbS (majority), Hb A_2: <3.5%, HbF (high), no HbA Severe disease with chronic hemolytic anemia
Sickle cell-β^0-thalassemia	Hb Sβ^0	HbS (majority), Hb A_2: >3.5%, Hb F (high), no HbA Low mean corpuscular volume and low man corpuscular hemoglobin; severe disease
Sickle cell-β^+-thalassemia	Hb Sβ^+/++	HbS (majority), Hb A_2: >3.5%, Hb F (high), Hb A: 5%−40% Variable mild-to-moderate sickle cell disease
Hemoglobin SC disease	Hb SC	Hb S: 50%, Hb C: 50% Moderate sickling disease but chronic chemolytic anemia may be present
Hemoglobin S/hereditary persistence of fetal hemoglobin		Hb S: 60%, Hb A_2: <3.5%, Hb F: 30%−40% Behaves as sickle cell trait

Diagnosis of sickle cell trait

- Hemoglobin electrophoresis demonstrates Hb A and HbS. The HbA is always greater than HbS. This statement is true for all other hemoglobin traits too. Hb A is approximately 50%−60% and Hb S is approximately 30%−40%.
- MCV and MCH values are not low, unless there is concomitant iron deficiency or anemia of chronic disease.
- There is no history of recent transfusion.

Diagnosis of sickle cell trait with concomitant α-thalassemia trait

- Hb A > HbS; however, the percentage of Hb S is less than 25%.
- MCV and MCH values are low without iron deficiency or anemia of chronic disease.
- There is no history of recent transfusion.

Diagnosis of sickle cell disease

- Hemoglobin electrophoresis reveals very high values of HbS. There is no HbA. Hb A2 values are less than 3.5%. Hb F levels may be high. In the Arab-Indian haplotype of sickle cell disease, Hb F maybe as high as 25%.

Diagnosis of sickle cell disease patient who has been transfused or has undergone red blood cell exchange

- Hemoglobin electrophoresis reveals Hb A and HbS. Hb A2 is less than 3.5%. Hb F is also present. If the HbS is greater than HbA, then this cannot be sickle cell trait. It must be sickle cell disease with RBC transfusion/exchange. However, if Hb A is greater than Hb S, then transfusion history is essential to differentiate from sickle cell trait.

Diagnosis of sickle cell/β-thalassemia (this in turn can be S/β 0- or S/β + - or S/β++ thalassemia)

- In any form of sickle cell/β-thalassemia Hb A2 needs to be greater than 3.5%. In addition, MCV and MCH should be low.
- In HbS/β0-thalassemia, there is no HbA. Hemoglobin electrophoresis will demonstrate Hb S, Hb F, and Hb A2 (>3.5%).
- In HbS/β +-thalassemia, Hb A is approximately 5%−15% with no history of RBC transfusion. HbS, HbF, and HBA2 (>3.5%) are all present.
- In HbS/β++-thalassemia, Hb A is approximately 20%−40% with no history of RBC transfusion. HbS, HbF, and HBA2 (>3.5%) are all present.

Diagnosis of sickle cell disease patient on hydroxyurea

- There are individuals with sickle cell disease or sickle cell β-thalassemia whose HbF is higher than expected. Such patients are placed on hydroxyurea to increase the percentage of HbF, which helps in reducing chance of sickling. Review of EMR should demonstrate hydroxyurea administration. In addition, hydroxyurea is antifolate resulting in macrocytic red cells.

Diagnosis of Hb SC disease

- Here the patient will have HbS and HbC roughly in equal amounts. MCV and MCH should be low (due to HbC not HbS). Patients will not have HbA, unless they have been transfused.

Diagnosis of HbS/hereditary persistence of fetal hemoglobin

- These patients clinically behave as sickle cell trait. Hemoglobin electrophoresis demonstrates HbS and high levels of HbF (20%−40%). CBC values are normal (hemoglobin, MCV, MCH). Clinically patients are asymptomatic.

Diagnosis of normal individual who has been transfused red blood cells from a sickle cell trait donor

- Normal individuals who have been donated with blood from a sickle cell trait donor will have HbS values typically <5%. There should be history of RBC transfusion.

  Diagnostic approach toward sickle cell disease is summarized in Table 4.9.

  Hemoglobin electrophoresis is useful in diagnosis of sickle cell disease by identifying Hb S. However, solubility test can also aid in diagnosis of sickle cell disease. When a blood sample containing Hb S is added to a test solution containing saponin (to lyse cells) and sodium hydrosulfite (to deoxygenate the solution), a cloudy turbid suspension is formed if Hb S is present. If no Hb S is present, the solution remains clear. False negative result may be observed if Hb S is <10% as often observed in infants younger than 3 months of age [9].

Hereditary persistence of fetal hemoglobin

In individuals with HPFH, significant amounts of fetal hemoglobin (Hb F) can be detected well into adulthood. In normal adults, Hb F represents less than 1% of total hemoglobin, but in HPFH, the percentage of HB F can be significantly elevated, but HbA_2 is also normal. HPFH is divided into two major groups: deletional and nondeletional. Deletional HPFH is caused by variable length deletion in β-globin gene cluster leading to decreased or absent β-globin synthesis and compensatory increase in γ-globin synthesis with a pancellular or homogenous distribution of Hb F in RBCs. Nondeletional HPFH is a broad category of related disorders with increased Hb F typically distributed heterocellularly. Heterocellular distribution is also seen in β-thalassemia and δβ-thalassemia.

Both homozygous and heterozygous HPFH are asymptomatics with no clinical or significant hematological change although individuals with homozygous HPFH may show up to 100% Hb F, whereas heterozygous typically shows 20%−28% HbF. If HPFH is associated with sickle cell, it can reduce the severity of disease. Compound heterozygotes for sickle hemoglobin (Hb S) and HPFH have high level of Hb F but these individuals experience few if any sickle cell disease−related complication [10]. If HPFH is associated with thalassemia, individuals also experience less severe disease.

Other hemoglobin variants

Hemoglobin D (hemoglobin D Punjab, also known as hemoglobin D Los Angeles) is formed due to substitution of glutamine for glutamic acid. Hb D Punjab is one of the most commonly observed abnormality worldwide not only found in Punjab region of India but also in Italy, Belgium, Austria, and Turkey. Hemoglobin D disease can occur in four different forms including heterozygous Hb D trait, Hb D-thalassemia, Hb SD disease, and very rarely homozygous Hb D disease. Heterozygous Hb D disease is a benign condition with no apparent illness, but when Hb D is associated with Hb S or β-thalassemia, clinical conditions such as sickling disease and moderate hemolytic anemia may be observed. Heterozygous Hb D is rare and usually presents with mild hemolytic anemia and mild-to-moderate splenomegaly [11].

Hemoglobin E is due to point mutation of β-globin, which results in substitution of lysine for glutamic acid in position 26. Individuals with HbE also produce less β-globin resulting in a thalassemia like phenotype. Hb E is unstable and can form Heinz bodies under oxidative stress. Hb E trait is associated with moderately severe microcytosis, but usually no anemia is present. However, individuals with Hb E homozygous present with modest anemia similar to thalassemia trait. However, when β-thalassemia is combined with Hb E, for example, in Hb E/β⁰ thalassemia, patients may have significant anemia requiring transfusion similar to patients with β-thalassemia intermedia.

Hemoglobin O-Arab (Hb O-Arab; also known as Hb Egypt) is a rare abnormal hemoglobin variant, where at position 121 of the β-globin, normal glutamic acid is replaced by lysine. Hb-O-Arab is found in people from Balkans, Middle East, and Africa. Patients who are heterozygous for Hb-O-Arab may experience mild anemia and microcytosis similar to patients with β-thalassemia minor. The homozygous form is extremely rare. Patients with Hb S/Hb O Arab may experience severe clinical symptoms similar to individuals with Hb S/S. Patients with Hb-O Arab/β-thalassemia may also experience severe anemia with hemoglobin level between 6 and 8 gm/dL and splenomegaly [12].

Hemoglobin Lepore is an unusual hemoglobin molecule that is composed of two α-chains and two δβ-chains as a result of fusion gene of δ and β genes. The δβ-chains have the first 87 amino acids of the δ-chain and 32 amino acids of the β-chain. There are three common variants of hemoglobin Lepore: Hb Lepore Washington, also known as Hb Lepore Boston, Hb Lepore Baltimore, and Hb Lepore Hollandia. Hemoglobin Lepore is seen in individuals from of Mediterranean descent. Individuals with HbA/Hb Lepore are asymptomatic with Hb Lepore representing 5%−15% of hemoglobin and slightly elevated Hb F (2%−3%). Homozygous Lepore individuals suffer from severe anemia similar to patients with β-thalassemia intermedia with Hb Lepore representing 8%−30% of hemoglobin and remainder hemoglobin is Hb F. Patients with Hb Lepore/β-thalassemia experience severe disease similar to patients with β-thalassemia major.

Hemoglobin G-Philadelphia (Hb G) is the commonest α-chain defect observed 1 in 5000 of African-Americans and is associated with α-thalassemia 2 deletions. Therefore, these individuals have only three functioning α gene and Hb G represents one-third of total hemoglobin. Hb S is the commonest β-chain defect often observed in African-American population, whereas Hb G is the commonest α-chain defect and again occurs most often in the African-American people. If an individual has one parent who has sickle cell trait (HbAS) and the other parent has Hb G trait (HbAG), then the individual will have the following globin chains:

- Normal α-chain
- Abnormal α-chain, represented as G
- Normal β-chain
- Abnormal β-chain, represented as S

This individual can then form the following hemoglobins:

- HbA (two normal α-chains and two normal β-chains)
- HbG (two abnormal α-G-chains and two normal β-chains)
- HbS (two normal α-chains and two abnormal β-S-chains)
- HbS/G (two abnormal α G chains and two abnormal β-S-chains)
- In addition, Hb G2 ($\alpha 2$, $\delta 2$) which is the counterpart of HbA_2 is also present.

In certain cases, HbA_2 variants may also be present. In such cases, the total HbA_2 (HbA_2 and HbA_2 variant) need to be considered for the diagnosis of β-thalassemia. HbA_2' is the most common of the known Hb A_2 variants reported in 1%−2% of African-Americans detected in heterozygous and homozygous states and in combination with other Hb variants and thalassemia. The major clinical significance of HbA_2' is that for the diagnosis or exclusion of β-thalassemia minor, the sum of HbA_2 and Hb A_2' must be considered. If the sum of HbA2 and HbA2' exceeds 3.5%, then β-thalassemia is a possibility. HbA_2' when present accounts for a small percentage (1%−2%) in heterozygotes and is difficult to detect by gel electrophoresis. It is, however, easily detected by capillary electrophoresis and HPLC. In HPLC, Hb A_2' elutes in the "S" window. In Hb AS trait and HB SS disease, Hb A_2' could be masked by the presence of Hb S. In Hb AC trait and Hb CC disease, glycosylated Hb C will also elute in the "S" window. In these conditions, Hb A_2' will remain undetected. Conversely, sickle cell patients on chronic transfusion protocol or recent efficient RBC exchange may result in a very small percentage of Hb S that the pathologist may interpret as Hb A2'. It has been documented that the HbA_2 concentration may be raised in HIV during treatment. Severe iron deficiency anemia can reduce HbA_2 levels, and this may obscure diagnosis of β-thalassemia trait. Hematological features of α- and β-thalassemia are given in Table 4.8 [13].

Individuals with high hemoglobin F

Increased levels of HbF are seen from time to time. The two broad categories of high HbF are

- Physiological and
- Pathological

Physiological causes of increased levels of HbF are young age and pregnancy. Newborns have very high levels of HbF, and values decrease rapidly with age. By the age of 6–9 months, values are close to adult values, i.e., <1%. Pregnancy can mildly increase Hb F values, typically up to 5%.

Pathological causes of high levels of Hb F include

- β-Thalassemia (individuals will have high HbA2 with low MCV and MCH)
- δβ-Thalassemia (normal HbA2 with low MCV and MCH; δβ-thalassemia trait patients have HbF ranging from 5% to 15%)
- HPFH (normal HbA2 with normal CBC)
- In association with HbS
- In association with Hb Lepore
- In association with certain hematologic conditions such as juvenile myelomonocytic leukemia (JMML), acute erythroid leukemia, aplastic anemia, and pernicious anemia
- In association with medication such as hydroxyurea, sodium valproate, and erythropoietin

Fetal hemoglobin induction is known to ameliorate the clinical complications of sickle cell disease [14]. Hb F quantification is useful in diagnosis of β-thalassemia and other hemoglobinopathies. But quantification of Hb F may be an issue when HPLC is used. Fast variants (e.g., Hb H or Hb Barts) may not be quantified as they may elute off the column before the instrument begins to integrate in many systems designed for adult samples. This will affect the quantity of Hb F. If an α-globin variant separates from Hb A, then there should be a Hb F variant that will often separate from normal Hb F, but it may not separate from other hemoglobin adducts present and then the total Hb F will not be adequately quantified. Hb F variants may also be due to mutation of the γ-globin chain, and again this may result in a separate peak and incorrect quantification. Some β-chain variants/adducts will not separate from Hb F, and this will lead to incorrect quantification. If Hb F appears to be greater than 10% on HPLC, its nature should be confirmed by an alternative method to exclude misidentification of Hb N or Hb J as Hb F.

Fast, unstable, and other rare hemoglobins

Fast hemoglobins are traditionally named as such because in the alkaline gel, they are found to travel beyond the last lane, which is the A lane. They produce a band found close to anode, farthest point from application of sample. Fast hemoglobins are

- Hb J (typically clinically insignificant)
- Hb I (typically clinically insignificant)
- Hb N (typically clinically insignificant)
- Hb H and
- Hb Barts

Hemoglobin J is characterized as a fast moving band in hemoglobin electrophoresis (band found close to anode, farthest point from application of sample), and more than 50 variants have been reported including Hb J Capetown and Hb J Chicago. However, heterozygous hemoglobinopathy involving Hb J is clinically insignificant. Hemoglobin I is due to a single α-globin substitution (substitution of lysine at position 16 for glutamic acid). Hb I is clinically insignificant unless in rare occasions when it is associated with α-thalassemia, where approximately 70% of hemoglobin is Hb I.

Examples of unstable hemoglobins (for these unstable hemoglobins, isopropanol test is positive) are

- Hemoglobin Koln
- Hemoglobin Hasharon, and
- Hemoglobin Zurich

Certain rarely reported hemoglobin variants, hemoglobin Malmo, hemoglobin Andrew, hemoglobin Minneapolis, hemoglobin British Columbia, and hemoglobin Kempsey are associated with erythrocytosis.

Important other hemoglobinopathies are summarized in Table 4.4.

Laboratory investigation of hemoglobinopathies

Multiple methodologies exist to detect hemoglobinopathies and thalassemias. Several methods that are routinely employed include gel electrophoresis, HPLC, capillary electrophoresis, and isoelectric focusing. If any one method detects an abnormality, a second method must be used to confirm the abnormality. In addition to relevant clinical history, review of the CBC and peripheral smear provides important correlation in the pursuit of an accurate diagnosis.

Table 4.4 Various other common hemoglobinopathies.

Diagnosis	Hemoglobin/Hematological	Comments
Hb C trait (Hb AC)	Hb A: 60%; Hb C: 40% Normal/microcytic	Hb C implies ancestry from West Africa; clinically insignificant
Hb CC disease	No Hb A, Hb C almost 100% Mild microcytic	Mild chronic hemolytic anemia
Hb C trait/ α-thalassemia	Hb A: major hemoglobin Hb C <30%	
Hb C/β-thalassemia	Microcytic, hypochromatic	Moderate-to-severe anemia with splenomegaly
Hb E trait (Hb AE)	Hb A major, Hb E: 30%−35% Normal/microcytic	No clinical significance, found in Cambodia, Laos, Thailand (Hb E triangle, where Hb E trait is 50%−60% of population). and South East Asia
Hb E disease	No Hb A; mostly Hb E Microcytic hypochromic red cells ± anemia	Usually asymptomatic
Hb E trait with α-thalassemia	Majority is A; Hb E <25%	
Hb O trait (Hb AO)	Majority is A; Hb O: 30%−40% Normal complete blood count (CBC)	Clinically insignificant but Hb S/O is a sickling disorder
Hb D trait (Hb AD)	Hb A > Hb D Normal CBC	Clinically insignificant; Hb S/D is a sickling disorder
Hb G trait (HbAG)	Hb A > Hb G Normal CBC	Clinically insignificant

Gel electrophoresis

In hemoglobin, electrophoresis red cell lysates are subjected to electric fields under alkaline (alkaline gel) and acidic (acid gel) pH. This can be carried out on filter paper, a cellulose acetate membrane, a starch gel, a citrate agar gel, or an agarose gel. Separation of different hemoglobins is largely but not solely dependent of the charge of the hemoglobin molecule. Change in the amino acid composition of the globin chains result in alteration of the charge of the hemoglobin molecule resulting in change of the speed of migration. In gel electrophoresis, different hemoglobins migrate at different speed. In the alkaline gel, we have the following main lanes, from *bottom of the gel moving toward the top*:

- C lane
- S lane
- F lane and
- A lane

A faint band which usually represents the enzyme carbonic anhydrase may be seen before the main C lane. Any band above the A lane represents fast hemoglobins. These can be HbJ, HbI, HbN, Hb Barts, and HbH. Hb H and HbI are able to move the furthest and are located in the H lane. The location of the H lane is the same distance from J as A is from J in the opposite direction. Patterns of various band in gel electrophoresis are summarized in Table 4.5.

Table 4.5 Migration of various hemoglobin bands in alkaline gel and acid gel electrophoresis.

Region	Hemoglobin present
Alkaline gel electrophoresis	
Top band (farthest from origin: H lane)	Hb H, Hb I Hb Barts and Hb N are between Hb J and Hb H
J lane	Hb J
A lane	Hb A
F lane	Hb F
S lane	Hb S, Hb D, Hb G, Hb Lepore
C lane	Hb C, Hb E, Hb O, Hb A2, Hb S/G hybrid
Carbonic anhydrase band (faint)	Hb G2, Hb A2', Hb CS
Acid gel electrophoresis	
Top band (fastest from origin) C lane	Hb C
S lane	Hb S Hb S/G hybrid; Hb O and Hb H are between S and A lane
A lane	Hb A, Hb E, Hb A2, Hb D, Hb G, Hb Lepore, Hb J, Hb I, Hb N, Hb H
F lane	Hb F, Hb Hope, Hb Barts

In the acid gel, the main lanes from top to bottom are

- C lane
- S lane
- A lane and
- F lane

Interpretation of alkaline and acid gels (please use Table 4.5 for reference):

We start with the alkaline gel, moving on to the acid gel. Normal individual will have a prominent band in the A lane in both the alkaline and acid gels. Very faint bands may be seen in the F lanes.

If we see a prominent band in the C lane in the alkaline gel, then we need to think of HbC or HbE or HbO. If there is a prominent band in the A lane in the same patient, then we may be dealing with C trait or E trait or O trait. If there is no band in the A lane, then we are probably dealing with homozygous forms of the hemoglobinopathies. Now, we need to look at the acid gel. If there is prominent band in the C lane then the band seen in the C lane of the alkaline gel is truly due to HbC. If in the acid gel there is no prominent band in the C lane or S lane or F lane, but there is only a band in the A lane, then the abnormal Hb is most likely HbE. If on the acid gel there is a band between the S and A lane, then the abnormal Hb is most likely HbO.

After we have formed a tentative diagnosis form the gels, a second method, either HPLC or capillary electrophoresis, must be performed to verify the abnormality.

Again, as a different example, we see a band in the S lane in the alkaline gel. Differential now includes Hb S, HbD, Hb G, and Hb Lepore. If this is true HbS, then we expect to see a band in the S lane in the acid gel. HbD, HbG, and Hb Lepore migrates in the A lane in the acid gel. Hb Lepore is usually underproduced. Bands due to Hb Lepore are thus faint. Other underproduced hemoglobins are

- Hb Constant Spring
- HbA2'

How should we differentiate HbD, HbG, and Hb Lepore? By capillary electrophoresis, they are all found in the same zone, zone 6. HPLC is best to differentiate these three abnormal hemoglobins. Hb Lepore has a retention time of approximately 3.7 min, the same window as HbA2 and HbE. Also, please note Hb Lepore is underproduced. Individuals with Hb Lepore also have minimally increased Hb F, which provides additional clue. Both HbD and HbG have the same retention time, approximately 3.9−4.2 min. However, individuals with HbG have an additional peak on HPLC at 4.6−4.7 min due to HbG2. This is not evident with HbD.

For diagnosis of Hb S/G hybrid on alkaline gel electrophoresis, one band is expected in the A lane, one band in the S lane (due to Hb S and Hb G), one band in the C lane (due to S/G hybrid), and one band in the carbonic anhydrase area (due to HbG2). Therefore, a total of four bands should be observed. If the band in the carbonic anhydrase is not prominent, at least three bands should be seen. On the acid gel electrophoresis, one band is expected in the A lane (due to Hb A, Hb G, and Hb G2) and one band in the S lane (due to Hb S and HbS/G hybrid). In capillary electrophoresis, a band should be seen in zone 5 (Hb S) and a band in zone 6 (Hb G). It is important to emphasize that for hemoglobinopathies, finding of gel electrophoresis must be confirmed by a second method, HPLC or capillary electrophoresis.

High-performance liquid chromatography

HPLC systems utilize a weak cation exchange column system and a sample of an RBC lysate in buffer is injected into the system followed by application of a mobile phase so that various hemoglobins can partition (interact) between stationary phase and mobile phase. The time required for different hemoglobin molecules to elute is referred to as retention time. The eluted hemoglobin molecules are detected by light absorbance. HPLC permits the provisional identification of many more variant types of hemoglobins that cannot be distinguished by conventional gel electrophoresis. When HPLC is used, a recognized problem is carryover of specimen from one to the next specimen. For example, if the first specimen belongs to a patient with sickle cell disease (Hb SS) then a small peak may be seen at the "S" window in the next specimen. This can lead to diagnostic confusion as well as the sample to be re-run. Approximate retention times of common hemoglobins in a typical HPLC analysis are summarized in Table 4.6.

Capillary electrophoresis

In capillary electrophoresis, a thin capillary tube made of fused silica is used. When an electric field is applied, the buffer solution within the capillary generates an electroendosmotic flow that moves toward the cathode. Separation of individual hemoglobins takes place due to differences in overall charges. Different hemoglobins are represented in different zones. Capillary zone electrophoresis has an advantage over HPLC in that hemoglobin adducts (glycated hemoglobins and the aging adduct Hb X1d) do not separate from the main hemoglobin peak in capillary electrophoresis and makes interpretation easier. Common hemoglobin zones in capillary electrophoresis are given in Table 4.7.

Other less commonly used methodologies include isoelectric focusing, DNA analysis, and mass spectrometry. It is important to note that hemoglobinopathies may interfere with measurement of

Table 4.6 Approximate retention times of various hemoglobins in high-performance liquid chromatography analysis.

Approximate retention time	Hemoglobin
0.7 min (peak 1)	Acetylated Hb F, Hb H, Hb Barts, Bilirubin
1.1 min	Hb F
1.3 min (peak 2)	Hb A1c, Hb Hope
1.7 min (peak 3)	Aged Hb A (HbA1d), Hb J, HbN, Hb I
2.5 min	Hb A, HbS1c
3.7 min	Hb A2, Hb E, Hb Lepore, Hb S1d
3.9−4.2 min	Hb D, Hb G
4.5 min	Hb S, HbA2′, Hb C1c Hb O Arab has a broad range from 4.5 to 5 min
4.6−4.7 min	Hb G2
4.9 min	Hb C (preceding the main peak is a small peak, Hb C1d), Hb S/G hybrid, Hb CS (3 peaks: 2%−3%)

Table 4.7 Various zones in which common hemoglobins appear in capillary electrophoresis.

Zone	Hemoglobin
Zone 1	HbA2′
Zone 2	Hb C, Hb CS
Zone 3	Hb A_2, Hb O-Arab
Zone 4	Hb E, Hb Koln
Zone 5	Hb S
Zone 6	Hb D-Punjab/Los Angeles/Iran, Hb G-Philadelphia
Zone 7	Hb F
Zone 8	Acetylated Hb F
Zone 9	Hb A
Zone 10	Hb Hope
Zone 11	Denatured Hb A
Zone 12	Hb Barts
Zone 13	
Zone 14	
Zone 15	Hb H

Table 4.8 Hematological features of α- and β-thalassemia.

Disease	Complete blood count	Hemoglobin electrophoresis
α-Thalassemia[a]		
Silent carrier	Hb: normal, man corpuscular hemoglobin (MCH) <27 pg	Normal
Trait	Hb: normal, MCH <26 pg, mean corpuscular volume (MCV) <75 fL	Normal
Hb H disease	Hb: 8–10 gm/dL, MCH <22 pg, MCV low	Hb H: 10%–20%
Hydrops fetalis	Hb < 6 gm/dL, MCH <20 pg	Hb Barts: 80%–90% Hb H <1%
β-Thalassemia[a]		
Minor	Hb: Normal or low, MCV: 55–75 fL#, MCH: 19–25 pg	Hb A_2 >3.5%
Intermedia	Hb: 6–10 gm/dL, MCV: 55–70 fL	Hb A_2: variable
Mild or compound	MCH: 15–23 pg	Hb F up to 100%
Heterozygous		
Major	Hb < 7 gm/dL, MCV: 50–60 fL MCH: 14–20 pg	Hb A_2: variable Hb F: high

#MCV: Abnormal: Adult <80 fL, Children [7–12]: <76, Children (6 months-6): <70.
[a]Mentzer index for children is < 13 for both α- and β-thalassemia.

Table 4.9 Diagnostic approach of sickle cell hemoglobinopathy.

Hemoglobin pattern	Diagnosis/Comments
A patient has Hb A and Hb S	Hb AS trait or Hb SS disease (posttransfusion) or Hb S/β + -thalassemia or a normal person transfused from a donor with Hb AS trait. Transfusion history is essential for diagnosis. For a patient with Hb AS trait, Hb A is majority and Hb S is 30%−40%; if donor was Hb S trait, then S% is usually between 0.8% and 14% of the total hemoglobin. In Hb S/β$^+$-thalassemia, Hb A$_2$ is expected to be high, and there should be microcytosis and hypochromia of the red cells. Hb A% is typically 5%−25% depending on severity of genetic defect.
A patient has Hb S but no Hb A	Hb SS disease; Hb S/β^0-thalassemia, Hb A2 is elevated with low mean corpuscular volume (MCV) and man corpuscular hemoglobin.
A patient has Hb S and high Hb F	Hb S/hereditary persistence of fetal hemoglobin and Hb SS disease while patient is on hydroxyurea. High MCV favors hydroxyurea; medication history will be required.

glycosylated hemoglobin (Hb A1C) providing an unreliable result. When an Hb A1C result is inconsistent with a patient's clinical picture, possibility of hemoglobinopathy must be considered. Depending on the methodology used for measurement of Hb A1C, such as HPLC, immunoassay, etc., results may be falsely elevated or lowered. Patients with Hb C trait particularly show variable results. In such case, a test which is not affected by hemoglobinopathy such as fructosamine measurement (represent average blood glucose: 2−3 weeks) may be used [15].

Reporting normal hemoglobin electrophoresis pattern

Normal hemoglobin electrophoresis pattern:

When hemoglobin electrophoresis reveals a normal pattern, i.e., HbF is <1%, HbA2 is <3.5% and the remainder is HbA, check the CBC values. If the MCV and MCH are not low, the case can be signed out as normal hemoglobin electrophoresis.

If the hemoglobin electrophoresis pattern is normal but the MCV and MCH are low, with or without anemia, we need to check the iron profile. If the iron profile is normal, then α-thalassemia trait cannot be ruled out. It should be signed out as such. Mentzer index (MCV/RBC count) is useful in differentiating thalassemia from iron deficiency anemia. In iron deficiency anemia, this ratio is usually greater than 13, but in thalassemia, this value is less than 13. However, for accurate diagnosis of α-thalassemia, genetic testing is essential.

Interpretations of various other hemoglobinopathies are given in Table 4.10.

Apparent hemoglobinopathy after blood transfusion

Blood transfusion history is essential in interpreting abnormal hemoglobin pattern because small peaks of abnormal hemoglobin may appear from blood transfusion. Apparent hemoglobinopathy after blood transfusion is rarely reported, but it may cause diagnostics dilemma resulting in repeated unnecessary

Table 4.10 Diagnostic approach to common hemoglobinopathies.

Diagnosis	Features
Diagnosis of Hb C	Band in the C lane in the alkaline gel; possibilities are C, E, or O Band in the C lane in the acid gel High-performance liquid chromatograph (HPLC) shows a peak around 5 min with a small peak just before this main peak (HbC1d). A small peak may also be observed at 4.5 min (HbC1c) Or capillary electrophoresis shows a peak in zone 2
Diagnosis of Hb E	Band in the C lane in the alkaline gel; possibilities are C, E, or O Band in A lane in acid gel HPLC show a peak at 3.5 min and is greater than 10% Or capillary electrophoresis shows a peak in zone 4
Diagnosis of Hb O	Band in the C lane in the alkaline gel; possibilities are C, E, or O Band between A and S lane in acid gel HPLC show a peak between 4.5 and 5 min Or capillary electrophoresis shows a peak in zone 3(O-Arab)
Diagnosis of Hb S	Band in the S lane in the alkaline gel; possibilities S, D, G, Lepore Band in the S lane in the acid gel HPLC show a peak at 4.5 min Or capillary electrophoresis shows a peak in zone 5
Diagnosis of Hb D	Band in the S lane in the alkaline gel; possibilities S, D, G, Lepore Band in the A lane in acid gel HPLC shows a peak at 3.9−4.2 min; no additional peak Or capillary electrophoresis shows a peak in zone 6
Diagnosis of Hb G	Band in the S lane in the alkaline gel; possibilities S, D, G, Lepore Band in the A lane in acid gel HPLC shows a peak at 3.9−4.2 min and a small additional peak (G2) Or capillary electrophoresis shows a peak in zone 6
Diagnosis of Hb Lepore[a]	Band (faint) in the S lane in the alkaline gel; possibilities S, D, G, Lepore Band in the A lane in acid gel HPLC shows a peak at 3.7 min (A_2 peak); quantity is lower than D or G or E. Small increase in HbF% Or capillary electrophoresis shows a peak in zone 6

[a]Hb Lepore band in the alkaline gel is faint.

testing. Kozarski et al. reported 52 incidences of apparent hemoglobinopathies out of which 46 were Hb C, 4 were Hb S, and 2 were Hb O-Arab. The percentage of abnormal hemoglobin ranged from 0.8% to 14% (median: 5.6%). The authors recommended identifying and notifying the donor in such event [16].

Universal newborn screen

Universal newborn screening for hemoglobinopathies are now required in all 50 states and District of Columbia. In addition, American College of Obstetricians and Gynecologist provides guideline for screening of couples who may be at risk of having children with hemoglobinopathy. Persons of Northern Europe, Japanese, Native Americans, or Korean descents are at low risk for hemoglobinopathies, but people with ancestors from Southeast Asia, Africa, or Mediterranean are at higher risk. A CBC should be done to accurately measure hemoglobin. If all parameters are normal and the couple belongs to a low risk group, no further testing may be necessary. For higher risk couples, hemoglobin analysis by electrophoresis or other method is recommended. Solubility test for sickle cell may be helpful. Genetic screening can help physicians to identify couple at risk of having children with hemoglobinopathy. Molecular protocols for hemoglobinopathies started in 1970s using Southern blotting and restriction fragment length polymorphism analysis for prenatal sickle cell disease. With development of polymerase chain reaction, molecular testing for hemoglobinopathies now requires much less DNA for analysis [17]. However, currently molecular testing for diagnosis of hemoglobinopathies is available in large academic medical centers and reference laboratories only. Universal screening for sickle cell disease in newborns is also strongly recommended [18].

Key points

- The normal hemoglobin (HbA) in adults contains two α-chains and two β-chains. Each α-chain contains 141 amino acids in length and each β-chain contains 146 amino acids. Hemoglobin A_2 (HbA$_2$) contains two α-chains and two δ-chains. The gene for the α-chain is located in chromosome 16 (two genes in each chromosome, a total of four genes), while genes for β- (one gene in each chromosome, a total of two genes), γ-, and δ-chains are located on chromosome 11.
- When hemoglobin is circulating with erythrocytes, glycosylation of the globin chains may takes place. These are referred to as X1c (X being any hemoglobin, e.g., HbA1c). When hemoglobin molecule is aging, glutathione is bound to cysteine at the 93rd position of the β-chain. This is HbAIII or HbA1d. Just like Hb A1c and HbA1d, there can exist HbC1c, HbC1d, Hb S1c, and HbS1d.
- Heme is synthesized in a complex way involving enzymes in both mitochondrion and cytosol.
- Hemoglobinopathies can be divided into three major categories.
- Quantitative disorders of hemoglobin synthesis: Production of structurally normal but decreased amounts of globin chains (thalassemia syndrome)
- Disorder (qualitative) in hemoglobin structure: Production of structurally abnormal globulin chains such as hemoglobin S, C, O, or E. Sickle cell syndrome is the most common example of such disease.
- Failure to switch globin chain synthesis after birth: HPFH (Hb F), a relatively benign condition, may coexist with thalassemia or sickle cell disease but with decreased severity of such diseases (protective effect).
- Hemoglobinopathies are transmitted in autosomal recessive fashion.
- Disorders due to β-chain defect such as sickle cell disease tend to manifest clinically after 6 months of age although diseases due to α-chain defect are manifested in utero or following birth.
- The hemoglobin variants of most clinical significance are hemoglobin S, C, and E. In West Africa, approximately 25% of individuals are heterozygous for hemoglobin S (Hb S) gene, which is related to sickle cell diseases. In addition, high frequencies of Hb S gene alleles are also found in people of Caribbean, South and Central Africa, Mediterranean, Arabian Peninsula, and East India. Hemoglobin C (Hb C) is found mostly in people living or originating from West Africa. Hemoglobin E (Hb E) is widely distributed between East India and Southeast Asia with highest

prevalence in Thailand, Laos, and Cambodia, but may be sporadically observed in parts of China and Indonesia. Thalassemia syndrome is not due to structural defect in globin chain but due to lack of sufficient synthesis of globin chain and is also a genetically inherited disease. Thalassemic syndrome can be divided into α-thalassemia and β-thalassemia. In general, β-thalassemia is observed in Mediterranean, Arabian Peninsula, Turkey, Iran, West and Central Africa, India, and other Southeast Asian countries, whereas α-thalassemia is commonly observed in parts of Africa, Mediterranean, Middle East, and throughout Southeast Asia.

α-Thalassemia occurs when there is a defect or deletion in one or more of four genes responsible for α-globin production. α-Thalassemia can be divided into four categories:

- The silent carriers: Characterized by only one defective or deleted gene but three functional genes. These individuals have no health problem. An unusual case of silent carrier is individuals carrying one defective Constant Spring mutation but three functional genes. These individuals also have no health problem.
- α-Thalassemia trait: Characterized by two deleted or defective genes and two functional genes. These individuals may have mild anemia.
- α-Thalassemia major (hemoglobin H disease): Characterized by three deleted or defective genes and only one functional gene. These patients have persistent anemia and significant health problems. When hemoglobin H disease is combined with hemoglobin Constant Spring, the severity of disease is more than hemoglobin H disease. However, if the child inherits one hemoglobin Constant Spring from mother and one from father, then the child has homozygous hemoglobin constant spring and severity of disease is similar to hemoglobin H disease.
- Hydrops Fetalis: Characterized by no functional α gene and these individuals have hemoglobin Bart. This condition is not compatible with life unless intrauterine transfusion is initiated.
- Hemoglobin Constant Spring (hemoglobin variant isolated from a family of ethnic Chinese background from Constant Spring district of Jamaica) is hemoglobin variant where mutation of α-globin gene produces an abnormally long α-chain (172 amino acids instead of normal 141 amino acids). Hemoglobin Constant Spring is due to nondeletion mutation of α gene which results in production of unstable α-globin. Moreover, this α-globin is produced in very low quantity (approximately 1% of normal expression level) and is found in people living or originating in Southeast Asia.
- β-Thalassemia can be broadly divided into three categories:
- β-Thalassemia trait: Characterized by one defective gene and one normal gene, and individuals may experience mild anemia but not transfusion dependent.
- β-Thalassemia intermedia: Characterized by two defective genes, but some β-globin production is still observed in these individuals. However, some individuals may have significant health problems requiring intermittent transfusion.
- β-Thalassemia major (Cooley's anemia): Characterized by two defective genes, but almost no function of either gene leading to no synthesis of β-globin. These individuals have severe form of disease requiring lifelong transfusion and may have shortened life span.
- Patients with β-thalassemia major have elevated HbA^2 and HbF (although in some individual HbF may be normal).
- In the heterozygous form (Hb AS), sickle cell trait protects from infection of *P. falciparum* malaria but not in the more severe form of homozygous sickle cell disease (Hb SS). The genetic defect

producing sickle hemoglobin is a single nucleotide substitution at codon 6 of the β-globin gene on chromosome 11 that results in a point mutation in β-globin chain of hemoglobin (substitution of valine for glutamic acid at sixth position).

- Double heterozygous states of Hb SC, Hb SD, and Hb SO Arab are important sickling states that should not be missed.
- Hemoglobin C is formed because of substitution of glutamic acid residue with a lysine residue at the sixth position of β-globin. Hemoglobin E is due to point mutation of β-globin, which results in substitution of lysine for glutamic acid in position 26
- Hemoglobin Lepore is an unusual hemoglobin molecule which is composed of two α-chains and two δβ-chains as a result of fusion gene of δ and β genes. The δβ-chains have the first 87 amino acids of the δ-chain and 32 amino acids of the β-chain.

Individuals with HbA/Hb Lepore are asymptomatic with Hb Lepore representing 5%−15% of hemoglobin, slightly elevated Hb F (2%−3%) and with MCV as well as MCH. However, homozygous Lepore individuals suffer from severe anemia similar to patients with β-thalassemia intermedia with Hb Lepore representing 8%−30% of hemoglobin and remainder hemoglobin is Hb F.

- Hemoglobin G-Philadelphia (Hb G) is the commonest α-chain defect observed 1 in 5000 of African-Americans and is associated with α-thalassemia 2 deletions.
- It is possible that an African-American individual may have Hb S/Hb G where hemoglobin molecule contains one normal α-chain, one α-G-chain, one normal β-chain, and one β-S-chain. This can results in detection of various hemoglobin in the blood including Hb A (α2, β2), Hb S (α2, β S2), Hb G (α G2, β2), and HbS/G (α G2, β S2). In addition, Hb G2 (α2, δ2) which is the counterpart of HbA2 is also present.
- Increase in fetal hemoglobin percentage is associated with multiple pathologic states. These include β-thalassemia, δβ-thalassemia, and HPFH. While β-thalassemia is associated with high HbA2, the latter two states are associated with normal HbA2 values. Hematologic malignancies are associated with increased hemoglobin F and include acute erythroid leukemia (AML, M6) and JMML. Aplastic anemia is also associated with an increase in percentage of Hb F. In elucidating the actual cause of high Hb F, it is important to consider the actual percentage of Hb F and HbA2 values as well as correlation with CBC and peripheral smear findings. It is also important to note that drugs (hydroxyurea, sodium valproate, erythropoietin) and stress erythropoiesis may also result in high Hb F. Hydroxyurea is used in sickle cell disease patients to increase the amount of Hb F, presence of which may help to reduce the clinical effects of the disease. Measuring the level of Hb F may be useful in determining the appropriate dose of hydroxyurea. In 15%−20% of cases of pregnancy, HbF may be raised to values as much as 5%.

References

[1] Rahimi Z. Genetic, epidemiology, hematological and clinical features of hemoglobinopathies in Iran. BioMed Res Int 2013;2013:803487.
[2] Meier ER, Rampersad A. Pediatric sickle cell disease: past successes and future challenges. Pediatr Res 2017;81:249−58.

[3] Kassim AA, Sharma D. Hematopoietic stem cell transplantation for sickle cell disease: the changing landscape. Hematol Oncol Stem Cell Ther 2017;10:259−66.

[4] Manning LR, Russell JR, Padovan JC, Chait BT, et al. Human embryonic, fetal and adult hemoglobins have different subunit interface strength. Correlation with lifespan in the red cell. Protein Sci 2007;16:164101658.

[5] Giordano PC. Strategies for basic laboratory diagnostics of the hemoglobinopathies in multi-ethnic societies: interpretation of results and pitfalls. Int J Lab Hematol 2013, Oct;35(5):465−79. https://doi.org/10.1111/ijlh.12037. Epub 2012 Dec 7.

[6] Rappaport VJ, Velazquez M, Willaims K. Hemoglobinopathies in pregnancy. Obstet Gynecol Clin N Am 2004;31:287−317.

[7] Sriiam S, Leecharoenkiat A, Lithanatudom P, Wannatung T, et al. Proteomic analysis of hemoglobin H constant spring (HB H-CS) erythroblasts. Blood Cells Mol Dis 2012;48:77−85.

[8] Kohne E. Hemoglobinopathies: clinical manifestations, diagnosis and treatment. Dtsch Arztebl Int 2011; 108:532−40.

[9] Lubin B, Witkowska E, Kleman K. Laboratory diagnosis of hemoglobinopathies. Clin Biochem 1991;24: 363−74.

[10] Ngo DA, Aygun B, Akinsheye I, Hankins JS. Fetal hemoglobin levels and hematological characteristics of compound heterozygous for hemoglobin S and deletional hereditary persistence of fetal hemoglobin. Br J Haematol 2012;156:259−64.

[11] Pandey S, Mishra RM, Pandey S, Shah V, et al. Molecular characterization of hemoglobin D Punjab traits and clinical-hematological profile of patients. Sao Paulo Med J 2012;130:248−51.

[12] Dror S. Clinical and hematological features of homozygous hemoglobin O-Arab [beta 121 Glu − LYS]. Pediatr Blood cancer 2013;60:506−7.

[13] Muncie H, Campbell JS. Alpha and beta thalassemia. Am Fam Physician 2009;339:344−71.

[14] Paikari A, Sheehan VA. Fetal haemoglobin induction in sickle cell disease. Br J Haematol 2018;180: 189−200.

[15] Smaldone A. Glycemic control and hemoglobinopathy: when A1C may not be reliable. Diabetes Spectr 2008;21:46−9.

[16] Kozarski TB, Howanitz PJ, Howanitz JH, Lilic N, et al. Blood transfusion leading to apparent hemoglobin C, S and O-Arab hemoglobinopathies. Arch Pathol Lab Med 2006;130:1830−3.

[17] Benson JA, Therell BL. History and current status of newborn screening for hemoglobinopathies. Semin Perinatol 2010;34:134−44.

[18] Colombatti R, Martella M, Cattaneo L, Viola G, et al. Results of a multicenter universal newborn screening program for sickle cell disease in Italy: a call to action. Pediatr Blood Canc 2019:e27657.

Benign white blood cell and platelet disorders

5

Introduction

Automated hematology analyzer can rapidly analyze whole blood specimens for the complete blood count (CBC). Results include red blood cell (RBC) count, white blood cell count (WBC), platelet count, hemoglobin concentration, hematocrit, RBC indices, and a leukocyte differential. A less sophisticated automated hematology analyzer in a physician's office setting may sometimes provide a limited CBC, using older technology for whole blood analysis (for example, impedance technology), which will generate only a three-part leukocyte differential. A three-part leukocyte differential will provide values for neutrophils, lymphocytes, and all other white cells together. More modern hematology analyzers are capable of analyzing all leukocytes using flow cytometry-based method, some in combination with cytochemistry or fluorescences or conductivity to count all the different types of WBC including neutrophils, lymphocytes, monocytes, basophils, and eosinophils (five-part differential). Nucleated RBCs are also detected. Leukocytosis or elevated WBC count is a common laboratory finding. Normal WBC differential also changes with age, and proper normal ranges must be established. For example, a WBC count of $30 \times 10\ 9/L$ (30,00/uL) is abnormal in an adult but normal in a newborn within the first few days of life. Leukocytosis is also a feature of leukemias, and thus distinguishing leukemias from other causes of leukocytosis is crucial. Examination of peripheral blood smear along with review of CBC analysis is essential for such differentiation, and if necessary, further analysis such as flow cytometry, molecular studies, and possible bone marrow examination must be undertaken [1]. Similarly, platelet disorder such as thrombocytopenia or thrombocytosis may be a benign condition or reflection of a serious condition such as severe thrombocytopenia observed in patients with acute leukemias. In this chapter, benign WBC and platelet disorders are reviewed.

Hereditary variation in white blood cell morphology

Various hereditary-mediated variations in WBC morphology have been described, which are mostly clinically benign except for Chediak—Higashi syndrome. Pelger-Huet anomaly is a rare disorder transmitted as autosomal dominant, where >75% of neutrophils are bilobed (neutrophils with hyposegmented nucleus). The characteristic morphology was described by Pelger in 1928, and in 1931, Huet identified it as an inherited disorder. This is a benign condition due to inherited defect of terminal neutrophil differentiation as a result of mutations in the lamin B receptor gene. Presence of occasional bilobed neutrophils is also seen in myelodysplastic syndrome (MDS) and is referred to as pseudo—Pelger-Huët cells. These have also been described in association with Filovirus infection. Distinguishing benign form of Pelger-Huet anomaly from acquired or pseudo—Pelger-Huet anomaly is important.

May—Hegglin anomaly, another rare disorder, is also transmitted as autosomal dominant and is characterized by thrombocytopenia, giant platelets, and leukocyte defect causing the appearance of very small rods (2—5 μm) in the cytoplasm (Dohle-like bodies). Actual Dohle bodies are seen with reactive or toxic polymorphonuclear neutrophils (PMNs). Reactive or toxic PMNs also have prominent azurophilic granules and vacuoles. These features are not seen in May—Hegglin anomaly. May—Hegglin anomaly is a benign condition as most patients do not appear to have any significant bleeding problem where treatment may be required.

Macrothrombocytopenia with neutrophilic inclusions (MYH9 disorders) is a group of disorders characterized by mutations in the MYH9 gene. This gene encodes the nonmuscle myosin heavy chain class IIA protein. These disorders are characterized by thrombocytopenia, large/giant platelet, and Dohle-like bodies in neutrophils. This group of disorders includes May—Hegglin anomaly (autosomal dominant), Sebastian syndrome, Fechtner syndrome (nephritis, ocular defects, and sensorineural hearing loss), and Epstein syndrome.

Chediak—Higashi syndrome is a rare multisystemic disorder transmitted as autosomal recessive characterized by hypopigmentation of the skin, eyes, and hair (silver hair), prolonged bleeding time, easy bruisability, and immunodeficiency. The pathological feature of this rare disease is the presence of massive lysosomal inclusion, which is formed through a combination of fusion, cytoplasmic injury and phagocytosis, in WBC. This abnormal inclusion may be responsible for most of the impaired leukocyte and other blood cell functions in these patients. In addition, dysfunction of natural killer cell is also observed. Approximately 85% of affected individuals develop the accelerated phase of this disease, a lymphoproliferative infiltration of bone marrow, and reticuloendothelial system mostly during childhood [2]. This could be a potentially fatal condition if not treated. Hematopoietic stem cell transplantation may be curative [3].

Alder—Reilly anomaly, a clinically benign rare condition, is transmitted as autosomal recessive and is characterized by large azurophilic granules (partially degraded protein—carbohydrate complexes known as mucopolysaccharides) in neutrophils and others granulocytes, monocytes, and lymphocytes. However, similar abnormalities are seen in mucopolysaccharidoses (MPS). Maroteaux—Lamy syndrome is when abnormal granulation of granulocytes and monocytes with lymphocyte occurs with vacuolation, as seen in MPS VI.

Changes in white cell counts

Changes in white cell count can be observed in various conditions where values may be increased or decreased. Leukocytosis is a common clinical observation where an increase in WBC count above two standard deviation of the mean is observed (above 11,000 per microliter). Leukocytosis is mostly a benign condition where elevated WBC reflects the normal response of bone marrow to infection, an inflammatory process, or drugs. However, leukocytosis may also be due to bone marrow abnormality related to leukemia or myeloproliferative disease. Widick and Winer summarized various causes of leukocytosis [4]. In this chapter, various changes in WBC count due to benign conditions will be discussed.

Neutrophilia

Neutrophilia is the most common cause of leukocytosis. Various causes of neutrophilia are summarized in Table 5.1. Most common cause is infection or inflammation, but there are congenital forms of neutrophilia, and such conditions are rarely encountered. For example, leukocyte adhesion deficiency is a rare autosomal recessive immunodeficiency disease characterized by severe recurrent bacterial infection due to inability of neutrophils to adhere to endothelial cell walls and migrating to the site of infection. Depending on the genetic effect, hematopoietic stem cell transplantation is often the only cure [5].

Various morphological changes are observed in reactive neutrophilia. Such morphologic changes seen in reactive neutrophils are summarized below:

- Dohle bodies (represent endoplasmic reticulum)
- Prominent 1st degree (azurophilic) granules in cytoplasm
- Vacuoles in cytoplasm

Significant neutrophilic leukocytosis may result in a picture resembling chronic myeloid leukemia (CML), which is referred to as leukemoid reaction. Basophilia and eosinophilia are not significant. The leukocyte alkaline phosphatase (LAP) score is high in such conditions. In contrast, the LAP score in chronic myeloid leukemia (CML) is low. Another cause of low LAP, although unrelated, is paroxysmal nocturnal hemoglobinuria.

To determine LAP score, PMNs are stained for alkaline phosphatase, and each neutrophil is given a score from 0 to 4. One hundred cells are counted. The total score is the actual score.

Eosinophilia and monocytosis

Eosinophils are WBCs that participate in immunological and allergic events. Eosinophilia is usually defined as eosinophil count greater than 500 cells per microliter (0.5 × 10^9/L). Parasitic infections are often responsible for eosinophilia in pediatric patients. Infections such as scarlet fever, chorea, and genitourinary infection may also cause eosinophilia. Skin rash, chronic inflammation such as rheumatoid arthritis and lupus erythematosus, and adrenal insufficiency such as Addison's disease may also cause eosinophilia. Pleural and pulmonary conditions such as Loffler's syndrome may also cause eosinophilia. Eosinophilia—myalgia is a disorder associated with dietary supplement tryptophan. Other causes of eosinophilia include malignancies affecting the immune system such as Hodgkin's lymphoma and non-Hodgkin's lymphoma [6].

Table 5.1 Various causes of neutrophilia.

- Infection
- Inflammation
- Any form of stress
- Splenectomy, hyposplenism
- Drugs: corticosteroids
- Acute hemorrhage
- Hemolytic anemia
- Congenital forms (rare): leukocyte adhesion deficiency, chronic idiopathic neutrophilia, hereditary neutrophilia

Hypereosinophilic syndrome is a rare and heterogenous group of hematological and systemic disorder characterized by eosinophil count above 1.5×10^9/L (1,500 cells per microliter) lasting over 6 months in absence of other known cause of eosinophilia. This syndrome may cause end organ damage, primarily heart, causing eosinophilic endomyocardial fibrosis. Patients with hypereosinophilic syndrome do not typically have asthma [7].

Monocytes are WBCs that give rise to macrophages and dendritic cells in the immune system. Causes of monocytosis include various infections (tuberculosis, brucellosis, typhoid, typhus, Rocky Mountain spotted fever, malaria, etc.), chronic myelomonocytic leukemia, juvenile myelomonocytic leukemia, Hodgkin lymphoma, acute myeloid leukemia (AML) M4/M5, and autoimmune diseases.

Basophilia

Basophils are inflammatory mediators of substance such as histamine, and along with mast cells, they have receptors for IgE. Basophilia is an uncommon situation. Causes include viral infections (varicella, chicken pox), inflammatory conditions (ulcerative colitis), CML or myeloproliferative disorder, myxoedema, and endocrinological causes (hypothyroidism, ovulation).

Neutropenia

Neutropenia is defined as an absolute neutrophil count that is more than two standard deviations below the normal neutrophil count. Mild-to-moderate neutropenia may not predispose an individual to an increased susceptibility to life-threatening infection, but patients with severe neutropenia (neutrophil count $<0.5 \times 10^9$/L) may be prone to severe even life-threatening infection. Severe neutropenia accompanied by fever of recent onset is a medical emergency requiring immediate investigation and treatment. Various causes of neutropenia (inherited and acquired) are summarized in Table 5.2. Most often neutropenia in adults is due to acquired causes such as drug-induced or postinfection. Felty's syndrome is a well-characterized clinical abnormality consisting of rheumatoid arthritis, splenomegaly, and severe neutropenia. The cause could be attributed to increased neutrophil margination and inhibition of granulopoiesis mediated by antibodies to neutrophils or by T cells.

Table 5.2 Various causes of selective neutropenia.

Acquired Causes	• Drug induced (e.g., methimazole, sulfasalazine, trimethoprim/sulfamethoxazole, Bactrim)
	• Postinfectious
	• Autoimmune disease (e.g., systemic lupus erythematosus)
	• Paroxysmal nocturnal hemoglobinuria
	• Felty's syndrome (triad of rheumatoid arthritis, splenomegaly, and neutropenia)
	• Splenomegaly
Inherited Forms	• Ethnic familial neutropenia
	• Cyclical neutropenia
	• Kostmann syndrome
	• Shwachman–Diamond syndrome

Neutropenia may also be inherited. Yemenite Jews and other population including Ethiopian Jews and Bedouins have low neutrophil counts, and this condition is called ethnic benign neutropenia because it is not associated with increased risk of infection. This condition may also be found in populations around the world including Africans, Africans-Americans, and African-Caribbean [8]. Cyclical neutropenia is a rare disorder caused by a stem cell regulatory defect characterized by a transient severe neutropenia occurring roughly every 21 days. The familial form seems to be inherited in an autosomal dominant pattern.

Shwachman–Diamond syndrome is transmitted as autosomal recessive and is characterized by exocrine pancreatic deficiency and neutropenia, thrombocytopenia, short stature, and mental retardation. This disorder may progress to MDS and AML.

Kostmann's syndrome is a group of inherited diseases, causing a congenital form of neutropenia, called severe congenital neutropenia (SCN). It was discovered in 1956 in a family in Sweden by the physician Rolf Kostmann. This disorder was originally thought to be due to point mutations in gene coding for granulocyte colony–stimulating factor (G-CSF). It has since been demonstrated that this mutation is an acquired somatic mutation and may be associated with the development of acute leukemia. The most common subtype, SCN1, is transmitted as autosomal dominant. Here, there are mutations in the gene for neutrophil elastase (ELANE, previously called ELA2). The initial recognized subtype, SCN3, is transmitted as autosomal recessive. Here, there are mutations in the HAX1 gene. X-linked inheritance may also be seen in SCN due to mutations in the WAS gene. Affected children develop frequent life-threatening infection due to severe neutropenia. There is increased risk of acute leukemia.

Lymphocytosis and infectious mononucleosis

Lymphocytosis occurs most commonly after viral infections (e.g., cytomegalovirus, mumps, varicella, influenza, rubella, etc.), lymphoid leukemias and lymphomas, and smoking. Lymphocytosis is rarely observed in bacterial infection, an exception being *Bordetella pertussis* infection. With lymphocytosis, reactive lymphocytes may be seen in the peripheral smear. Reactive lymphocytes are also referred to as Downey cells. There are three types of Downey cells:

- Type I: small cells with minimum cytoplasm, indented nucleus/irregular nuclear membrane, and condensed chromatin
- Type II: larger cells with abundant cytoplasm, the lymphocyte cytoplasm seem to hug the red cells. This is the most common type of Downey cells.
- Type III: cells with large moderate basophilic cytoplasm and nucleus with coarse chromatin. Nucleoli are apparent.

In infectious mononucleosis caused by Epstein–Barr virus, there is lymphocytosis with characteristic large atypical lymphocyte in the blood. The virus infects B lymphocytes, and T lymphocytes attack the virally infected B lymphocytes. Reactive lymphocytes are the T lymphocytes. Features of infectious mononucleosis from the peripheral blood include

- white cells of 50% or more are mononuclear cells
- At least 10% of the lymphocytes exhibit reactive changes
- There is lymphocytic morphologic heterogeneity (this means different types of reactive lymphocytes are seen)

Lymphocytopenia

Lymphocytes consist of T lymphocytes, B lymphocytes, and natural killer cells. The term lympho-cytopenia (lymphopenia) refers to less than 1000 lymphocytes per microliter of blood in adults or less than 3000 lymphocytes per microliter of blood in children. Severe combined immunodeficiency is a heterogenous disorder characterized by severe deficiency of T and B lymphocytes as well as natural killer cells. The X-linked inherited form is most commonly characterized by the absence of T lymphocytes and natural killer cells but poorly functioning B cells. A deficiency of adenosine deaminase underlies 30%−40% of autosomal recessive form of this inherited disorder. The different forms of severe combined immunodeficiency are clinically indistinguishable and observed in early infancy manifested by severe infection. Treatment includes correction of the defect by stem cell transplant or enzyme replacement with adenosine deaminase. Lymphocytopenia may also be acquired, for example, in patients with HIV infection [9].

Platelet disorders

Common platelet disorders include thrombocytopenia, thrombocytopathia, and thrombocytosis. Bolton-Maggs et al. reviewed various inherited platelet disorders with guidelines for their management [10]. Thrombocytopenia is often discovered incidentally during office visit of a patient when CBC is ordered along with other tests. However, severe thrombocytopenia may be a reflection of a severe disease. Similarly thrombocytosis is a common finding during a routine blood test and may represent a benign condition. However, like severe thrombocytopenia, severe thrombocytosis may be a reflection of a serious clinical condition requiring further investigation.

Normal platelet physiology and pathology leading to disorders of coagulation are discussed in Chapter 15.

Thrombocytopenias

Thrombocytopenia is defined as platelet count below 150,000 per microliter of blood (150 × 10^9/L), but patients even with a platelet count of 50,000 per microliter or more may be asymptomatic. However, counts from 10,000 to 30,000 per microliter may be associated with bleeding, and especially patients with platelet count below 10,000 per microliter are very sensitive to spontaneous bleeding.

Before discussing thrombocytopenia, it is important to mention pseudothrombocytopenia. This is when the platelet count appears low but is not truly so. Causes of pseudothrombocytopenia are

- Platelet clumping (may occur when blood is collected in EDTA; blood needs to be recollected in blue or green top)
- Platelet satellitism (when platelets adhere to neutrophils, again seen when blood is collected in EDTA)
- Presence of numerous large platelets (platelets are falsely counted as red cells)
- Traumatic venipuncture (platelets are activated and cause them to aggregate)

True thrombocytopenia may be due to decreased platelet production (congenital or acquired), increased destruction of platelets, increased platelet consumption, or sequestration. Various causes of

Table 5.3 Various causes of thrombocytopenia.

Decreased Production (any cause of bone marrow suppression/ failure)	• Bone marrow failure (aplastic anemia, paroxysmal nocturnal hemoglobinuria, etc.) • Bone marrow suppression due to medication, chemotherapy, or radiation therapy • Infection (cytomegalovirus, HIV, parvovirus B19, hepatitis C, etc.) • Myelodysplastic syndrome • Neoplastic marrow infiltration • Inherited forms are summarized in Table 5.4
Increased Platelet Consumption	• Disseminated intravascular coagulation • Thrombotic thrombocytopenic purpura • Hemolytic uremic syndrome
Increased Platelet Destruction	• Immune thrombocytopenic purpura • Mechanical destruction, e.g., mechanical valves, extracorporeal bypass
Thrombocytopenias due to Sequestration	• Sequestration in hemangiomas (Kasabach−Merritt syndrome)

thrombocytopenias, except inherited forms, are given in Table 5.3. Inherited forms of thrombocytopenia are summarized in Table 5.4.

Congenital thrombocytopenia can be broadly divided into three groups namely cytopenia with small platelets, cytopenia with normal platelets, and cytopenia with large platelets.

Kasabach−Merritt syndrome is a rare locally aggressive vascular tumor characterized by a rapidly enlarging vascular anomaly, consumption coagulopathy, thrombocytopenia, prolonged bleeding time, hypofibrinogenemia, presence of D dimer, and fibrin split products with or without microangiopathic hemolytic anemia. Prognosis is poor due to availability of few treatment options [11].

Table 5.4 Various types of congenital thrombocytopenia.

Congenital thrombocytopenia with small platelets	Congenital thrombocytopenia with normal-sized platelets	Congenital thrombocytopenia with large platelets
• Wiskott−Aldrich syndrome (WAS): X-linked recessive related to mutation of WASP gene • X-linked thrombocytopenia: isolated thrombocytopenia related also to mutation of WASP gene, but disease is milder than WAS • Inherited microthrombocytes: transmitted as autosomal dominant with normal platelet function	• Thrombocytopenia with absent radii (TAR syndrome) • Amegakaryocytic thrombocytopenia due to mutation of MPL gene • Fanconi's anemia: autosomal recessive	• Bernard−Soulier syndrome: autosomal recessive • May−Hegglin anomaly: autosomal dominant • Sebastian syndrome: autosomal dominant • Epstein syndrome: autosomal dominant • Fechtner syndrome: autosomal dominant • Gray platelet syndrome: autosomal recessive • DiGeorge and velocardiofacial syndrome: autosomal dominant

Congenital thrombocytopenia with small platelets could be related to Wiskott–Aldrich syndrome (WAS), which is characterized by eczema, immunodeficiency, and thrombocytopenia. This disease is transmitted as X-linked recessive due to inheritance of the WASP gene, located at Xp11 encoding WAS protein (WASP). Platelets are dysfunctional. X-linked thrombocytopenia, a congenital disorder characterized by isolated thrombocytopenia and small platelet but in general without other complications as seen in WAS, is a mild allelic variant also caused by mutation of WASP gene [12]. Inherited microthrombocytes is a disorder transmitted as autosomal dominant with normal platelet function.

Congenital thrombocytopenia with normal-sized platelets can be related to thrombocytopenia with absent radii (TAR syndrome), amegakaryocytic thrombocytopenia which is due to mutation of MPL gene that encodes thrombopoietin receptor (thrombopoietin is required for maturation of megakaryoblasts to megakaryocytes), or Fanconi's anemia which is transmitted in autosomal recesssive pattern.

Congenital thrombocytopenia with large platelets can be related to Bernard–Soulier syndrome, which is transmitted as autosomal recessive pattern. In early phase of primary hemostasis, platelets adhere to damaged vessel walls by binding via the platelet glycoprotein (GP)Ib-V-IX complex to von Willebrand factor exposed on the subendothelium. In this disorder, the (GP)Ib-V-IX complex is abnormal. Platelet aggregation studies show impaired aggregation to ristocetin. Some cases of Bernard–Soulier syndrome are due to defects of the GpIbbeta gene located on chromosome 22. This gene may be affected in velocardiofacial syndrome or DiGeorge syndrome associated with deletion of 22q11.2. Mhawech et al. reviewed various inherited giant platelet disorders [13].

Congenital thrombocytopenia with large platelets can also be due to May–Hegglin anomaly, another inherited disorder is transmitted in autosomal dominant pattern. This disorder is due to defective myosin heavy chain 9 gene at 22q11. Neutrophils have Dohle-like bodies. Sebastian syndrome is transmitted as autosomal dominant pattern and is due to defective myosin heavy chain 9 gene at 22q11. Epstein syndrome is transmitted as autosomal dominant pattern and is related to defective myosin heavy chain 9 gene at 22q11. Patients have Alport's-like syndrome with features of nephritis, sensorineural deafness, and cataract. Fechtner syndrome is transmitted as autosomal dominant and also due to defective myosin heavy chain 9 gene at 22q11. Patients have Alport's-like syndrome with features of nephritis, sensorineural deafness, cataract and Dohle-like bodies in neutrophils. Gray platelet syndrome, a rare congenital autosomal recessive bleeding disorder, is due to hypogranular platelets, which are dysfunctional. DiGeorge and velocardiofacial syndrome are transmitted as autosomal dominant with loss of function of GP1BB gene at 22q11. In this disorder, cardiac, parathyroid, and thymus abnormalities may be observed.

Thrombocytosis

Thrombocytosis is defined as platelet count exceeding 450,000 per microliter (450×10^9/L). This abnormality is termed as primary thrombocytosis if platelet increase is related to alterations targeting the hematopoietic cells in the bone marrow. Examples of such states are essential thrombocythemia or thrombocytosis seen in other myeloproliferative disorders [14]. The disorder is considered as secondary (also called reactive thrombocytosis) if platelet increase is due to external cause such infection, inflammation, neoplasms, or iron deficiency. Secondary thrombocytosis can also be due to redistribution such as observed postsplenectomy.

There are several examples of primary thrombocytosis which are inherited disorders. Familial thrombocytosis is transmitted in autosomal dominant pattern and is due to mutation in the thrombopoietin receptor gene (myeloproliferative leukemia virus oncogene MPL gene; first identified from the murine myeloproliferative leukemia virus that was capable of immortalizing bone marrow hematopoietic cells from different lineages). Several mutations of this gene causing thrombocytopenia have been reported. Inherited primary thrombocytopenia may also due to mutation of the promoter of thrombopoietin gene that encodes thrombopoietin. Thrombopoietin is the primary humoral regulator of platelet production.

Thrombocytopathia

Thrombocytopathia is any of several hematological disorders characterized by dysfunctional platelets (thrombocytes) which lead to prolonged bleeding time, defective clot formation, and a tendency for hemorrhage. Thrombocytopathia may be congenital or acquired.

Congenital causes of thrombocytopathia include

- Disorders of platelet adhesion: von Willebrand disease (VWD), Bernard—Soulier syndrome
- Disorders of platelet activation: storage pool disorders, Chediak—Higashi syndrome, and Hermansky—Pudlak syndrome
- Disorders of platelet aggregation: Glanzmann's syndrome

Acquired causes of thrombocytopathia include

- Drugs: Aspirin, NSAIDs (nonsteroidal antiinflammatory drugs)
- Uremia
- Acquired VWD
- Myeloproliferative diseases
- Antiplatelet antibodies

Key points

- Pelger—Huet anomaly is transmitted as autosomal dominant pattern where >75% of neutrophils are bilobed. Presence of occasional bilobed neutrophils is also seen in MDS and is referred to as Pseudo—Pelger-Huët cells.
- May—Hegglin anomaly is transmitted as autosomal dominant pattern and is characterized by thrombocytopenia, giant platelets, prominent Dohle-like bodies.
- Chediak—Higashi syndrome is transmitted as autosomal recessive pattern and is characterized by giant neutrophilic granules (due to fusion of lysosomes), defective chemotaxis and phagocytosis, immunodeficiency, and oculocutaneous albinism.
- Alder—Reilly anomaly is transmitted as autosomal recessive pattern and is characterized by large azurophilic granules in neutrophils and others granulocytes, monocytes, and lymphocytes; the condition is clinically benign.
- Kostmann's syndrome is an example of congenital neutropenia, which inherited as autosomal recessive pattern due to point mutations in gene coding for G-CSF.

- Shwachman–Diamond syndrome is another example of congenital neutropenia, which is also transmitted as autosomal recessive pattern. This disorder is characterized by exocrine pancreatic deficiency and neutropenia, thrombocytopenia, short stature, and mental retardation; this may progress to MDS and AML.
- Reactive lymphocytes are also referred to as Downey cells. There are three types of Downey cells: Type I: small cells with minimum cytoplasm, indented nucleus/irregular nuclear membrane, and condensed chromatin; Type II: larger cells with abundant cytoplasm; the lymphocyte cytoplasm seem to hug the red cells. This is the most common type of Downey cells; and Type III: cells with large moderate basophilic cytoplasm and nucleus with coarse chromatin. Nucleoli are apparent.
- Features of infectious mononucleosis, from the peripheral blood; 50% or more of the white cells are mononuclear cells; at least 10% of the lymphocytes exhibit reactive changes, and there is lymphocytic morphologic heterogeneity (this means different types of reactive lymphocytes are seen)
- Wiskott–Aldrich syndrome is characterized by eczema, immunodeficiency, and thrombocytopenia. This disorder is transmitted as X-linked recessive due to inheritance of the WASP gene, located at Xp11. Platelets are dysfunctional.
- Bernard–Soulier syndrome is transmitted as autosomal recessive. The GpIb/IX/V complex is abnormal. Platelet aggregation studies show impaired aggregation to ristocetin. Some cases of Bernard–Soulier syndrome are due to defects of the GpIbbeta gene located on chromosome 22. This gene may be affected in velocardiofacial syndrome or DiGeorge syndrome associated with deletion of 22q11.2
- Causes of thrombocytopenia with giant platelets and associated with defective myosin heavy chain 9 gene at 22q11 are Bernard–Soulier syndrome, May Hegglin anomaly, Sebastian syndrome, Epstein syndrome, and Fechtner syndrome.

References

[1] George TI. Malignant or benign leukocytosis. Hematol Am Soc Hematol Educ Prog 2012;2012:475–84.
[2] Roy A, Kar R, Basu D, Srivani S, et al. Clinico-hematological profile of Chediak-Higashi syndrome: experience from a tertiary care center in South India. Indian J Pathol Microbiol 2011;54:547–51.
[3] Nagai K, Ochi F, Maeda M, Ohga S, et al. Clinical characteristics and outcomes of Chediak-Higashi syndrome: a nationwide survey in Japan. Pediatr Blood Cancer 2013;60:1582–6.
[4] Widick P, Winer ES. Leukocytosis and leukemia. Prim Care 2016;43:575–87.
[5] Van de Vijver E, van den Berg TK, Kuijpers TW. Leukocyte adhesion deficiencies. Hematol Oncol Clin N Am 2013;27:101–16.
[6] Abramson N, Melton B. Leukocytosis: basics of clinical assessment. Am Fam Physician 2000;62:2053–60.
[7] Kahn JE, Bletry O, Guillevin L. Hypereosinophilic syndrome. Best Pract Res Clin Rheumatol 2008;22: 863–82.
[8] Paz Z, Nails M, Ziv E. The genetics of benign neutropenia. Isr Med Assoc J 2011;13:625–9.
[9] Stock W, Hoffman R. White blood cell 1: non-malignant disorders. Lancet 2000;355:1351–7.
[10] Bolton-Maggs PH, Chalmers EA, Collins PW, Harrison P, et al. A review of inherited platelet disorders with guidelines for their management on behalf of the UKHCDO. Br J Haematol 2006;135:603–33.

[11] Acharya S, Pillai K, Francis A, Criton S, et al. Kasabach—Merritt syndrome: management with interferon. Indian J Dermatol 2010;55:281—3.

[12] Imai K, Morio T, Zhu Y, Kin Y, et al. Clinical course of patients with WASP gene mutation. Blood 2004;103: 456—64.

[13] Mhawech P, Saleem A. Inherited giant platelet disorders. Classification and literature review. Am J Clin Pathol 2000;113:176—90.

[14] Vannucchi AM, Guglielmelli P, Tefferi A. Polycythemia vera and essential thrombocythemia: algorithmic approach. Curr Opin Hematol 2018;25:112—9.

Myeloid neoplasms

6

Introduction

Mature blood cells (white blood cells [WBCs], red blood cells [RBCs], and platelets) are created from one of two types of hematopoietic stem cells present in the bone marrow, lymphoid stem cells, and myeloid stem cells. Lymphoid stem cells mature to lymphocytes, and myeloid stem cells mature into granulocytes, monocytes, RBC, and platelets. In patients with leukemia, the bone marrow produces either abnormal lymphoid cells or abnormal myeloid cells. This chapter focuses on myeloid neoplasms only.

The term "myeloid" includes all cells belonging to the granulocyte (neutrophils, eosinophils, and basophils), monocyte/macrophage, erythroid, megakaryocyte, and mast cell lineages. Myeloid neoplasms (myeloid malignancies) are clonal diseases of hematopoietic stem or progenitor cells due to genetic or epigenetic factors that perturb key process of normal development of such cells. The disease state could be acute such as in acute myeloid leukemia (AML) or chronic such as myelodysplastic syndrome (MDS). Mutations responsible for these disorders occur in several genes whose encoded proteins principally belong to five classes including signaling pathway proteins, transcription factors, epigenetic regulators, tumor suppressors, and components of spliceosome [1].

Classification of myeloid neoplasms

The 2017 update to the World Health Organization (WHO) categorizes myeloid malignancies into six primary types:

- Myeloproliferative neoplasm (MPN): These are clonal hematopoietic stem cell disorders which typically follow a chronic clinical course and manifest as elevated peripheral blood counts for one or more lineages. The bone marrow is usually hypercellular due to proliferation of one or more cell lineage. Dysplasia is typically absent. Blasts are typically not increased in either the blood or bone marrow; if blasts are increased it may be indicative of progression of disease.
- Mastocytosis: It is a clonal proliferation of mast cells that accumulate in the skin or various organs. Because of its unique features, mastocytosis has been separated from the other MPNs and is now within its own diagnostic category in the 2017 update to the WHO classification.
- Myeloid/lymphoid neoplasms associated with eosinophilia and gene rearrangements: These are chronic diseases defined by the presence of any of the following fusion genes that result in the expression of an abnormal tyrosine kinase: platelet-derived growth factor receptor, alpha

polypeptide (PDGFRA), platelet-derived growth factor receptor, alpha polypeptide (PDGFRB), or fibroblast growth factor receptor 1 ([FGFR1). They are typically accompanied by eosinophilia. This category now also includes the provisional entity myeloid/lymphoid neoplasms with t(8;9)(p22;p24.1) *PCM1-JAK2*. Otherwise, it remains largely unchanged since 2008.

- MDS/MPN, mixed features of MDS and MPN: This is typically a chronic process. In the peripheral blood, white cell counts are usually increased. Bone marrow is hypercellular. Dysplasia is present.

- MDS: Another chronic process characterized by observation of cytopenia in peripheral blood and the bone marrow is typically hypercellular. Dysplasia is a dominant feature. In some subtypes, blasts may be significantly increased.

- AML: It is an acute illness where myeloblasts, monoblasts/promonocytes, and/or megakaryoblasts are 20% or higher; however, less than 20% blasts may be encountered in AML if it is associated with disease defining chromosomal translocations, including t(15;17), t(8;21), t(16;16), or inv 16.

Myeloproliferative neoplasm

The MPNs can be further divided into chronic myeloid leukemia, *BCR-ABL* positive (CML), chronic neutrophilic leukemia, polycythemia vera (PV), primary myelofibrosis (PMF), essential thrombocythemia (ET), chronic eosinophilic leukemia (CEL), not otherwise specified (NOS), and MPN, unclassifiable (MPN-U).

Chronic myeloid leukemia, *BCR-ABL1* positive

CML is the most common form of the myeloid neoplasms where median onset is observed in the fifth or sixth decade of life. This type of leukemia is defined by presence of Philadelphia chromosome and/or *BCR/ABL1* fusion gene. Philadelphia chromosome was first described by Peter C. Nowell in 1960, a faculty member of University of Pennsylvania, Philadelphia, as an unusual small chromosome present in leukocytes of patients with CML [2]. As methodologies for cytogenetics improved, this chromosomal abnormality was determined to be due to reciprocal translocation involving the long arms of chromosome 9 and 22 t(9;22)(q34.1;q11.2). This translocation results in the formation of the *BCR-ABL1* fusion gene (*ABL1* gene from chromosome 9 fuses with *BCR* gene on chromosome 22; *BCR* stands for breakpoint cluster region and *ABL1* is an oncogene known as Abelson murine leukemia viral oncogene, which encodes and enhances tyrosine kinase activity). The abnormal fusion gene is present in all myeloid lineages as well in some lymphoid cells. This disease has three clinical phases: chronic phase, accelerated phase, and blast crisis. Most often, this disease is diagnosed in chronic phase. The patient may present with weakness and organomegaly, but about 50% of new cases are asymptomatic and present with an elevated WBC count detected by routine complete blood count. The chronic phase can last for years before transforming to accelerated phase or directly to blast crisis. The Philadelphia chromosome is present in 90%–95% of cases of CML [3]. In the Philadelphia chromosome–negative cases, *BCR-ABL1* fusion gene can be detected by fluorescence in situ hybridization (FISH), polymerase chain reaction (PCR), or Southern blot techniques. A diagnosis of CML requires the presence of Philadelphia chromosome or documentation of the presence of *BCR-ABL1* fusion gene. Discovery of imatinib mesylate as a tyrosine kinase inhibitor has dramatically improved the clinical course of patients suffering from CML.

The site of the breakpoint in the *BCR* gene may vary, and this may also have an influence on the phenotype of the disease. In the vast majority of cases, the breakpoint is referred to as the major breakpoint cluster region (M-BCR). The resultant abnormal fusion protein, p210, has increased tyrosine kinase activity. If the breakpoint is in the region referred to as the minor region, then the fusion protein is p190, which is most frequently associated with Philadelphia chromosome positive acute lymphoblastic leukemia (ALL). If present in CML, patients with the p190 fusion protein may have increased number of monocytes and can resemble chronic myelomonocytic leukemia (CMML). If the breakpoint is in the region referred to as mu region, then the fusion protein is p230. These patients have prominent neutrophilic maturation and/or thrombocytosis.

Characteristics of peripheral blood in chronic phase of CML include

- Leukocytosis: peak in myelocytes and polymorphonuclear neutrophils (PMNs), absolute basophilia (almost invariably present), absolute eosinophilia (commonly present), and absolute monocytosis (with normal %). Orderly differentiation is preserved. Significant dysplasia is not present, and blasts are not increased (usually <2% of WBCs).
- Platelets: normal or increased (low platelets uncommon)
- RBC: nucleated red blood cells (NRBCs) may be seen

Characteristic features of bone marrow examination (chronic phase) include

- Increased M:E ratio (myeloid cells to erythroid cells ratio), usually greater than 10:1; blast population up to 9%
- Paratrabecular cuff of immature myeloid cells (5−10 cells thick; normal 2−3)
- Megakaryocytes: small, hypolobated ('dwarf' megakaryocytes)
- Increased reticulin fibrosis
- Pseudo-Gaucher cells and sea blue histiocytes (seen because of increased cell turn over)
- No significant dysplasia

Accelerated phase of CML shows following features:

- Persistent or increasing WBC (>10,000/mm3 or >10,000/microliter) and/or persistent splenomegaly unresponsive to therapy (specific for CML)
- Persistent thrombocytosis (>1 million/µL) despite therapy
- Persistent thrombocytopenia (<100,000/µL) unrelated to therapy
- Clonal cytogenetic evolution (additional chromosomal abnormalities)
- Basophilia 20% or higher
- Myeloblasts 10%−19% in peripheral blood or bone marrow

In addition, response (resistance or failure) to therapy (tyrosine kinase inhibitors [TKIs] or stem cell transplant) are also now being included as provisional criteria for accelerated phase [AP].

Blast crisis of CML shows the following features:

- Blasts in the peripheral blood or bone marrow are 20% or higher
- Extramedullary blast proliferation or cluster of blasts in the bone marrow

In two-thirds of cases, the blast lineage is myeloid, and in 20%−30% of cases the blasts are lymphoblasts.

Chronic neutrophilic leukemia

Chronic neutrophilic leukemia is a rare MPN with approximately 150 reported cases. This disease is disease is difficult to diagnose because diagnosis is contingent on exclusion of underlying causes of reactive neutrophilia, particularly if evidence of clonality is lacking. Philadelphia chromosome or *BCR-ABL1* fusion gene must also be absent, thus excluding the diagnosis of CML. Recent discovery of specific oncogenic mutation in the colony-stimulating factor 3 receptor gene (*CSF3R*) in these patients has opened a new era for diagnosis and treatment [4,5]. There may also be accompanying *SETBP1* or *ASXL1* (which confers a worse prognosis) mutations. The neutrophils often show toxic granulation. Clinically, there may be hepatosplenomegaly, and there is an also an association with multiple myeloma.

Diagnostic criteria of chronic neutrophilic leukemia include

- Persistent peripheral blood leukocytosis (WBC > 25,000 with blasts <1%, bands and PMNs >80%, and immature cells [promyelocytes, myelocytes, metamyelocytes] < 10%) without any other causes of neutrophilia. Monocytes are not increased, and there is no dysgranulopoiesis.
- Hypercellular marrow with increased neutrophils which have normal maturation and blasts <5%.
- Does not meet criteria for *BCAR-ABL* positive CML, PV, ET, or PMF.
- No rearrangement in *PDGFRA*, *PDFRB*, or *FGFR1* and no *PCM1-JAK2* fusion.
- *CSF3R* T618I or other activating *CSF3R* mutation or persistent neutrophilia for over 3 months with no reactive cause identified and no plasma cell neoplasm.

Polycythemia vera, primary myelofibrosis, and essential thrombocythemia

PV, PMF, and ET are three Philadelphia chromosome—negative clonal stem cell disorders.

PV is characterized by increase in red cells and thus hemoglobin (>16.5 g/dL in men and >16 g/dL in women). As a result of the increased hemoglobin levels, erythropoietin levels are often significantly decreased. The bone marrow is hypercellular, and proliferation of all cell lines (panmyelosis) may be seen. Megakaryocytes are usually increased in number. In almost all patients, there occurs *JAK2* (Janus Kinase 2) gene gain of function somatic mutation. In the vast majority (>95%), *JAK2* V617F mutation is documented, but in about 3% of cases of PV, *JAK2* exon 12 mutations are seen instead. In the late stages of PV, reticulin and collagen fibrosis may develop in the bone marrow, known as post-PV myelofibrosis (MF). There may also be teardrop-shaped RBCs in the peripheral blood. At this point, erythropoiesis may actually decrease (known as "spent phase"), and there may be cytopenias (from ineffective medullary hematopoiesis); the spleen may further increase in size (due to extramedullary hematopoiesis). If there are also ≥20% accompanying blasts, this is known as blast-phase post-PV MF.

In PMF, there occurs megakaryocytic and also granulocytic proliferation with ultimate fibrosis formation. The megakaryocytes have an abnormal morphology, including hypolobulated, "bulbous" or "cloud-like" forms. The disease typically progresses from an early/prefibrotic phase (with MF grade 0−1) to the overt or classic PMF (MF grade 2−3) [6]. In the fibrotic stage, there is significant reticulin or collagen fibrosis in the bone marrow, and the peripheral blood classically shows leukoerythroblastosis (left-shifted granulocytic lineage and nucleated RBCs, with accompanying teardrop-shaped RBCs). Extramedullary hematopoiesis may thus occur in the liver and spleen, with resultant enlargement of these organs. *JAK2* V617F mutations are seen in about 50%−60% of cases; *CALR* mutations in about 30%; and *MPL* mutations in about 8%; approximately 12% of cases are negative for all three mutations.

In ET, the proliferation is typically restricted to megakaryocytes. Platelets counts are thus increased ($\geq 450 \times 10^9$/L). In the bone marrow, there is a marked proliferation of megakaryocytes with hypersegmented ("staghorn") nuclei. Granulocytic hyperplasia is highly uncommon, and there is no increase in blasts or myelodysplasia. *JAK2* V617F mutations are seen in 50%−60%; *CALR* mutations in 30%; and *MPL* mutations in 3%; and 12% of cases are negative for all three mutations.

All of the three conditions (PV, PMF, and ET) may ultimately result in bone marrow fibrosis, as well as convert to acute leukemia, especially AML. The *JAK2* (Janus Kinase 2) gene and *MPL* gene (myeloproliferative leukemia virus oncogene) play an important role in cell signaling and proliferation. Mutations in these genes confer activation of *JAK-STAT* pathway (STAT: signal transducer and activator of transcription-a gene regulatory protein) and other pathways promoting differentiation and proliferation of various lineages. The *JAK2* V617 mutation is present in approximately 95% of patients with PV, in 58% patients with PMF, and in 50% patients with ET [7]. Various characteristics and diagnostic features of these diseases are summarized in Table 6.1.

Chronic eosinophilic leukemia

CEL is a chronic myeloproliferative disease with unknown etiology, where a clonal proliferation of eosinophilic precursor leads to increased eosinophils in blood, bone marrow, or peripheral tissues. Eventually, this leads to eosinophilic filtration and functional damage of peripheral organs with

Table 6.1 Key features of polycythemia vera, primary myelofibrosis, and essential thrombocythemia.

Polycythemia vera	Primary myelofibrosis	Essential thrombocythemia
Genetics Almost all cases have *JAK2* mutation; most common mutation is *JAK2* V617F (>95%); JAK2 exon 12 mutations in about 3% of cases.	**Genetics** Approximately 58% cases with *JAK2* mutation, CALR mutations in 30%; and MPL mutations in 8%; 12% of cases are negative for all three mutations.	**Genetics** Approximately 50% cases with *JAK2* mutation CALR mutations in 30%; and MPL mutations in 3%; and 12% negative for all three mutations.
Characteristics Characterized by increased erythropoiesis. There may be proliferation of granulocytes and megakaryocytes. Three phases of disease: prodromal, clinically overt, and postpolycythemic fibrosis. Clinical features include: plethora, hypertension, and thrombosis	**Characteristics** Extramedullary hematopoiesis with resultant hepatosplenomegaly is present. Thrombocytosis may be a feature.	**Characteristics** Patients may be asymptomatic. Unexplained thrombocytosis with or without thrombosis or hemorrhage.
Diagnosis Diagnostic criteria include hemoglobin >18.5 g/dL for male and > 16.5 g/dL for female (or increased red cell mass); low erythropoietin; splenomegaly; hypercellular marrow (all cell lines)	**Diagnosis** Peripheral blood: leucoerythroblastic anemia with teardrop red cells. Bone marrow: increased megakaryocytes with atypia (hypolobated, bulbous); increased fibrosis; osteosclerosis; intrasinusoidal hematopoiesis	**Diagnosis** In peripheral blood, platelets >450,000/mm^3 (>450,000 per microliter) Bone marrow: increased megakaryocytes with hyperlobulated ("staghorn") nuclei and increased mature cytoplasm

associated signs and symptoms of that particular organ. Organs that can be involved include heart (endomyocardial fibrosis, restrictive cardiomyopathy, valvular dysfunction), neurological (peripheral neuropathy, CNS dysfunction), lung infiltrates, and diarrhea. Although a very rare disease, tyrosine kinase inhibitor such as imatinib can be used for patient management.

Diagnostic criteria of CEL include

- Marked eosinophilia defined as eosinophilia (>1500/mm3 or 1.5×10^9/L of blood and persisting for more than 6 months) with evidence of clonality of eosinophils or increase in myeloblasts in peripheral blood or bone marrow.
- Exclude *BCR-ABL*−positive CML, PV, ET, PMF, chronic neutrophilic leukemia (CNL), CMML, and atypical (*BCR-ABL* negative) CML.
- No rearrangement of *PDFGRA*, *PDGFRB*, or *FGFR1*, and no *PCM1-JAK2*, *ETV6-JAK2*, or *BCR-JAK2* fusion.
- Blasts <20% in the peripheral blood (PB) or bone marrow (BM) and no other diagnostic rearrangements for AML such as t(16;16) or t(8;21).
- There is a clonal cytogenetic or molecular genetic abnormality or blasts ≥2% in the PB or ≥5% in the BM.

All other causes of eosinophilia, including neoplasms such as Hodgkin lymphoma, ALL, T-cell lymphomas, and systemic mastocytosis, must be excluded as well. If clonality of eosinophils cannot be proved or there are no increased myeloblasts, then the diagnosis is idiopathic hypereosinophilic syndrome, where eosinophilia is present for at least 6 months with features of organ involvement and dysfunction [8, 9].

Myeloproliferative neoplasm, unclassifiable

This diagnostic category is reserved exclusively for cases that have clinical, pathologic, laboratory, or molecular features consistent with an MPN but do not meet diagnostic criteria for any one in particular. The presence of *BCR-ABL* fusion or rearrangements in *PDFGRA*, *PDGFRB* or *FGFR1*, *or PCM1-JAK2* fusion automatically excludes a diagnosis of MPN-U. However, the nonspecific mutations *JAK2*, *CALR*, or *MPL* support the diagnosis of MPN.

Mastocytosis

This is a clonal neoplastic disorder of mast cells involving one or more organs. In these involved organs, multifocal clusters or aggregates of abnormal mast cells are present. Because of its unique characteristics, mastocytosis is now listed as its own diagnostic category and is no longer broadly listed under MPN. Classifications of mastocytosis include cutaneous mastocytosis (most common, usually observed in children and with good prognosis), systemic mastocytosis, and mast cell sarcoma. Cutaneous mastocytosis consists of three variants: urticaria pigmentosa/maculopapular cutaneous mastocytosis (most common form), diffuse cutaneous mastocytosis, and mastocytoma of the skin (single lesion, almost exclusively in children). Systemic mastocytosis almost always involves the bone marrow and consists of six variants: indolent systemic mastocytosis, bone marrow mastocytosis (no skin lesions), smoldering mastocytosis, systemic mastocytosis with an associated hematological neoplasm (meets criteria for both mastocytosis and a separate hematologic neoplasm), aggressive systemic mastocytosis, and mast cell leukemia ($\geq$20% mast cells in BM aspirate smears). Mast cell sarcoma is extremely rare and is a localized destructive tumor of highly atypical mast cells.

Histological and immunohistochemical evaluation of a bone marrow trephine biopsy is used for diagnosis and staining of mast cells with Giemsa and chloroacetate esterase (CAE) is helpful. Moreover, almost all mast cells regardless of their origin, stage of maturation, and activation of sublineage status express tryptase. Other mast cell−related antigens such as chymase, CD117, and for neoplastic mast cells, CD2 and CD25 can be used for diagnosis. The diagnosis of systemic mastocytosis requires the presence of the major criterion and one minor criterion, or in the absence of the major criterion, three minor criterions to be present.

- Major criterion: Multifocal, dense infiltrates of mast cells ($\geq$15 mast cells in aggregates) in the bone marrow or extracutaneous organs should be present.
- Minor criterion: (a) In the bone marrow or extracutaneous organs, >25% of the mast cells are spindle shaped or have atypical morphology, (b) mast cells are positive for CD2 and/or CD25, (c) serum total tryptase is high, and (d) point mutation at codon 816 of *KIT* is detected.

In aggressive systemic mastocytosis, features that may be present include cytopenias, hepatomegaly with liver dysfunctions, and splenomegaly with hypersplenism, skeletal osteolytic lesions, and gastrointestinal infiltration with malabsorption.

Myeloid/lymphoid neoplasms with eosinophilia and gene rearrangement

Myeloid and lymphoid neoplasms with eosinophilia and abnormalities of platelet-derived growth factor receptor alpha (*PDGFRA*), platelet-derived growth factor receptor beta (*PDGFRB*), and fibroblast growth factors receptor-1 (*FGFR1*) are a group of hematological neoplasms resulting from abnormal fusion genes that encode constitutively activated tyrosine kinase. Patients with any of the three diseases may present present similar to an MPN. Conventional cytogenetics (karyotyping) or FISH will detect the majority of abnormalities involving *PDGFRB* and *FGFR1*. However, molecular studies, such as reverse trancriptase -polymerase chain reaction (RT-PCR), may be needed to detect abnormalities involving *PDGFRA*, namely the cryptic del(4)(q12) that results in the *FIP1L1-PDGFRA* fusion gene. [10]. The 2018 update to the WHO added a provisional entity *PCM1-JAK2* fusion.

Various features of these disorders include

- Patients with *PDGFRA* rearrangement usually present as CEL, but they can also present as AML or T-lymphoblastic leukemia with eosinophilia. Patients respond to tyrosine kinase inhibitors (first line therapy).
- Patients with *PDGFRB* rearrangement present more often as CMML with eosinophilia. Again, tyrosine kinase inhibitors mesylate is the first line therapy.
- For a patient with *FGFR1* rearrangement, lymphomatous presentation is common especially with T-lymphoblastic lymphoma with eosinophilia. Unfortunately, these patients do not respond to therapy with imatinib mesylate and carry a poor prognosis [11].
- Myeloid/lymphoid neoplasms with *PCM1-JAK2* is a new provisional entity based on its shared characteristics with other entities in this group. It typically involves a t(8;9)(p22; p24.1) resulting in *PCM1-JAK2* fusion. Its features are usually those of MPN or MDS/MPN with eosinophilia.

In addition, there are often left-shifted erythroid predominance, lymphoid aggregates, and myelofibrosis.

Myelodysplastic/myeloproliferative neoplasms

MDS/MPNs can be classified under four categories:

- CMML
- Atypical CML, *BCR-ABL1* negative
- Juvenile myelomonocytic leukemia (JMML)
- MDS/MPN-U

CMML is characterized by monocytosis and dysplasia. Peripheral blood monocytosis is the hallmark and, by definition, must be $\geq 1 \times 10^9$/L and $\geq 10\%$ of the WBC. The bone marrow is usually hypercellular, and dysplasia is present. CMML can be further classified into various subtypes: CMML-0 (blasts $<2\%$ in blood, $<5\%$ in bone marrow), CMML-1 (blasts 2%–4% in blood, 5%–9% in bone marrow, and no Auer rods), and CMML-2 (blasts 5%–19% in blood or 10%–19% in bone marrow or with Auer rods). As expected, blasts must be less than 20%, otherwise this would be classified acute leukemia. Also, Philadelphia chromosome should be absent, and *BCR-ABL* gene rearrangement test should also be negative. In cases of CMML with eosinophilia, rearrangements of *PDGFRA*, PDGFRB, *FGFR1*, and *PCM1-JAK2* must be excluded. The most common cytogenetic abnormalities include gain of chromosome 8 and loss of chromosome 7. The most frequently mutated genes in CMML include *ASXL1* (40%), *TET2* (58%), and *SRSF2* (46%); they can be used to establish clonality. Other less frequently mutated genes include *RUNX1*, *NRAS*, and *CBL*.

Atypical chronic myeloid leukemia (aCML) is characterized by peripheral blood leukocytosis ($\geq 13 \times 10^9$/L) and expansion of the neutrophil precursors (promyelocytes, myelocytes, and metamyelocytes $\geq 10\%$ of WBC), resembling CML. However, dysplasia (dysgranulopoiesis) is present, and basophilia is absent or minimal ($<2\%$ in PB). Unlike CMML, monocytosis is not a feature (10% in PB). Blasts must be $<20\%$. By definition, Philadelphia chromosome is absent and tests for *BCR-ABL1* gene rearrangement are negative. There is also no *PDGFRA*, *PDGFB*, or *FGFR1* rearrangement or *PCM1-JAK2*. *SETBP1* and *ETNK1* mutations are relatively common in aCML.

JMML is characterized by leukocytosis and monocytosis (1×10^9/L) in individuals less than 14 years of age with features of dysplasia. The hemoglobin F (HbF) percentage may be increased in JMML. JMML is associated with neurofibromatosis 1.

MDS/MPN with ring sideroblasts and thrombocytosis ($\geq 450 \times 10^9$/L) has $<1\%$ blasts in the PB and dyserythropoiesis, $<5\%$ blasts, and $\geq 15\%$ ring sideroblasts in the bone marrow. There is an association with *SF3B1* (60%–90%) mutations, often found co-occurring with *JAK2* V617F ($>60\%$) or less commonly with *MPL* W515 or *CALR* ($<10\%$). Anemia is present.

Myelodysplastic syndromes

In MDS, the bone marrow produces cells that are not up to the adequate quality. The poor quality cells are not good enough to be released into the circulation and thus are destroyed within the bone marrow, and the process is called ineffective hematopoiesis. In a normal marrow, there can be up to 10%

ineffective hematopoiesis, but in MDS, this percentage is significantly higher. As a result, the bone marrow becomes hypercellular with cytopenia observed in the peripheral blood. Cytopenia in at least one lineage is required for a diagnosis of MDS. As expected, in MDS, morphologic abnormalities (dysplasia) are usually present in cells in both the peripheral blood and in the bone marrow. Major features of MDS are that it is a clonal stem cell disease occurring principally in older adults with a median age of 70 years. MDS is characterized by cytopenia in the peripheral blood with a hypercellular marrow, and morphological evidence of dysplasia is present. In addition, there are certain cytogenetic abnormalities associated with MDS. Patients with MDS are also at an increased risk of developing AML.

MDS is classified according to the 2017 WHO classification as

- Myelodysplastic syndrome with single lineage dysplasia (MDS-SLD)
- Myelodysplastic syndrome with ringed sideroblasts (MDS-RS)
- Myelodysplastic syndrome with multilineage dysplasia (MDS-MD)
- Myelodysplastic syndrome with excess of blasts (MDS-EB)
- MDS associated with isolated del(5q)
- Myelodysplastic syndrome, unclassified (MDS-U)
- Childhood MDS
- Refractory cytopenia of childhood (RCC)

Features of dysplasia in red cells, erythroid precursors, granulocytes and megakaryocytes

Features of dysplasia in red cells and erythroid precursors include dimorphic red cells, macrocytic red cells, nuclear budding, internuclear bridging, bi/multinucleated cells, karyorrhexis, megaloblastoid changes, cytoplasmic fraying, nuclear cytoplasmic dyssynchrony, cytoplasmic vacuolization, ringed sideroblasts, and PAS (periodic acid−Schiff) positivity.

It is important to note that there are many causes of dimorphic and macrocytic red cells, and these are considered soft signs. Cytoplasmic vacuolization can also be seen with prolonged alcohol intake. Vitamin B_{12} and folate deficiency may result in features of dyserythropoiesis listed above. Therefore, a diagnosis of MDS requires exclusion of folate and vitamin B_{12} deficiency.

Features of dysplasia in granulocytes include nuclear hypolobulation (pseudo−Pelger-Huët cells), hypersegmented PMNs, abnormally small or large granulocytes, hypogranulation, and presence of Auer rods.

When an individual has majority of the neutrophils (usually 75% or more) with two nuclear segments, this is referred to as hereditary autosomal dominant Pelger-Huët anomaly (caused by *LBR* gene mutation). This is inherited as autosomal dominant and is clinically insignificant. In MDS, pseudo−Pelger-Huët cells have two nuclear segments are hypogranular, and a small number of cells have this morphology.

Features of dysplasia in megakaryocytes include micromegakaryocytes, nuclear hypolobulation, and multinucleation.

However, it is important to note that small megakaryocytes are also seen in CML. When dealing with a case of MDS, it is important to review the peripheral smear and bone marrow slides. Flow cytometry may or may not contribute to the diagnosis, but cytogenetic study is essential.

Arriving at a diagnosis of MDS and subclassifying MDS

Approach to assessment of peripheral smear and bone marrow in patients with suspected MDS:

*   Is unicytopenia or bicytopenia or pancytopenia present in the peripheral blood?
*   What is the blast count in the peripheral blood and or in the bone marrow?
*   Are Auer rods present in the peripheral blood and/or in the bone marrow?
*   In the bone marrow, dysplasia should be $\geq$10% in the affected cell line(s).
*   In the bone marrow, how many cell lines are affected by dysplasia?
*   Are more than 15% of ringed sideroblasts (as a percentage of all erythroid cells) present? This will be evident in the bone marrow aspirate smear stained for iron.

It is important to note that if the blast count is 20% or more, this is not MDS but AML.

MDS-EB is an MDS based on increased blasts in either the peripheral blood or bone marrow. The degree of cytopenias or how many cell lines are affected by dysplasia does not matter. The presence of ringed sideroblasts does not matter. However, the blast count must be below 20%. MDS-EB can be divided into two subcategories, MDS-EB-1 and MDS-EB-2. In MDS-EB-1, the blast count in the peripheral blood is 2%–4% and in the bone marrow is 5%–9%. In MDS-EB-2, the blast count in the peripheral blood is 5%–19% and in the bone marrow is 10%–19%. If there are Auer rods in the peripheral blood or bone marrow, then this is MDS-EB-2, regardless of blast count.

MDS-EB and erythroid predominance encompasses cases in which there are increased myeloblasts and maturing erythroblasts account for $\geq$50% of total bone marrow cells. In the previous 2008 WHO classification, there was a category of erythroleukemia (the 20% blast cutoff was based on nonerythroid nucleated cells), but now the blast percentage is based on all nucleated marrow cells (including the erythroid cells). As such, most of those cases would now be classified as MDS-EB.

In about 10%–15% of cases of MDS, evidence of significant fibrosis is present, referred to as MDS with fibrosis (MDS-F). MDS-F is often associated with increased blast counts and falls into the MDS-EB range (MDS-EB-F). The presence of significant fibrosis is also associated with an aggressive course regardless of blast count.

If there is unicytopenia or bicytopenia in the peripheral blood with <1% blasts and unilineage dysplasia (with $\geq$10% cells dysplastic), <5% blasts this is MDS-SLD. If dysplasia is present in at least two cell lines with <1% blasts in peripheral blood and <5% in the bone marrow, then this is MDS-MLD; $\geq$15% of ringed sideroblasts may or may not be present.

If there is anemia, with erythroid dysplasia and $\geq$15% of ringed sideroblasts, then this is MDS-RS. MDS-RS can be further subclassified into MDS-RS-SLD or MDS-RS-MLD, if there is single lineage (erythroid lineage only) or multilineage dysplasia, respectively. Similar to MDS/MPN with ring sideroblasts and thrombocytosis, MDS-RS is also closely associated with mutations in *SF3B1*. A diagnosis of MDS-RS can be made with as few as $\geq$ 5% ring sideroblasts in the bone marrow if an *SF3B1* mutation is also present.

MDS with isolated del(5q) is characterized by

*   Increased incidence in females
*   Anemia, with normal or increased platelet counts
*   Splenomegaly
*   Normal or increased megakaryocytes with hypolobated nuclei

- Blasts less than 1% in the peripheral blood and <5% in the bone marrow
- del(5q) cytogenetic abnormality

Cases of MDS with additional chromosomal abnormalities may still be classified as MDS with del(5q), unless the abnormality is deletion of chromosome 7 or 7q.

MDS-U is diagnosed in any of the following three scenarios:

- Pancytopenia with other features of MDS-SLD
- Features suggestive of MDS but the blast count in the peripheral blood is 1% on at least two separate occasions
- Less than 10% dysplasia in the cell line/s but there is a cytogenetic abnormality that is considered as presumptive evidence of MDS

It is important to note that in all of the above the blast count in peripheral blood should be <1% and in the bone marrow <5% with the exception of MDS-EB-1 and MDS-EB-2. Thus, it is the percentage of blasts and Auer rods that dictate MDS-EB-1 or -2. Once MDS-EB-1 and 2 are excluded, then a key point is whether there is dysplasia in the bone marrow affecting one or two cell lines. If two cell lines are affected, then it is MDS-MLD. If only one cell line is affected, then this is MDS-SLD. MDS-RS is unique in the sense that the erythroid cell line is affected with ringed sideroblasts. Ringed sideroblasts may also be seen in MDS-MLD. MDS with del(5q) is also unique with features mentioned earlier.

Sideroblasts are red cell precursors with iron. Both of these cells are physiological. Ringed sideroblasts are red cell precursors with iron distributed in a ring manner around the nucleus. It does not have to be a complete ring. One-third of the circumference of the nucleus is enough to fulfill the criteria. The iron is present within the mitochondria.

MDS is subdivided into three risk groups: The risk of AML is highest with MDS-EB-2 and lowest with MDS-SLD, and MDS with isolated del(5q).

- Low Risk: MDS-SLD, MDS-RS-SLD, and MDS with isolated del(5q)
- Intermediate risk: MDS-MLD and MDS-RS-MLD
- High risk: MDS-EB

Previously, MDS was classified according to the FAB (French–American–British) classification into five classes including refractory anemia (RA), refractory anemia with ringed sideroblasts, refractory anemia with excess of blasts (RAEB), refractory anemia with excess of blasts in transformation (RAEB-t) and CMML. However, CMML is now part of the group MPN/MDS as per the WHO classification of MPNs.

RAEB-t was used with blasts count between 20% and 30% as previously AML required a blast count of 30% or higher. Now, with AML being diagnosed with blast count of 20% or higher, RAEB-t is an obsolete term.

Abnormal localization of immature precursors

This term is used when immature myeloid cell clusters (five to eight cells) are present in the central portion of marrow, away from usual locations (paratrabecular or perivascular) of myeloid cell maturation. Three or more such foci fulfill the criteria for ALIP. ALIP is frequently present in cases of MDS-EB and is associated with a rapid evolution to AML. If ALIP is found in other subtypes of MDS, the case should be reevaluated.

Cytogenetic abnormalities associated with myelodysplastic syndrome

Approximately 40%−50% patients with primary MDS have clonal cytogenetic abnormalities, whereas 80% of therapy related MDS cases have cytogenetic abnormalities. Therefore, cytogenetic study is useful in diagnosis of MDS and prediction of prognosis of individual patients using the International Prognostic Scoring System [12,13]. Many cytogenetic abnormalities have been described in patients with MDS but common abnormalities include

- Unbalanced: −7 or del(7q), −5 or del(5q), −20 or del(20q), −13 or del(13q), del(11q), del(12p) or t(12p), del(9q), idic(X)(q13), i(17q) or t(17p), +8, -Y.
- Balanced: t(11;16)(q23.3; p13.3), t(3;21)(q26.2;q22.1), inv(3)(q21.3;q26.2)/t(3;3)(q21.3;q26.2), t(1;3)(p36.3;q21.2), t(2;11)(p21;q23.3), t(6;9)(p23;q34.1).

However, there are many other abnormalities. Interestingly, MDS patients with +8 (trisomy 8) as sole aberration are predominately male. An isolated cytogenetic abnormality of −Y, +8, or del(20q) in the absence of morphologic support is not considered diagnostic for MDS. However, the other cytogenetic abnormalities, such as del(7q), are considered presumptive evidence for MDS. Certain aberrations are associated with good prognosis, whereas with other with intermediate or poor prognosis. Some examples are given below:

- Karyotypes are associated with good prognosis: Normal, −Y, del(5q), and del(20q)
- Karyotypes associated with intermediate prognosis: Trisomy 8 (+8) and more than one abnormality
- Karyotypes associated with poor prognosis: Complex (three or more abnormalities) or chromosome 7 abnormalities.

In additional to cytogenetic findings, recurrent somatic mutations in over 50 genes (including *SF3B1*, *TET2*, *ASXL1*, *SRSF2*, *DNMT3A*, and *RUNX1*) have been found in a great majority of MDS cases (80%−90%).

Unusual situations in myelodysplastic syndrome

In about 10% of cases, the bone marrow in MDS can be hypocellular, referred to as hypoplastic MDS; the differential diagnosis in such situations includes aplastic anemia.

Childhood MDS is rare (<5% of hematologic neoplasms in children <14 years old). Some cases may be due to aplastic anemia, a bone marrow failure syndrome, or an MDS with germline predisposition. Criteria for subcategorization of childhood MDS is similar to that of adults. However, because it may be challenging to successfully diagnose low-grade MDS in children, the provisional entity RCC was added to the 2017 WHO. RCC is the most common type of childhood MDS (50% of cases). It shows persistent cytopenia and <5% BM blasts and <2% PB blasts. Most (80%) children with RCC have a hypocellular bone marrow for age.

Myeloid neoplasms with germline predisposition

Myeloid neoplasms, including MDS and AML, occurring in association with inherited or de novo germline mutations in specific genes are identified by molecular testing. This possibility should be considered when there is a long-standing history of thrombocytopenia or bleeding tendencies, if there is a first- or second-degree relative with a hematological disorder or a solid tumor with a germline predisposition or syndromic symptoms (i.e., congenital limb abnormalities). A complete family history (including bleeding history) is essential for genetic counseling purposes and to avoid inadvertent stem cell transplant from a family member with the same mutation (which could result in donor-derived leukemia in the recipient).

Subdivided into three major groups:

1. Myeloid neoplasms with germline predisposition without a preexisting disorder or organ dysfunction
   - AML with germline *CEBPA* mutation: often biallelic (germline mutation on $5'$ end of the gene on one allele and a somatic mutation on the $3'$ end of the other allele acquired at the time of progression to AML). Only biallelic mutations are associated with a good prognosis, and only AML with biallelic *CEBPA* mutations are considered a specific subtype of AML.
   - Myeloid neoplasms with germline *DDX41* mutation: many also are biallelic, with one mutation being germline and the other somatic; longer latency than other myeloid neoplasms with germline predisposition (age of malignancy onset ~60s).
2. Myeloid neoplasms with germline predisposition and preexisting platelet disorders: all three are autosomal dominant
   - Myeloid neoplasms with germline *RUNX1:* familial platelet disorder with predisposition to AML. Young age of onset of MDS/AML.
   - Myeloid neoplasms with germline *ANKRD26*: aka thrombocytopenia 2, with moderate thrombocytopenia and only mild bleeding tendency.
   - Myeloid neoplasms with germline *ETV6*: aka thrombocytopenia 5; mild to moderate bleeding tendency
3. Myeloid neoplasms with germline predisposition and other organ dysfunction
   - Myeloid neoplasms with germline *GATA2* mutation: immunodeficiency, infections, reduced monocyte, B cells and NK cells, lymphedema.
   - Myeloid neoplasms associated with bone marrow failure syndromes (i.e., Fanconi anemia, severe congenital neutropenia, Shwachman–Diamond syndrome, Diamond–Blackfan anemia); often diagnosed in childhood due to bone marrow failure or systemic symptoms.
   - Myeloid neoplasms associated with telomere biology disorders (i.e., dyskeratosis congenita, and syndromes related to *TERC* and *TERT*).
   - JMML associated with neurofibromatosis, Noonan syndrome, or Noonan syndrome–like disorders.
   - Myeloid neoplasms associated with Down syndrome

Acute leukemia

In acute leukemia, immature cells proliferate and fail to mature within the bone marrow. The bone marrow cannot produce normal cells, and when approximately one trillion (one million x one million) leukemic cells accumulate in the bone marrow, these cells start to spill over into the peripheral blood. The immature leukemic cells are thus now present in the peripheral blood. The term used for the leukemic process before the leukemic cells are found in the peripheral blood is subleukemic leukemia. In acute leukemia, patients typically present with blasts in the peripheral blood. Blasts are not a normal finding in the peripheral blood.

Causes of blasts in the peripheral blood include

- Acute leukemia
- MDS
- Neupogen effect
- Neonates with a reactive condition
- Leukoerythroblastic blood picture
- Leukemoid reaction
- CML

In acute leukemia, typically anemia, thrombocytopenia and leukocytosis with increased blasts are observed. For diagnosis of acute leukemia, it is important to determine if the patient is suffering from AML or ALL. If the patient is suspected to have AML, then it is important to decide whether it is acute promyelocytic leukemia (APL, a subtype of AML where accumulation of immune granulocytes called promyelocytes accumulate) or not, because APL patients are especially prone to disseminated intra-vascular coagulation (DIC), and therapy with all trans retinoic acid (ATRA, also known as tretinoin) should be started as soon as possible. To make a diagnosis, age of patient should be considered because ALL is seen most often in children, whereas AML is seen in all ages. Moreover, morphology of blasts should be carefully observed and also presence of Auer rods also should be investigated.

Blasts

Any blasts have a high nuclear to cytoplasmic ratio, fine chromatin, and with or without nucleoli. Myeloblasts tend to be larger when compared with lymphoblasts, the chromatin finer and nucleoli, if present, tend to be more in number (more than two). The cytoplasm may have fine azurophilic granules. Sometimes these granules condense to form Auer rods. There are three types of myeloblasts: Type I (rare granules), Type II (few granules, 5—20), and Type III (abundant granules). Auer rods are mostly seen with type III myeloblasts. Atypical promyelocytes (seen in APL, discussed later), mon-oblasts, promonocytes, and megakaryoblasts are considered blast equivalents.

Lymphoblasts are smaller blasts, and the chromatin is not as fine as that of myeloblasts. Nucleoli are absent or if present are <2. Typically, lymphoblasts have no granules.

The presence of Auer rods and/or expression of myeloperoxidase (MPO) is useful in establishing myeloid lineage to the blasts. When a conclusive diagnosis of AML or ALL cannot be made, ancillary tests such as cytochemistry and flow cytometry are extremely helpful. Typically, in all cases of acute leukemia flow cytometry, cytogenetic and molecular studies should be done. These studies can be performed form the peripheral blood and the bone marrow specimen.

Cytochemistry

Various staining methods can be used for cytochemistry analysis for diagnosis. These various stains are listed below:

- MPO: stains myeloblasts in a Golgi distribution; monoblasts are negative, promonocytes or monocytes may have cytoplasmic stain. Importantly, a lack of MPO expression alone does not exclude myeloid lineage. Very early or minimally differentiated myeloblasts (i.e., AML with minimal differentiation, M0), monoblasts, and megakaryoblasts show little to no MPO expression. MPO is a peroxidase enzyme stored in the cytoplasmic granules of neutrophils and other myeloid cells. Its main function is to produce hypochlorous acid in the oxidative burst of neutrophils to kill bacteria and other pathogens

- Nonspecific esterase (e.g., butyrate) stain: stains cytoplasm of cells of monocytic lineage
- CAE: stains cells of neutrophilic lineage and mast cells and also stains abnormal eosinophils (seen in M4Eo)
- Sudan Black: parallels MPO but less specific and thus not used.
- PAS: stains FAB M6 and M7 subtypes of AML; may also stain ALL. Thus, it is not useful to reliably differentiate AML and ALL.

Classification of acute myeloid leukemia and diagnosis

The current classification of AML is according to the WHO classification. The immediate previous classification is known as the FAB (Table 6.2).

For diagnosis of AML, the presence of 20% or more blasts in the bone marrow is required. If the blast count is lower, diagnosis of MDS should be considered. An exception to the 20% blast count rule is AML with certain recurrent genetic abnormalities, namely t(15;17), t(8;21), t(16;16), and inv 16. If an individual has any of these cytogenetic abnormalities, then they have AML irrespective of their blast count.

Acute myeloid leukemia with recurrent genetic abnormalities

Of all the leukemia related prognostic factors, the karyotype of the leukemic cells is the most predictive for response to therapy and survival.

The genetic abnormality of t(15;17)(q24.1;q21.2) *PML-RARA* results from fusion of the retinoic receptor α gene (*RARA*) on chromosome 17q12 with promyelocytic leukemia (*PML*) gene on

Table 6.2 World Health Organization (WHO) and FAB (French–American–British) classification of acute myeloid leukemia (AML).

WHO classification of AML	FAB classification of AML
<ul><li>AML with recurrent genetic abnormalities</li><li>AML with myelodysplastic syndrome (MDS)–related changes: history of MDS; AML with MDS-related cytogenetics; AML with multilineage dysplasia (50%)</li><li>Therapy-related myeloid neoplasms (alkylating agents, topoisomerase II inhibitors, radiation)</li><li>AML (not otherwise specified): AML with minimal differentiation, AML without maturation, AML with maturation, acute myelomonocytic leukemia, acute monoblastic and monocytic leukemia, pure erythroid leukemia, acute megakaryoblastic leukemia, acute basophilic leukemia, acute panmyelosis with myelofibrosis</li><li>Myeloid sarcoma</li><li>Myeloid proliferations related to Down syndrome</li></ul>	<ul><li>M0: AML with minimum differentiation</li><li>M1: AML without maturation—here promyelocytes, myelocytes, metamyelocytes, and bands/PMNs are <10%</li><li>M2: AML with maturation—here promyelocytes, myelocytes, metamyelocytes, and bands/PMNs are >10%</li><li>M3: acute promyelocytic leukemia (classical and microgranular)</li><li>M4: acute myelomonocytic leukemia, where cells of monocytic lineage are >20%; M4 has a variant called M4Eo</li><li>M5: acute monocytic leukemia, where cells of monocytic lineage are >80% (subdivided into M5a and M5b)</li><li>M7: acute megakaryoblastic leukemia</li></ul>

chromosome 15q22 and causes APL. In the WHO classification, it is included in the category of recurrent genetic abnormalities. In the FAB classification, it is M3. APL is characterized by proliferation and maturation arrest of abnormal promyelocytes. In the setting of AML, the abnormal promyelocytes may be counted as blasts. APL typically presents de novo, with an abrupt onset in younger patients (most cases present before the sixth decade of life). Patients are at increased risk of DIC. However, they respond well to ATRA, anthracycline based chemotherapy, and arsenic trioxide. It is possible to see atypical promyelocytes and detect *PML-RARA* transcripts by RT-PCR in the bone marrow for several weeks after the initiation of treatment, but these findings do not appear to affect prognosis. In contrast, detecting *PML-RARA* transcripts, a patient who has completed induction chemotherapy and achieved complete remission is worrisome for impending relapse.

There are two types of APL: classical and microgranular. In the classical case, abnormal promyelocytes with abundant azurophilic granules are seen. Cells with short and thick Auer rods are easily apparent. Cells with multiple Auer rods are referred to as Faggot cells. The morphologic clue is the appearance of the nuclei of the abnormal promyelocytes. The nuclei are highly irregular having bilobed appearance, sometimes referred to as "apple-core appearance." Cytochemical staining with MPO will be strongly positive, characteristically obscuring almost the entire nucleus. In the microgranular variant, the granules of the abnormal promyelocytes are there but not visible by light microscopy. The WBC count in microgranular variant of APL is also usually very high. In addition, a *FLT3*-ITD mutation in APL is associated with higher WBC counts and the microgranular variant and the breakpoint region 3 (BCR 3) of the *PML* gene. Some cases of APL have translocations resulting in fusion of RARA (chr 17q12) with a gene other than PML. These include *ZBTB16* (chr 11q23), *NUMA1* (chr 11q13), *NPM1* (chr 5q35), and *STAT5B* (chr 17q11.2). APL with *ZBTB16-RARA* and *STAT5B-RARA* are resistant to ATRA.

AML with t(8;21)(q22;q22.1) *RUNX1-RUNX1T1*: The morphology is generally similar to AML M2 (FAB classification). There may be salmon-colored granules, perinuclear clearing (hof), single long and slender Auer rods, and cytoplasmic vacuoles. Immunophenotypically, they often express B-cell markers such as CD19, cytoplasmic CD79a, or PAX5. Some cases may also express CD56 and may confer a worse prognosis. Finally, rare cases of "hypoproliferative" disease or cases initially presenting as myeloid sarcoma may be encountered. The peripheral blood may show leukopenia and <20% blasts, but detection of t(8;21) establishes the diagnosis of AML.

AML with t(16;16)(p13.1q22) *CBFB-MYH11* or inv(16)(p13.1q22) results from the fusion of the *CBFB* gene at chromosome 16q22 to the *MYH11* gene at chromosome 16p13.1. It may have the M4 Eo subtype of AML (FAB classification). These patients have eosinophilia with presence of abnormal eosinophils. These abnormal eosinophils have basophilic granules and stain with CAE.

Acute myeloid leukemia with t(8;21)(q22; q22.1) and AML with inv(16)(p13.1q22) or t(16;16)(p13.1q22) are commonly referred to as the "core-binding factor leukemias." Core-binding factor (CBF) is a transcription factor necessary for normal hematopoiesis, and disruption of function leads to leukemogenesis. The t(8;21), which involves the *RUNX1* gene (also known as CBF-α), affects the alpha subunit of CBF. The inv(16) or t(16;16), which involves the *CBFB* gene, affects the beta subunit of CBF. The CBF leukemias have an increased propensity for monocytic differentiation and extramedullary manifestation (myeloid sarcoma). Overall, they have a good prognosis due to responsiveness to chemotherapy. However, a concomitant mutation in *KIT* gene (found in 20%−25% of cases) confers a worsened prognosis. As such, all cases of CBF leukemia should be tested for *KIT*.

AML with t(9;11)(p21.3;q23.3) *KMT2A-MLLT3* can have monocytic features analogous to AML M4 (FAB). The *KMT2A* gene is located on chromosome 11q23 and was previously known as the *MLL* gene. *KMT2A* rearrangement can also be seen in ALL, therapy-related AML, mixed phenotype acute leukemia (MPAL), and AML with myelodysplasia-related changes (AML-MRC).

AML with t(6;9)(p23;q34.1) *DEK-NUP214* often has accompanying basophilia and multilineage dysplasia. It frequently occurs in children and young adults, often presents as pancytopenia, and has a poor prognosis.

AML with inv(3) or t(3;3)(q21.3q26.2) *GATA2-MECOM* often has dysplastic megakaryocytes (hypolobated) or multilineage dysplasia. It can present with normal or increased platelet counts. The *MECOM* gene used to be called the *EVI1* gene. It also has a poor prognosis.

AML with t(1;22)(p13.3; q13.1) *RBM15-MKL1* is an acute megakaryoblastic leukemia analogous to AML M7 in the FAB classification. It almost exclusively occurs in infants, but in those who do *not* have Down syndrome (as opposed to myeloid leukemia associated with Down syndrome).

AML with *BCR-ABL1* is a de novo AML with the classic t(9;22)(q34.1; q11.2) and usually the p210 fusion protein. However, there no other evidence of an underlying CML. Cases of MPAL with *BCR-ABL* are excluded from this category. There is usually leukocytosis predominantly composed of blasts. In contrast to the blast phase of CML, there is a lower frequency of splenomegaly and less basophilia (typically <2%) in the peripheral blood.

AML with mutated *NPM1* is usually a de novo AML with normal karyotype and myelomonocytic or monocytic features. *NPM1* is one of the most common recurrent mutations in AML, occurring between 27% and 35% of all adult cases and usually involving exon 12 of the gene. In cases with a normal karyotype and in the absence of a *FLT3* mutation, AML with mutated *NPM1* has a generally good prognosis. The presence of "cup"-like or "fish mouth"—like nuclear indentation in the AML blasts is a morphologic clue for the presence of *NPM1* (and/or *FLT3*) mutations.

AML with biallelic mutation of *CEBPA* is usually a de novo AML with a normal karyotype and generally good prognosis. Of note, the favorable prognosis applies only to biallelic mutations of *CEBPA*, meaning that the mutation is present in both (paternal and maternal) alleles. It has morphology similar to that of AML M1 or M2 (FAB).

AML with mutated *RUNX1* is another de novo AML and is essentially a diagnosis of exclusion. The diagnosis is not made in the presence of other recurrent genetic abnormalities, therapy-related AML, or AML-MRC.

AML-MRC requires >20% blasts and either a history or MDS or MDS/MPN, an MDS related cytogenetic abnormality, or multilineage dysplasia by morphology. Multilineage dysplasia in this context means at least two bone marrow cell lines are dysplastic and at least 50% of each cell line is dysplastic.

Therapy-related myeloid neoplasms are seen in patients who have received chemotherapy (such as alkylating agents, topoisomerase II inhibitors) or radiation. They are often associated with abnormalities in chromosome 5 or 7 and rearrangements involving 11q23 including t(9;11).

Acute myeloid leukemia, not otherwise specified

AML (NOS) according to the WHO classification corresponds to the FAB classification with the exception of acute basophilic leukemia and acute panmyelosis with myelofibrosis, which do not exist in the FAB classification. Also, APL does not belong in the AML, NOS category as it is part of AML with recurrent genetic abnormalities. Features of various subtypes of AML according to FAB classifications are summarized below:

- AML with minimum differentiation, M0: blasts with no granules or few granules are predominately present. Auer rods are not seen. Special staining for MPO is negative. Thus, it is not possible to provide a decision whether the blasts are myeloblasts or lymphoblasts. Flow cytometry is crucial in this case. The blasts express myeloid antigens and thus labeled as AML.
- AML without maturation, M1: myeloblasts dominate the picture (here promyelocytes, myelocytes, metamyelocytes, and bands/PMNs are <10%). However, it is possible to designate the blasts as myeloblasts by positive staining for MPO in ≥3% of the blasts.
- AML with maturation, M2 promyelocytes, myelocytes, metamyelocytes, and bands/PMNs are ≥10%.
- APL, M3 was discussed above
- Acute myelomonocytic leukemia, M4: here cells of monocytic lineage (monoblasts, promonocytes, and monocytes) are ≥20% but less than 80%. A variant of M4 is M4 Eo. This has been discussed above.
- Acute monocytic leukemia, FAB M5: here cells of monocytic lineage are ≥80%. If the majority of the cells are monoblasts (typically 80% or more), then this may be categorized as M5a (acute monoblastic leukemia). In M5b, the majority of the monocytic cells are promonocytes or monocytes. It is important to note that in the setting of AML, promonocytes are still counted as blasts.
- Pure erythroid leukemia involves >80% erythroid cells with ≥30% erythroblasts. The erythroid cells are typically dysplastic and also may have vacuoles. These vacuoles may be positive for PAS stain. Cases previously diagnosed as acute erythroid leukemia (FAB M6) now have the blast percentage based on total cells and as such most fall under the category of MDS-EB (discussed earlier in the chapter) or AML-MRC.
- Acute megakaryoblastic leukemia, M7: here at least half of the blast population is of megakaryocytic lineage. Megakaryoblasts typically have cytoplasmic blebs. Features of dysplasia may also be present in the myeloid cell lines.
- Acute basophilic leukemia: here the predominant differentiation of the blasts is to basophils.
- Acute panmyelosis with myelofibrosis: here there is panmyelosis with ≥20% blasts as well as fibrosis.

Myeloid sarcoma

The blasts form a mass outside of the bone marrow (extramedullary tumor). It is considered equivalent to a diagnosis of bone marrow AML. The immunophenotyped reflects the type of blast composing the tumor (i.e., myeloid, monoblastic, or even megakaryoblastic). Myeloid sarcoma can also have recurrent genetic abnormalities such as t(8;21), inv(16) or *KMT2A* rearrangement.

Myeloid proliferations associated with Down syndrome

Patients with Down syndrome have an increased risk (10–100×) for leukemia, most commonly acute megakaryoblastic leukemia. It usually presents at <5 years of age. Some (~10%) neonates with Down syndrome may have transient abnormal myelopoiesis, mimicking megakaryocytic leukemia; however, this condition resolves spontaneously over the first 3 months of life. *GATA1* mutations are present in the blasts (pathognomonic). The prognosis is actually quite good in young children with AML associated with Down syndrome and *GATA1* mutation.

Acute myeloid leukemia and flow cytometry

Individuals with AML should have flow blasts positive for myeloid markers such as CD13, CD33, and CD117. Cells that denote immaturity are CD117, human leukocyte antigen-DR isotype (HLA-DR), and CD34. Thus, in AML with minimum differentiation (FAB classification: M0), positivity for CD13, CD33, CD117, HLA-DR, and CD34 should be observed but MPO should be negative. In AML without maturation (FAB classification: M1), positivity for the above mentioned markers and positivity for MPO should be observed.

In AML with maturation (FAB classification: M2), positivity for CD13, CD33, CD117, HLA-DR, CD34, and MPO should be observed and positivity for mature markers such as CD11b, CD15 and CD65.

In APL, the abnormal promyelocytes are CD34 negative, completely or partially negative for HLA-DR, CD11b, and CD15, but positive for CD13 (heterogenous) and CD33 (homogenous and bright) and CD117. The abnormal promyelocytes lack CD10 and CD16, which are mature myeloid markers. It has been reported that combination of absence of CD11b, CD11c, and HLA-DR identifies 100% of APLs [5]. CD2 positivity is a feature of APL.

In acute myelomonocytic leukemia (FAB classification: M4), positivity for mature myeloid markers such as CD15 and CD65 should be seen. Positivity for monocytic markers such as CD4, CD11b, CD11c, CD14, CD36, or CD64 should also be observed, and at least 20% myeloblasts positive for myeloid markers should also be present. In acute monocytic leukemia (FAB classification: M5), positivity for monocytic markers will be seen.

Pure erythroid leukemia typically shows expression of glycophorin and hemoglobin A (HbA) as well as CD71, CD117, CD235a and CD36. MPO, human leukocyte antigen (HLA-DR), CD34 and CD64 are usually negative. CD36 positivity can be seen in both monocytes and erythroid cells; however, if the cells in question are negative for CD64 (which is another monocytic marker) we can conclude that these cells are actually erythroid cells.

In acute megakaryoblastic leukemia (FAB classification: M7), positivity for CD41 and CD61 is expected. HLA-DR and CD34 may again be negative. CD41 is the glycoprotein IIb/IIIa antigen, and CD61 is the glycoprotein IIIa antigen.

Molecular and acute myeloid leukemia

AML with gene mutations include mutations of fms-related tyrosine kinase 3 (*FLT3*), nucleophosmin (*NPM1*), and less commonly mutations of *CEBPA* gene, *KIT*, *MLL*, *WT1*, *NRAS*, and *KRAS*.

FLT3 is located at 13q12 and encodes a tyrosine kinase receptor involved in hematopoietic stem cell differentiation and proliferation. *FLT3* mutations are seen in AML and MDS. In AML, FLT3 mutations are most commonly seen in AML with normal karyotype, APL, and AML with t(6;9). *FLT3* mutations impart adverse outcome. Mutations of *NPM1* and biallelic *CEBPA* in absence of *FLT3* mutation in AML with normal karyotype indicate a more favorable prognosis of the disease.

The genetic mutations can be broadly divided into two classes: class I and class II. Class I mutations are likely later events in leukemogenesis and promote proliferation and survival but do not affect differentiation. They frequently involve receptor tyrosine kinases and affect prognosis but do not themselves define a distinct subtype of AML. Class I mutations include *FLT3*-ITD, *KIT*, *RAS*, and *JAK2*. In contrast, Class II mutations are likely initiating events, impair differentiation, and define distinct AML subtypes. Class II mutations include *NPM1*, *CEBPA*, *KMT2A*, *PML-RARA*, *RUNX1-RUNX1T1*, and *CBFB-MYH11*.

Cytogenetics and acute myeloid leukemia

In addition to the genetic abnormalities mentioned in the category of recurrent genetic abnormalities, other genetic abnormalities in AML may also be seen. Therefore, cytogenetics analysis is useful for diagnosis. Common abnormalities include

- t(1;22): usually associated with megakaryocytic lineage
- t(3) or inv 3: usually associated with abnormal megakaryocytes, multilineage dysplasia, and increased platelets
- t(6;9): usually associated with monocytic lineage, basophilia, and multilineage dysplasia
- t(9;11): usually associated with monocytic lineage.

Recall that the karyotype of the leukemic cells is the most predictive for response to therapy and survival. However, the most common leukemic karyotype (~50%) is a normal karyotype, and it is considered to be an intermediate prognosis (Table 6.3).

Key points

- CML: Defined by presence of Philadelphia chromosome and/or *BCR-ABL1* fusion gene. The abnormal fusion gene is present in all myeloid lineages as well in some lymphoid cells.
- Philadelphia (Ph) chromosome (seen in 90%−95% of cases of CML) is due to reciprocal translocation resulting in t(9;22)(q34.1;q11.2). *ABL1* gene from chromosome 9 fuses with *BCR* gene on chromosome 22. In the Philadelphia chromosome−negative cases, *BCR-ABL1* fusion gene can be detected by FISH, PCR, or Southern blot techniques.
- CML has three clinical phases: chronic phase, accelerated phase, and blast crisis.

Table 6.3 Prognostic implication of cytogenetic abnormalities in acute myeloid leukemia.

Favorable	Intermediate	Adverse
t(8;21)(q22;q22.1) *RUNX1-RUNX1T1*	Normal karyotype	inv(3)(q21.3q26.2) or t(3;3)(q21.3q26.2) *GATA2-MECOM*
Inv(16) (p13.1q22) or t(16;16)(p13.1q22) *CBFB-MYH11*	Normal karyotype with mutated *NPM1* and *FLT3*-ITD	t(6;9)(p23;q34.1) *DEK-NUP214*
t(15;17)(q24.1;q21.2) *PML-RARA*	Normal karyotype with wild-type *NPM1* and *FLT3*-ITD	t(v;11q23.3) *KMT2A* rearranged
Normal karyotype with mutated *NPM1* without *FLT3*-ITD	t(9;11)(p21.3;q23.3) *KMT2A-MLLT3*	−5 or del(5q)
Normal karyotype with biallelic mutated *CEBPA*		−7
		Complex karyotype (≥3 chromosomal abnormalities)

- Most cases of CML involve the M-BCR, resulting in the abnormal fusion protein, p210, which has increased tyrosinase kinase activity. If the breakpoint is in the minor region, then the fusion protein is p190 and is most frequently associated with Ph positive ALL. If p190 is present in CML, these patients have increased number of monocytes and can resemble CMML. If the breakpoint is in the mu region then the fusion protein is p230, and these patients have prominent neutrophilic maturation and/or thrombocytosis.
- The chronic phase of CML is characterized by peripheral blood: leukocytosis: peak in myelocytes and PMNs, absolute basophilia (invariably present), absolute eosinophilia (commonly present), and absolute monocytosis (with normal %); orderly differentiation is preserved; platelets: normal or increased (low platelets uncommon) and RBC: NRBCs may be seen.
- The chronic phase of CML is characterized by bone marrow: increased M:E ratio, usually greater than 10:1, paratrabecular cuff of immature myeloid cells (5–10 cells thick; normal 2–3), megakaryocytes: small, hypolobated (micromegakaryocyte), increased reticulin fibrosis, pseudo-Gaucher cells and sea blue histiocytes (seen due to increased cell turn over). No significant dysplasia is present.
- Accelerated phase of CML is characterized by persistent or increasing WBC ($>$10,000/mm^3) and/or persistent splenomegaly unresponsive to therapy (specific for CML), persistent thrombocytosis ($>$1million/mm^3) despite therapy, persistent thrombocytopenia ($<$100,000/mm3) unrelated to therapy, clonal cytogenetic abnormalities (additional chromosomal defects), and basophilia 20% or higher. Myeloblasts should represent 10%–19% in peripheral blood or bone marrow.
- Blast crisis of CML is characterized by $\geq$20% blasts in the peripheral blood or bone marrow, or an extramedullary blast proliferation; in two third of cases the blast lineage is myeloid, but in 20%–30% of cases the blasts are lymphoblasts.
- PV: almost all cases have *JAK2* mutation; most common mutation is *JAK2* V617F and diagnostic criteria include hemoglobin $>$ 16.5 g/dL(male) or $>$ 16.0 g/dL (female) (or increased red cell mass); low erythropoietin; splenomegaly; hypercellular marrow (all cell lines aka panmyelosis).
- PMF: 58% with *JAK2* mutation and key features include leukoerythroblastic anemia with teardrop red cells (peripheral blood), increased megakaryocytes with atypia (hypolobated, "bulbous," or "cloud-like" nuclei), increased fibrosis, osteosclerosis, and intrasinusoidal hematopoiesis in the bone marrow.
- ET: 50% with *JAK2* mutation and key features include platelets $>$450,000/mm^3 with increased megakaryocytes with hyperlobulated ("staghorn") nuclei and increased mature cytoplasm.
- Mastocytosis is a clonal neoplastic disorder of mast cells involving one or more organs. In involved organs, there will be presence of multifocal clusters or aggregates of abnormal mast cells. Mast cells can be stained with Giemsa, CAE, tryptase/chymase, CD117, and, for neoplastic mast cells, CD2 and CD25. The diagnosis of systemic mastocytosis requires the presence of the major criterion and one minor criterion or, in the absence of the major criterion, three minor criterions to be present.
- CMML is characterized by monocytosis and dysplasia. Naturally blasts have to be less than 20% as otherwise this would be acute leukemia. Also, Philadelphia chromosome is absent. A test for *BCR-ABL* gene rearrangement is negative.
- Atypical CML is characterized by leukocytosis resembling CML. However, dysplasia is present, and basophilia is not present. Philadelphia chromosome is absent or a test for *BCR-ABL1* gene rearrangement is negative.

- JMML is characterized by leukocytosis and monocytosis in individuals less than 14 years of age with features of dysplasia. The HbF percentage may be increased in JMML.
- Key features of MDS include the following: MDS is a clonal stem cell disease, MDS occur principally in older adults with a median age of 70 years, MDS is characterized by cytopenia in the peripheral blood with a hypercellular marrow, morphological evidence of dysplasia is present, and there are certain cytogenetic abormalities associated with MDS. There is also an increased risk of AML.
- Subclassification of MDS: MDS-EB-1, MDS-EB-2, MDS-SLD, MDS-MLD, MDS-RS.
- MDS associated with isolated del(5q) is characterized by increased incidence in females, anemia, with normal or increased platelet counts, splenomegaly, normal or increased megakaryocytes with hypolobated nuclei, blasts less than 1% in the peripheral blood, and <5% in the bone marrow, isolated del(5q) cytogenetic abnormality.
- The following are common cytogenetic abnormalities seen in MDS: Unbalanced: -7 or del(7q), -5 or del(5q), -20 or del(20q), -13 or del(13q), del(11q), del(12p) or t(12p), del(9q), idic(X)(q13), i(17q) or t(17p), +8, -Y; Balanced: t(11;16) (q23.3; p13.3), t(3;21) (q26.2;q22.1), inv(3)(q21.3;q26.2) /t(3;3) (q21.3;q26.2), t(1;3)(p36.3;q21.2), t(2;11)(p21;q23.3), t(6;9)(p23; q34.1).
- In isolation; +8, del(20q), or -Y are not considered presumptive evidence of MDS; however, in the setting of persistent cytopenias, the other cytogenetic abnormalities listed are considered presumptive evidence of MDS.
- The following karyotypes are associated with good prognosis of MDS: normal, $-Y$, del(5q), and del(20q). The following are associated with poor prognosis of MDS: complex (three or more abnormalities) or chromosome 7 abnormalities. All other abnormalities impart intermediate prognosis.
- Diagnosis of AML requires the presence of 20% or more blasts in the bone marrow. An exception to the 20% blast count rule is AML with recurrent genetic abnormalities, specifically t(15;17), t(8; 21), t(16;16), and Inv 16 (if an individual has any of these cytogenetic abnormalities, then they have AML irrespective of their blast count).
- The cytogenetic abnormality t(15;17) results in APL. Patients with APL are at increased risk of DIC but respond to ATRA therapy.
- Individuals with t(8;21) may have typical FAB M2 morphology. Typical M2 morphology includes blasts with salmon-colored granules, slender Auer rods, and cytoplasmic vacuoles. B-cell markers, such as CD19, may be positive by flow cytometry.
- Individuals with t(16;16) or inv16 may have FAB M4 Eo subtype of AML. Here, there is eosinophilia with presence of abnormal eosinophils. These abnormal eosinophils have basophilic granules and stain with CAE.
- AML with MDS-related changes and therapy-related myeloid neoplasms are easy to understand. If a patient with AML has prior history of MDS or has MDS-related cytogenetics or has AML with multilineage dysplasia, then he should be diagnosed as AML with MDS-related changes. Multilineage dysplasia in this context means at least two bone marrow cell lines are dysplastic and at least 50% of each cell line is dysplastic. Therapy-related myeloid neoplasms are seen in patients who have received chemotherapy (e.g., alkylating agents, topoisomerase II inhibitors) or radiation.

- AML, NOS, according to the WHO classification, roughly corresponds to the FAB classification with the exception of acute basophilic leukemia and acute panmyelosis with myelofibrosis, which do not exist in the FAB classification. Also, APL does not belong in the AML, NOS category as it is part of AML with recurrent genetic abnormalities.
- AML with minimum differentiation, FAB M0: here blasts are predominately present. The blasts typically have no granules or few granules. Auer rods are not seen. Special staining for MPO is negative. Thus, it is not possible to provide a decision whether the blasts are myeloblasts or lymphoblasts. Flow cytometry is crucial in this case. The blasts express myeloid antigens and thus the case is labeled as AML.
- AML without maturation, FAB M1 (FAB classification: M1): again myeloblasts dominate the picture (here promyelocytes, myelocytes, metamyelocytes, and bands/PMNs are $<10\%$). However, it is possible to designate the blasts as myeloblasts with MPO staining (positive staining for MPO in $\geq 3\%$ of the blasts).
- AML with maturation, FAB M2: here promyelocytes, myelocytes, metamyelocytes, and bands/PMNs are $\geq 10\%$.
- Acute myelomonocytic leukemia, FAB M4: here cells of monocytic lineage are $>20\%$ but less than 80%. A variant of M4 is M4 Eo. This has been discussed above.
- Acute monocytic leukemia, FAB M5: here cells of monocytic lineage are $\geq 80\%$. If the majority of the cells are monoblasts (typically 80% or more), then this may be categorized as M5a (acute monoblastic leukemia). In M5b, the majority of the monocytic cells are promonocytes or monocytes. Pease note in the setting of AML, promonocytes are counted as blasts.
- Acute erythroid leukemia, M6: previously, here 50% or more of the bone marrow cells are crythroid. Of the nonerythroid population, at least 20% are myeloblasts. The erythroid cells are typically dysplastic and also may have vacuoles. These vacuoles may be positive for PAS. However, acute erythroid leukemia is now classified as MDS, and the blast percentage is taken out of total cells. Pure erythroid leukemia is characterized by numerous proerythroblasts.
- Acute megakaryoblastic leukemia, M7: here at least half of the blast population is of megakaryocytic lineage. Megakaryoblasts typically have cytoplasmic blebs. Features of dysplasia may also be present in the myeloid cell lines.

References

[1] Murati A, Brecqueville B, Devillier R, Mozziconacci MJ, et al. Myeloid malignancies: mutations, models and management. BMC Cancer 2012;12:304.
[2] Koretzky G. The legacy of the Philadelphia chromosome. J Clin Investig 2007;117:2030−2.
[3] Keung YK, Beaty M, Powell B, Molnar I, et al. Philadelphia chromosome positive myelodysplastic syndrome and acute myeloid leukemia-retrospective study and review of literature. Leuk Res 2004;28:579−86.
[4] Elliott MA, Tefferi A. The molecular genetics of chronic neutrophilic leukemia: defining a new era in diagnosis and therapy. Curr Opin Hematol 2014;21:148−54.
[5] Elliott MA, Tefferi A. Chronic neutrophilic leukemia: 2018 update on diagnosis, molecular genetics and management. Am J Hematol 2018;93:578−87.
[6] Thiele J, Kvasnicka HM, Beelen DW, Flucke U, et al. Megakaryopoiesis and myelofibrosis in chronic myeloid leukemia after allogeneic bone marrow transplantation: an immunohistochemical study of 127 patients. Mod Pathol 2001;14:129−38.

[7] Dos Santoa LC, Riberio JC, Silva NP, Cerutti NP, et al. Cytogenetics, JAK2 and MPL mutations in polycythemia vera, primary myelofibrosis and essential thrombocythemia. Rev Bras Hematol Hemoter 2011; 33:417—24.

[8] Wang SA, Hasserjian RP, Tam W, Tsai AG, et al. Bone marrow morphology is a strong discriminator between chronic eosinophilic leukemia, not otherwise specified and reactive idiopathic hypereosinophilic syndrome. Haematologica 2017;102:352—1360.

[9] Wang SA. The Diagnostic Work-Up of Hypereosinophilia. Pathobiology 2019;86:39—52.

[10] Gotlib J. World Health Organization defined eosinophilic disorders: 2014 updates on diagnosis, risk stratification and management. Am J Hematol 2014;89:325—37.

[11] Savage N, George TI, Gotlib J. Myeloid neoplasms associated with eosinophilia and rearrangement of PDGFRA, PDGFRB and FGFR1: a review. Int J Lab Hematol 2013;35:491—500.

[12] Kawankar N, Vundinti BR. Cytogenetic abnormalities in myelodysplastic syndrome: an overview. Hematology 2011;16:131—8.

[13] Wall M. Recurrent cytogenetic abnormalities in myelodysplastic syndrome. Methods Mol Biol 2017;1541: 209—22.

Monoclonal gammopathies and their detection

Introduction

Monoclonal gammopathy is present in a patient when a monoclonal protein (also referred as M protein or paraprotein) is identified in a patient's serum, urine, or both. Monoclonal gammopathy may be due to plasma cell dyscrasia or a B-cell lymphoproliferative disorder producing this paraprotein. Monoclonal gammopathy may present as monoclonal gammopathy of undetermined significance (MGUS), which requires no specific treatment to frank multiple myeloma. Multiple myeloma, a malignant disorder of the bone marrow, is the most common form of myeloma. This disease is called multiple myeloma because it affects multiple organs in the body. In multiple myeloma, plasma cells that proliferate at a low rate become malignant with a massive clonal expression resulting in a high rate of production of monoclonal immunoglobulin in the circulation. MGUS was first described in 1978 and is a precancerous condition affecting approximately 3% of people over 50 years of age [1]. This condition may progress to multiple myeloma among 1% of these individuals every year. Circulating microRNAs may be a potential biomarker for distinguishing normal people from patients with multiple myeloma or related conditions [2]. The risk of malignant transformation of MGUS into multiple myeloma is higher in females than males and is also higher in individuals showing the presence of IgA paraprotein as compared with individuals with IgG paraprotein [3].

The paraprotein can be an intact immunoglobulin, only light chains (light chain myeloma, light chain deposition disease, or amyloid light chain amyloidosis), or rarely found only as heavy chains (heavy chain disease [HCD]). Paraproteins can be detected in the serum and can also be excreted into the urine. Sometimes, if the paraprotein is only light chain (light chain disease), then this is detected in the urine alone but not in the serum. The serum may paradoxically exhibit only hypogammaglobulinemia. It is important to note that the presence of paraprotein in serum, urine, or both indicates monoclonal gammopathy and not necessarily the presence of multiple myeloma in a patient. Multiple myeloma is only one of the causes of monoclonal gammopathy. Transient monoclonal gammopathy may be observed in an immunocompromised patient suffering from infection due to an opportunistic pathogen such as cytomegalovirus [4]. Monoclonal gammopathy is usually observed in patients over 50 years of age and is rare in children. Gerritsen et al. studied 4000 pediatric patients over a 10-year period and observed monoclonal gammopathy only in 155 children, but such gammopathies were found most frequently in patients suffering from primary and secondary immunodeficiency, hematological malignancies, autoimmune disease, and severe aplastic anemia. Follow-up analysis revealed that most of these monoclonal gammopathies were transient [5].

Hematology and Coagulation. https://doi.org/10.1016/B978-0-12-814964-5.00007-3

Diagnostic approach to monoclonal gammopathy using electrophoresis

Serum protein electrophoresis (SPEP), urine electrophoresis, serum immunofixation, and urine immunofixation are all performed primarily to investigate suspicion of monoclonal gammopathy in a patient. In addition, serum light chain and serum heavy chain assays are employed. Lastly, mass spectrometry (MS) is also a method that uses high-resolution molecular mass measurements to accurately identify and classify M proteins in the serum. In addition to being more accurate than immunofixation, initial reports suggest that, unlike other methods, MS is also able to accurately discriminate between therapeutic monoclonal antibodies and endogenous M proteins. However, MS is not yet a standard part of the evaluation of M proteins in most medical centers. Agarose gel electrophoresis and capillary electrophoresis are two principal methods employed in screening for paraproteins. Both methods are applicable for both serum and urine specimens. Once a paraprotein is detected, confirmation and the isotyping of paraprotein are essential, which is usually achieved by immunofixation. For urine immunofixation, the best practice is to utilize a 24-h urine specimen that has been concentrated; such technique allows for detection of even a faint band.

Serum protein electrophoresis

SPEP is an inexpensive, easy-to-perform screening procedure for initial identification of monoclonal bands. Monoclonal bands are usually seen in the gamma zone but may be seen in proximity of the beta band or rarely in the alpha-2 region. Blood can be collected in a tube with clot activator, and after separation from blood components, serum is then placed on special paper treated with agarose gel followed by exposure to an electric current in the presence of a buffer solution (electrophoretic cell). Various serum proteins are then separated based on charge. After a predetermined time of exposure to an electric field, the special paper is removed, dried, and placed on a fixative to prevent further diffusion of specimen components followed by staining to visualize various protein bands. Coomassie Brilliant Blue is a common staining agent to visualize bands in SPEP. Then using a densitometer, each fraction is quantitated. The serum protein components are separated into five major fractions:

- Albumin
- Alpha-1 globulins (alpha-1 zone)
- Alpha-2 globulins (alpha-2 zone)
- Beta globulins (beta zone often splits into beta one and beta two band)
- Gamma globulins (gamma zone)

Albumin and globulins are two major fractions of electrophoresis pattern. Albumin, the largest band, lies closest to the positive electrode (anode) and has a molecular weight of approximately 67 kDa (67,000 Da).

Reduced intensity of this band is observed in

- inflammation,
- liver dysfunction,
- uremia,
- nephrotic syndrome, and
- other conditions that lead to hypoalbuminemia, such as critically illness and pregnancy.

A smear observed in front of the albumin band may be due to hyperbilirubinemia or due to presence of certain drugs.

A band in front of the albumin band may be due to prealbumin (a carrier for thyroxine and vitamin A), which is commonly seen in cerebrospinal fluid specimens or serum specimens in patients with malnutrition.

Two, rather than one, albumin bands may represent bisalbuminemia. This is a familial abnormality with no clinical significance. However, occurrence of bisalbuminemia is rare [6].

Analbuminemia is a genetically inherited metabolic disorder first described in 1954, but this disorder is a rare disorder affecting less than one in 1 million births. The condition is benign as low albumin levels are compensated by high levels of nonalbumin proteins and circulatory adaptation. Hyperlipidemia is usually observed in these patients. Pseudo-analbuminemia due to the presence of a slow-moving albumin variant appearing in the alpha-1 region of SPEP has also been reported [7].

Moving toward the negative portion of the gel (cathode), the alpha zone is the next band after albumin. The alpha zone can be subdivided into two zones:

- the alpha-1 band and
- alpha-2 band.

The alpha-1 band mostly consists of

- alpha-1-antitrypsin (AT) (90%),
- alpha-1-chymotrypsin, and
- thyroid-binding globulin.

Alpha-1-antitrypsin is an acute-phase reactant, and its concentration is increased in inflammation and other conditions. The alpha-1-antitrypsin band is decreased in patients with alpha-1-antitrypsin deficiency or decreased production of globulin in patients with severe liver disease.

At the leading edge of this band, a haze due to high-density lipoprotein (HDL) may be observed. The alpha-2 band consists of

- alpha-2-macroglobulin,
- haptoglobin, and
- ceruloplasmin.

Because both haptoglobin and ceruloplasmin are acute-phase reactants, this band is increased in inflammatory states. Alpha-2-macroglobulin is increased in nephrotic syndrome and cirrhosis of liver.

The beta zone may consist of two bands:

- beta-1 and
- beta-2.

Beta-1 is mostly composed of transferrin and low-density lipoprotein (LDL). An increased beta-1 band is observed in iron deficiency anemia due to increased level of free transferrin. This band may also be elevated in pregnant women.

Very low–density lipoprotein usually appears in the pre-beta zone.

The beta-2 band is mostly composed of complement proteins. If two bands are observed in the beta-2 region, it implies either electrophoresis of plasma specimen (fibrinogen band) instead of serum specimen or IgA paraprotein.

Much of the clinical interest of SPEP is focused on the gamma zone of the SPEP because immunoglobulins mostly migrate to this region. Usually the C-reactive protein band is found between the beta and gamma region. SPEP is most commonly ordered when multiple myeloma is suspected and observation of a monoclonal band (M band, paraprotein) indicates that monoclonal gammopathy may be present in the patient. If M band or paraprotein is observed in SPEP, the following steps are performed:

- The monoclonal band is measured quantitatively using densitometric scan of the gel.
- Serum and/or urine immunofixation is conducted to confirm the presence of the paraprotein as well as determining the isotype of the paraprotein.
- A serum light chain assay is conducted or recommended to the ordering clinician.

In about 5% of cases, two paraproteins may be detected. This is referred to as biclonal gammopathy.

A patient may also have nonsecretory myeloma, as in the case of a plasma cell neoplasm in which the clonal cells are not either producing or secreting M proteins.

The most commonly observed paraprotein is IgG followed by IgA, light chain, and rarely IgD.

When a monoclonal band is identified using SPEP, serum immunofixation and 24-h urine immunofixation is typically recommended. There are certain situations where a band may be apparent, but in reality, it is not a monoclonal band. Examples include the following:

- Fibrinogen is seen as a discrete band when electrophoresis is performed on plasma instead of serum specimen. This fibrinogen band is seen between the beta and gamma regions. If the electrophoresis is repeated after the addition of thrombin, this band should disappear. In addition, immunofixation study should be negative.
- Intravascular hemolysis results in release of free hemoglobin in circulation which binds to haptoglobin. The hemoglobin−haptoglobin complex may appear as a large band in the alpha-2 area. Serum immunofixation studies should be negative in such cases.
- In patients with iron deficiency anemia, concentrations of transferrin may be high, which may result in a band in the beta region. Again, immunofixation should be negative.
- Patients with nephrotic syndrome usually show low albumin and total protein, but this condition may also produce increased alpha-2 and beta fractions. Bands in either of these regions may mimic a monoclonal band.
- When performing gel electrophoresis, a band may be visible at the point of application. Typically, this band is present in all samples performed at the same time using same agarose gel support material.

Common problems associated with interpretation of SPEP are summarized in Table 7.1. A low concentration of a paraprotein may not be detected by serum electrophoresis. There are also certain situations where a false negative interpretation could be made on serum electrophoresis. These situations include

- A clear band is not seen in cases of alpha-HCD. This is presumably due to the tendency of these chains to polymerize or due to their high carbohydrate content. HCDs are rare B-cell lymphoproliferative neoplasm characterized by production of a monoclonal component consisting of monoclonal immunoglobulin heavy chain without associated light chain.
- In mu HCD, a localized band is found in only 40% of cases. Panhypogammaglobulinemia is a prominent feature in such patients.

Table 7.1 Common problems associated with serum protein electrophoresis (SPEP).

- SPEP performed using plasma instead of serum produces an addition distinct band between beta and gamma zone due to fibrinogen but such band is absent in subsequent immunofixation study.
- A band may be seen at the point of application. Typically, this band is present in all samples performed at the same time.
- If concentration of transferrin is high (for example, iron deficiency), a strong band in the beta region is observed.
- In nephrotic syndrome, prominent bands may be seen in alpha-2 and beta regions, which are not due to monoclonal proteins.
- Hemoglobin—haptoglobin complexes (seen in intravascular hemolysis) may produce a band in the alpha-2 region.
- Paraproteins may form dimers, pentamers, polymers, or aggregates with each other resulting in abroad smear rather than a distinct band.
- In light chain myeloma, light chains are rapidly excreted in the urine and no corresponding band may be present in SPEP.

- In occasional cases of gamma HCD, a localized band may not be seen.
- When a paraprotein forms dimers, pentamers, polymers, or aggregates with each other or when forming complexes with other plasma components, a broad smear may be visible instead of a distinct band.
- Some patients may produce only light chains that are rapidly excreted in the urine, and no distinct band may be present in the SPEP. Urine protein electrophoresis is more appropriate for diagnosis of light chain disease. When light chains cause nephropathy and result in renal insufficiency, excretion of the light chains is hampered, and a band may be seen in serum electrophoresis.
- In some patients with IgD myeloma, the paraprotein band may be very faint.

Hypogammaglobulinemia may be congenital or acquired. Amongst the acquired causes are multiple myeloma and primary amyloidosis. Panhypogammaglobulinemia can occur in about 10% of cases of multiple myeloma. Most of these patients have a Bence Jones protein in the urine but lack intact immunoglobulins in the serum. Bence Jones proteins are monoclonal free kappa or lambda light chains in the urine. Detection of Bence Jones protein may be suggestive of multiple myeloma or Waldenström's macroglobulinemia. Panhypogammaglobulinemia can also be seen in 20% of cases of primary amyloidosis. It is important to recommend urine immunofixation studies when panhypogammaglobulinemia is present in SPEP.

Although monoclonal gammopathy is the major reason for performing SPEP, polyclonal gammopathy may be observed in some patients. Monoclonal gammopathies are associated with a clonal process that is malignant or potentially malignant. However, polyclonal gammopathy, in which there is a nonspecific increase in gamma globulins, is not associated with malignancies. SPEP may also exhibit changes that imply specific underlying clinical conditions other than monoclonal gammopathy.

Common features of SPEP in various disease states other than monoclonal gammopathy include the following:

- Inflammation: Increased intensity of alpha-1 and alpha-2 with sharp a leading edge of alpha-1 may be observed, but with chronic inflammation the albumin band may be decreased with increased gamma zone due to polyclonal gammopathy.

- Nephrotic syndrome: In nephrotic syndrome, the albumin band is decreased due to hypoalbuminemia. In addition, the alpha-2 band may be more distinct.
- Cirrhosis or chronic liver disease: A low albumin band due to significant hypoalbuminemia with a prominent beta-2 band and beta-gamma bridging is characteristics features of liver cirrhosis or chronic liver disease. In addition, polyclonal hypergammaglobulinemia is observed.

Urine electrophoresis

Urine protein electrophoresis is analogous to the SPEP and is used to detect monoclonal proteins in the urine. Ideally, it should be performed on a 24-h urine sample (concentrated 50−100 times). Molecules less than 15 kDa are filtered through a glomerular filtration process and are excreted freely into urine. In contrast, only selected molecules with molecular weight between 16 and 69 KDa can be filtered by the kidney and may appear in the urine. Albumin is approximately 67 kDa. Therefore, trace albumin in urine is physiological.

Molecular weight of the protein, concentration of the protein in the blood, charge, and hydrostatic pressure all regulate passage of a protein through the glomerular filtration glomerular process.

Proteins that pass through glomerular filtration include albumin, alpha-1 acid glycoprotein (orosomucoid), alpha-1 microglobulin, beta-2 microglobulin, retinol-binding protein, and trace amounts of gamma globulins. However, 90% of these are reabsorbed and only a small amount may be excreted in the urine. Normally, total urinary protein is < 150 mg/24 h and consists of mostly albumin and Tamm−Horsfall protein (secreted from ascending limb of loop of Henle). The extent of proteinuria can be assessed by quantifying the amount of proteinuria and expressing as protein to creatinine ratio. Normal protein to creatinine ratio is <0.5 in children 6 months to 2 years of age, <0.25 in children above 2 years, and <0.2 in adults.

Proteinuria with minor injury (typically only albumin is lost in urine) can be related to vigorous physical exercise, congestive heart failure, pregnancy, alcohol abuse, or hyperthermia. Overflow proteinuria can be seen in patients with myeloma or massive hemolysis of crush injury (myoglobin in urine). In addition, beta-2 microglobulin, eosinophil-derived neurotoxin, and lysozyme can produce bands in urine electrophoresis. Therefore, immunofixation studies are required to document true M proteins and rule out presence of other proteins in urine electrophoresis.

Proteinuria can be classified as glomerular, tubular, or combined proteinuria. Glomerular proteinuria can be subclassified as selective glomerular proteinuria (urine will have albumin and transferrin bands) or nonselective glomerular proteinuria (urine will have presence of all different types of proteins). In glomerular proteinuria, the dominant protein present is always albumin. In tubular proteinuria, albumin is a minor component. The presence of alpha-1 microglobulin and beta-2 microglobulin are indicators of tubular damage.

Immunofixation studies

In immunofixation, electrophoresis of one specimen from a patient suspected of monoclonal gammopathy is performed using five separate lanes. Then, each sample is overlaid with different monoclonal antibodies: anti-gamma (to detect gamma heavy chain), anti-mu (to detect mu heavy chain), anti-alpha (to detect alpha heavy chain), anti-kappa (to detect kappa light chain), and anti-lambda (to detect lambda light chain). Antigen antibody reaction should take place. After washing to remove unbound antibodies, the gel paper is stained, which allows identification of specific isotope of the monoclonal protein. A normal SPEP does not exclude diagnosis of myeloma because approximately 11% myeloma patients may have normal SPEP. Therefore, serum and urine immunofixation studies should be performed regardless of serum electrophoresis results if clinical suspicion is high. It is also important to note that an M band or paraprotein in SPEP may not be a true band unless it is identified by using serum or urine immunofixation as these tests are more sensitive than SPEP. In addition, immunofixation technique can also determine the particular isotype of the monoclonal protein. However, immunofixation technique cannot estimate the quantity of the M protein. In contrast, SPEP is capable of estimating the concentration of an M protein.

In multiple myeloma, sometimes only free light chains are produced. The concentration of the light chains in serum may be so low that these light chains remain undetected using SPEP and even serum immunofixation. In such cases, immunofixation on a 24-h urine sample is useful. Another available test is detection of serum free light chains by immunoassay, which allows for calculating the ratio of kappa to lambda free light chains. This test is more sensitive than urine immunofixation.

One source of possible error in urine immunofixation study is the "step ladder" pattern. Here, multiple bands are seen in the kappa (more often) or lambda lanes and are indicative of polyclonal spillage rather than monoclonal spillage into the urine. During urine immunofixation, five or six faint, regular diffuse bands with hazy background staining between bands may be seen. This is more often seen in the kappa lane than the lambda lane. This is referred to as the step ladder pattern and is a feature of polyclonal hypergammaglobulinemia with spillage into the urine.

Capillary zone electrophoresis

Capillary zone electrophoresis is an alternative method of performing protein electrophoresis. Protein stains are not required, and a point of application is not observed. It is considered to be faster and more sensitive compared with agarose gel electrophoresis where a classic case of monoclonal gammopathy produces a peak, typically in the gamma zone. However, subtle changes in the gamma zones may also represent underlying monoclonal gammopathy. Interpretation can be subjective, and a relatively high percentage of cases may be referred for ancillary studies such as immunofixation depending on preference of the pathologist interpreting the result. However, disregarding a subtle change in capillary zone electrophoresis may potentially result in missing a case. However, because of increased analytical sensitivity of capillary electrophoresis over traditional protein electrophoresis, capillary electrophoresis may detect additional irregularities that are suspicious for a monoclonal component. This is most noticeable in the beta-1, beta-2, and gamma globulin fractions. The causes of nonmonoclonal irregularities are manifold but are rarely reported back to the ordering physician. Recently, Regeniter A and Siede reviewed the basic concepts to correctly identify typical nonmonoclonal irregularities according to their electrophoresis fractions and their possible clinical implications [8].

Free light chain assay

Patients with monoclonal gammopathy may have negative SPEP and serum immunofixation studies. This may be due to very low levels of paraproteins and light chain gammopathy in which the light chains are very rapidly cleared from the serum by the kidneys. Because of this, urine electrophoresis and urine immunofixation is part of the workup for cases where monoclonal gammopathy is a clinical consideration. Urine electrophoresis and urine immunofixation studies are also performed to document the amount (if any) of potentially nephrotoxic light chains being excreted in the urine in a case of monoclonal gammopathy.

Quantitative serum assays for kappa and lambda free light chain disease have increased the sensitivity of serum testing strategies for identifying monoclonal gammopathies, especially the light chain diseases. Cases that may appear as nonsecretory myeloma can actually be cases of light chain myeloma. Free light chain assays allow disease monitoring and provide prognostic information for MGUS, smoldering myeloma.

The rapid clearance of light chains by the kidney is reduced in renal failure. Levels may be 20−30 times higher than normal in end-stage renal disease (ESRD). In addition, the kappa/lambda ratio may be as high as 3:1 in renal failure (normal 0.26−1.65). Therefore, patients with renal failure may be misdiagnosed as kappa light chain monoclonal gammopathy. If a patient has lambda light chain monoclonal gammopathy, with the relative increase in kappa light chain in renal failure, the ratio may become normal. Thus, a case of lambda light chain monoclonal gammopathy may be missed.

Heavy/light chain assay

One of the great diagnostic benefits of sFLC analysis is the κ/λ ratio. This is because (1) it provides a quantitative assessment of clonality; (2) the clinical ranges are enhanced due to immunosuppression of the nontumour FLCs; and (3) there is automatic compensation for variable metabolism. These same advantages can be exploited for intact immunoglobulins if the different light chain types are measured (e.g., to produce a ratio of IgGκ/IgGλ). Raising polyclonal antibodies specific for the unique junctional epitopes, spanning the heavy and light chain immunoglobulin constant regions, is a significant challenge, but reagents have now been developed for the three main immunoglobulins (i.e., IgGκ, IgGλ, IgAκ, IgAλ, IgMκ, and IgMλ). Turbidimetric/nephelometric immunoglobulin heavy/light chain (HLC) assays were made available for general use in 2009−2010 with the trade name of Hevylite.

HLC assays typically have a greater clinical sensitivity than SPE (for monoclonal immunoglobulin) and can exceed that of IFE in some instances. In addition, the assays can be particularly helpful for monitoring patients with IgA myeloma if their monoclonal immunoglobulin comigrates with other proteins in the β-region of SPE gels. Preliminary studies have also indicated that suppression of the uninvolved HLC pair (e.g., IgGλ in an IgGκ patient) may be more informative than assessment of general immunoparesis, and HLC analysis has provided prognostic information in both MGUS and MM.

Paraprotein interferences in clinical laboratory tests

Interference of paraprotein can produce both false positive and false negative test results depending on the analyte and the analyzer. However, the magnitude of interference may not correlate with the amount of paraprotein present in serum. The most common interferences include falsely low HDL cholesterol, falsely high bilirubin, and altered values of inorganic phosphate. Other tests in which altered results may occur include LDL cholesterol, C-reactive protein, creatinine, glucose, urea nitrogen, inorganic calcium, and blood count. There is a poor correlation between concentration or type of paraprotein and likelihood of interference [9].

Plasma cell disorders

Plasma cell neoplasm is characterized by the clonal expansion of terminally differentiated B-lymphoid cells that have undergone somatic hypermutation that results in production of monoclonal immunoglobulin protein. Plasma cell neoplasm encompasses a spectrum of diseases and includes

Monoclonal gammopathy of undetermined significance (MGUS)
Idiopathic Bence Jones proteinuria
Monoclonal gammopathy of renal significance (MGRS)
POEMS syndrome
AL amyloidosis, light and heavy chain deposition disease
Solitary plasmacytoma
Multiple and smoldering myeloma

Monoclonal gammopathy of undetermined significance

MGUS, a disorder of plasma cell, could be benign but in some patients may progress to a malignant disorder. MGUS occurs in over 3% of the general Caucasian population over 50 years of age. The diagnostic criteria are M protein concentration <30 g/L, plasma cells in bone marrow <10%, and absence of organ or tissue damage (CRAB: C for hypercalcemia, R for renal dysfunction, A for anemia, and B for bone lesions). Risk of progression is approximately 1%/yr. There are three distinct subtypes:

IgM MGUS: This may progress to lymphoplasmacytic lymphoma/Waldenström macroglobulinemia
IgM MGUS: This may progress to myeloma
Light chain MGUS: This may progress to idiopathic Bence Jones proteinuria or light chain myeloma

Idiopathic Bence Jones proteinuria

This is also known as smoldering light chain multiple myeloma.

Monoclonal gammopathy of renal significance

MGRS represents a group of disorders in which a monoclonal immunoglobulin secreted by a nonmalignant or premalignant B-cell or plasma cell clone results in renal damage. These disorders do not meet diagnostic criteria for overt symptomatic multiple myeloma or a lymphoproliferative disorder. The term MGRS was proposed in 2012 by the International Kidney and Monoclonal Gammopathy Research Group to collectively describe patients who would otherwise meet the criteria for MGUS but demonstrate renal injury attributable to the underlying monoclonal protein.

In MGRS, the renal lesions are primarily caused by the abnormal deposition or activity of monoclonal proteins in the kidney. Deposition of the monoclonal proteins may occur within the glomeruli, tubules, vessels, or interstitium of the kidney, depending on the specific biochemical characteristics of the pathogenic light and/or heavy chains involved.

MGRS-associated kidney diseases encompass a wide spectrum of renal pathology and include such lesions as immunoglobulin-associated amyloidosis, the monoclonal immunoglobulin deposition diseases (MIDDs; light chain deposition disease, heavy chain deposition disease, and light and heavy chain deposition disease), proliferative glomerulonephritis with monoclonal immunoglobulin deposits, C3 glomerulopathy with monoclonal gammopathy, light chain proximal tubulopathy (Fanconi syndrome), amongst others.

Patients with MGRS frequently develop progressive kidney disease and ESRD. MGRS-related kidney diseases recur in most patients after kidney transplantation and can lead to rapid allograft loss.

POEMS

POEMS (P for polyneuropathy, O for organomegaly, E for endocrinopathy, M for myeloma, and S for skin changes) syndrome is also known as osteosclerotic myeloma.

Plasmacytoma

Solitary plasmacytoma of bone and extraosseous plasmacytoma (most often in upper respiratory tract).

Monoclonal immunoglobin deposition disease (MIDD)

In this disease, tissue deposition of monoclonal Ig, Ig chain, or fragment takes place. In Ig-related primary amyloidosis, Congo red staining of amyloid protein in tissue can be used for diagnostic purpose. Under electron microscopy, amyloid deposits appears to be composed of linear, non-branching, aggregated fibrils that are 7.5−10 nm thick of indefinite length arranged in a loose meshwork. MIDD can be related to systemic light chain or heavy chain deposition disease.

Plasma cell myeloma

In classical or symptomatic plasma cell myeloma, M protein is found in serum or urine (typically >30 g/L of IgG or >25 g/L of IgA or >1g/24 h of urine light chain), bone marrow shows clonal plasma cells (usually >10% of nucleated cells in bone marrow), or there is evidence of plasmacytoma. In addition, there is evidence of organ or tissue damage (CRAB). Other variants include (a) smoldering (lab criteria met but no CRAB; the plasma cells in the bone marrow are greater than 10% but less than 60%), (b) nonsecretory (85% nonsecretors have Ig in the cytoplasm but are not secreting them and 15% nonproducers); and (c) plasma cell leukemia (>2000 plasma cells/mm3 or 20% by difference).

Morphology of plasma cells in myeloma

In myeloma, plasma cells are found as interstitial clusters or focal nodules or in diffuse sheets in the bone marrow. The cells may appear mature or immature, plasmablastic, or pleomorphic. The following descriptive terms have been used but none are diagnostic for myeloma:

- Mott cell: plasma cell with grape-like cluster of cytoplasmic inclusion
- Russell bodies: plasma cells with cherry red cytoplasmic inclusions
- Dutcher bodies: plasma cell with intranuclear inclusions

- Flame-shaped cells: plasma cells vermillion staining glycogen-rich IgA
- Thesaurocytes: plasma cells with ground glass cytoplasm

Immunophenotype of neoplastic plasma cells

Plasma cells (normal and neoplastic) express CD38 (bright) and CD138. CD138 is more specific but is less sensitive. Normal peripheral blood plasma cells are CD45+. In bone marrow, there are two subsets of plasma cells: one major subset positive for CD45 and a smaller negative one. Aberrant CD56 expression is identified in most patients with myeloma. Bright CD38 expression with coexpression of CD56 is used to detect abnormal population of plasma cells by flow cytometry. Abnormal plasma cells also are CD19−, in contrast to normal plasma cells which are CD19+. Normal and abnormal plasma cells are negative for CD20.

The abnormal population of plasma cells once identified should also exhibit cytoplasmic light chain restriction.

Cytogenetics in myeloma diagnosis

Conventional cytogenetics documents abnormality in approximately 30% of cases; with FISH (fluorescence in situ hybridization), this increases to 90%. 14q32 translocations involving the heavy chain locus are the most frequent abnormality.

Genetic abnormalities associated with an unfavorable prognosis are

- Hypodiploidy, del 13, del 17p, t (4;14), t(14;16)

Genetic abnormalities associated with a favorable prognosis are

- Hyperdiploidy, t(11;14)

Therefore, cytogenetics and molecular profiling have value in determining prognosis of multiple myeloma [10,11]. It is interesting to note that high serum beta-2 microglobulin and low serum albumin is also associated with less favorable prognosis.

Key points

- SPEP, urine electrophoresis, serum immunofixation, and urine immunofixation are all performed primarily to investigate suspicion of monoclonal gammopathy in a patient. Monoclonal gammopathy is present in a patient when a monoclonal protein (also called paraprotein or M protein) is identified in a patient's serum, urine, or both. The paraprotein can be an intact immunoglobulin, only light chains (light chain myeloma, light chain deposition disease, or amyloid light chain amyloidosis), or rarely found only as heavy chains (HCD).
- Agarose gel electrophoresis and capillary electrophoresis are two principal methods employed in screening for paraproteins. Both methods are applicable for both serum and urine specimens. Paraproteins are seen usually in the gamma region of the electrophoresis but also may be present in beta or rarely in the alpha-2 region.

- Once a paraprotein is detected, confirmation and the isotyping of paraprotein is essential, which is usually achieved by immunofixation. In about 5% of cases, two paraproteins may be detected. This is referred to as biclonal gammopathy. A patient may also have nonsecretory myeloma, as in the case of a plasma cell neoplasm in which the clonal cells are not either producing or secreting M proteins. The most commonly observed paraprotein is IgG followed by IgA, light chain, and rarely IgD. A normal SPEP does not exclude diagnosis of myeloma because approximately 11% myeloma patients may have normal SPEP. Therefore, serum and urine immunofixation studies should be performed regardless of serum electrophoresis results if clinical suspicion is high.
- Other components of SPEP include albumin, alpha-1 globulins (alpha-1 zone), alpha-2 globulins (alpha-2 zone), beta globulins (beta zone often splits into beta one and beta two band), and gamma globulins (gamma zone).
- Reduced intensity of the albumin band is observed in inflammation, liver dysfunction, uremia, nephrotic syndrome, and other conditions that lead to hypoalbuminemia. A smear observed in front of the albumin band may be due to hyperbilirubinemia or due to presence of certain drugs. A band in front of the albumin band may be due to prealbumin (a carrier for thyroxine and vitamin A), which is commonly seen in cerebrospinal fluid specimens or serum specimens in patients with malnutrition. Two, rather than one, albumin bands may represent bisalbuminemia. This is a familial abnormality with no clinical significance.
- The alpha-1 band mostly consists of alpha-1-antitrypsin (AT) (90%), alpha-1-chymotrypsin, and thyroid-binding globulin. Alpha-1-antitrypsin is an acute-phase reactant, and its concentration is increased in inflammation and other conditions. The alpha-1-antitrypsin band is decreased in patients with alpha-1-antitrypsin deficiency or decreased production of globulin in patients with severe liver disease. At the leading edge of this band, a haze due to HDL may be observed. The alpha-2 band consists of alpha-2-macroglobulin, haptoglobin, and ceruloplasmin. Because both haptoglobin and ceruloplasmin are acute-phase reactants, this band is increased in inflammatory states. Alpha-2-macroglobulin is increased in nephrotic syndrome and cirrhosis of liver.
- The beta zone may consist of two bands, beta-1 and beta-2. Beta-1 is mostly composed of transferrin and LDL. An increased beta-1 band is observed in iron deficiency anemia due to increased level of free transferrin. This band may also be elevated in pregnant women. Very low–density lipoprotein usually appears in the pre-beta zone. The beta-2 band is mostly composed of complement proteins. If two bands are observed in the beta-2 region, it implies either electrophoresis of plasma specimen (fibrinogen band) instead of serum specimen or IgA paraprotein.
- There are certain situations where a band may be apparent, but in reality, it is not a monoclonal band. For example, fibrinogen is seen as a discrete band between beta and gamma region when electrophoresis is performed on plasma instead of serum specimen. If the electrophoresis is repeated after the addition of thrombin, this band should disappear. In addition, immunofixation study should be negative. Intravascular hemolysis results in release of free hemoglobin in circulation which binds to haptoglobin. The hemoglobin–haptoglobin complex may appear as a large band in the alpha-2 area. Serum immunofixation studies should be negative in such cases.
- In patients with iron deficiency anemia, concentrations of transferrin may be high, which may result in a band in the beta region. Again, immunofixation should be negative.
- Patients with nephrotic syndrome usually show low albumin and total protein, but this condition may also produce increased alpha-2 and beta fractions. Bands in either of these regions may mimic a monoclonal band.

- Hypogammaglobulinemia may be congenital or acquired. Amongst the acquired causes are multiple myeloma and primary amyloidosis. Panhypogammaglobulinemia can occur in about 10% of cases of multiple myeloma. Most of these patients have a Bence Jones protein in the urine but lack intact immunoglobulins in the serum.

Common features of SPEP in various disease states other than monoclonal gammopathy include the following:

- Inflammation: Increased intensity of alpha-1 and alpha-2 with sharp a leading edge of alpha-1 may be observed, but with chronic inflammation, the albumin band may be decreased with increased gamma zone due to polyclonal gammopathy.
- Nephrotic syndrome: In nephrotic syndrome, the albumin band is decreased due to hypoalbuminemia. In addition, the alpha-2 band may be more distinct.
- Cirrhosis or chronic liver disease: A low albumin band due to significant hypoalbuminemia with a prominent beta-2 band and beta-gamma bridging is characteristics features of liver cirrhosis or chronic liver disease. In addition, polyclonal hypergammaglobulinemia is observed.

Proteinuria can be classified as glomerular, tubular, or combined proteinuria. Glomerular proteinuria can be subclassified as selective glomerular proteinuria (urine will have albumin and transferrin bands) or nonselective glomerular proteinuria (urine will have presence of all different types of proteins). In glomerular proteinuria, the dominant protein present is always albumin. In tubular proteinuria, albumin is a minor component. The presence of alpha-1 microglobulin and beta-2 microglobulin is a indicator of tubular damage.

- One source of possible error in urine immunofixation study is the "step ladder" pattern. Here, multiple bands are seen in the kappa (more often) or lambda lanes and are indicative of polyclonal spillage rather than monoclonal spillage into the urine. During urine immunofixation, five or six faint, regular diffuse bands with hazy background staining between bands may be seen. This is more often seen in the kappa lane than the lambda lane. This is referred to as the step ladder pattern and is a feature of polyclonal hypergammaglobulinemia with spillage into the urine.
- The rapid clearance of light chains by the kidney is reduced in renal failure. Levels may be 20–30 times higher than normal in ESRD. In addition, the kappa/lambda ratio may be as high as 3:1 in renal failure (normal 0.26–1.65). Therefore, patients with renal failure may be misdiagnosed as kappa light chain monoclonal gammopathy. If a patient has lambda light chain monoclonal gammopathy, with the relative increase in kappa light chain in renal failure, the ratio may be come normal. Thus, a case of lambda light chain monoclonal gammopathy may be missed.
- Plasma cell dyscrasias are MGUS, POEMS, plasmacytoma, Ig deposition diseases, and plasma cell myeloma.
- Plasma cells (normal and neoplastic) express CD38(bright) and CD138. CD138 is more specific but is less sensitive. Normal peripheral blood plasma cells are CD45+. In bone marrow, there are two subsets of plasma cells: one major subset positive for CD45 and a smaller negative one. Bright CD38 expression with coexpression of CD56 is used to detect abnormal population of plasma cells by flow cytometry. Abnormal plasma cells also are CD19−, in contrast to normal plasma cells which are CD19+. Normal and abnormal plasma cells are negative for CD20.

References

[1] Agarwal A, Ghobrial IM. Monoclonal gammopathy of undetermined significance and smoldering multiple myolemma: a review of current understanding of epidemiology, biology, risk stratification, and management of myeloma precursor disease. Clin Cancer Res 2013;19:985–94.

[2] Jones CI, Zabolotskaya MV, King AJ, Stewart HJ, et al. Identification of circulating micro RNAs as diagnostic biomarkers for use in multiple myeloma. Br J Cancer 2012;107:1987–96.

[3] Gregersen H, Mellemkjaer L, Ibsen JS, Dahlerup JF, et al. The impact of M component type and immunoglobulin concentration on the risk of malignant transformation in patients with monoclonal gammopathy of undetermined significance. Haematologica 2001;86:1172–9.

[4] Vodopick H, Chaskes SJ, Solomon A, Stewart JA. Transient monoclonal gammopathy associated with cytomegalovirus infection. Blood 1974;44:189–95.

[5] Gerritsen E, Vossen J, van Tol M, Jol-van der Zijde C, et al. Monoclonal gammopathies in children. J Clin Immunol 1999;9:296–305.

[6] Agarwal P, Parkash A, Tejwani N, Mehta A. Bisalbuminemia: a rare finding on serum electrophoresis. Indian J Hematol Blood Transfus 2018;34:558–9.

[7] Gras J, Padros R, Marti I, Gomez-Acha JA. Pseudo-analbuminemia due to the presence of a slow albumin variant moving into the alpha1 zone. Clin Chim Acta 1980;104:125–8.

[8] Regeniter A, Siede WH. Peaks and tails: evaluation of irregularities in capillary serum protein electrophoresis. Clin Biochem 2018;51:48–55.

[9] Roy V. Artifactual laboratory abnormalities in patients with paraproteinemia. South Med J 2009;102:167–70.

[10] Sawyer JR. The prognostic significance of cytogenetics and molecular profiling in multiple myeloma. Cancer Genet 2011;204:3–12.

[11] Tarigopula A, Chandrashekar V, Govindasamy P. Cytogenetic profiling of myelomas, association with complete blood count: study of 180 patients. Lab Med 2017;49:68–74.

Application of flow cytometry in diagnosis of hematological disorders

---

Introduction

Flow cytomatery (cyto: single cell, meter: measurement) is a laser-based biophysical method capable of providing both qualitative and quantitative measurements of multiple characteristics of a single cell or any other particle by measuring the optical and fluorescence characteristics of that particle. Flow cytometry can provide information on cell size (forward scatter) and cytoplasmic complexity or granularity (side scatter), as well as a wide range of membrane-bound and intracellular proteins. Fluorescent dyes, called fluorochromes, are conjugated to specific antibodies specific which then bind to the corresponding target proteins on the cell membrane or inside the cell. When the labeled cells are passed through a light source, the fluorescent molecules are excited to a higher energy state, and upon returning to their ground states, the fluorochromes emit light that is measured by the flow cytometer. The use of several fluorochromes each with similar excitation wavelengths but different emission wavelengths allows measuring antigen expression on various cells and or cell parameters at the same time. Emitted light is measured and is at a higher wavelength than excitation wavelength. Commonly used fluorochromes include phycoerythrin (PE), fluorescein isothiocyanate (FITC), allophycocyanin (APC), and pacific blue (PB), but many other dyes are commercially available including tandem dyes with internal fluorescence energy transfer capability [1,2]. Flow cytometry can even provide information on DNA and RNA content with certain fluorochromes, such as propidium idoide (PI), which can bind or intercalate with DNA or RNA.

The first impedance-based flow cytometric device was discovered by Wallace H. Coulter, and the first fluorescence-based flow cytometer was developed by Wolfgang Gohde from Germany. Other than hematology, flow cytometry is also applied for diagnosis in immunology (histocompatibility cross-matching), oncology, blood banking, and diagnosis of certain genetic disorders. This technique is also used outside medicine for research, including in marine biology, for example.

The cluster of differentiation (also known as cluster of designation) is abbreviated as CD (CD nomenclature) and is a protocol for identification of cell surface molecules providing target for immunophenotyping of cells. These cell surface molecules are glycoproteins with complex functions, including acting as receptors and cell signaling. Precise functions of many cell surface glycoproteins are still unknown. The CD nomenclature was established by the first International Workshop and Conference on Human Leukocyte Differentiation Antigen (HLDA) in 1984 for the classification of many monoclonal antibodies generated by different research laboratories worldwide against epitopes of these surface glycoproteins molecules on leukocytes. Since then, its application has been extended to many other cell types, and more than 300 unique clusters or subclusters have been identified.

Hematology and Coagulation. https://doi.org/10.1016/B978-0-12-814964-5.00008-5

The proposed surface molecule is assigned a unique number once two monoclonal antibodies are shown to bind with the surface antigen. The ninth HLDA International Congress was the most recent congress, which was held in 2010 in Barcelona, Spain, where 64 new antibodies were tested [3].

There are few factors that should be considered before making a conclusion from flow cytometry:

- Are the cells studied viable?
- Are the negative controls within acceptable limits?
- Are we sure that the cells that are considered as debris (mainly red cells) actually so?
- For acute leukemia, are the gated cells blasts?

These important issues should be addressed also keeping in mind following points:

- Dead cells can be identified by the using a specific dye. During flow cytometry analysis, one such dyes that is used to assess viability is 7-aminoactinomycin D (7-AAD).
- An isotype control (i.e., where an antibody that has the same Ig isotype as the test antibody used, but a different specificity which is known to be irrelevant to the sample being analyzed) is used to determine whether fluorescence that is observed is due to nonspecific binding of the fluorescent antibody. This can be checked with CD71, marker for erythroid cells. The cells being analyzed should be negative for CD71.
- Typically, a plot of side scatter versus CD45 is used. Blasts should have low side scatter (side scatter represents complexity/granularity, and blasts are cells with a large nucleus and thus represent a low complex cell population) and weak CD45. Lymphocytes have brighter CD45 expression than blasts.

The diagnostic approach to flow cytometry is critical to successfully identify and diagnose both normal and abnormal populations. It takes skill and practice and requires understanding of the normal expression patterns in the sample being examined to correctly identify any abnormal populations. Appropriate gating (i.e., selection of populations) is also critical. In general, panels are typically carefully and meticulously designed for specific diseases with a series of markers that will separate out populations and allow the identification any abnormal population. The normal pattern should be readily identifiable, and any deviation from the normal pattern can be gated or selected for further classification and identification of its expression profile.

Index of commonly used flow cytometric markers:

- CD1a: Expressed in immature T-cell and therefore positive in T-cell acute lymphoblastic anemia (T-ALL).
- CD2: Expressed in T and NK cells; may be aberrantly expressed in acute myeloid leukemia (AML).
- sCD3 (surface antigen CD3): Expressed in more mature T-cells but in not NK cells; may be lost in T-cell lymphomas.
- cCD3 (cytoplasmic antigen CD3): Expressed in all T-cells and also NK cells.
- CD4: Expressed in subset of T-cells and monocytes (in monocytic leukemias, there is disproportionate increase in CD4 positive cells).
- CD5: Expressed in T-cells; may be lost or decreased in T-cell lymphomas; may be aberrantly expressed in some B-cell lymphomas, most notably chronic lymphocytic leukemia/small lymphocytic lymphoma (CLL/SLL) and mantle cell lymphoma (MCL).
- CD7: Expressed in T and NK cells; may be lost or decreased in T-cell lymphomas.
- CD8: Expressed in cytotoxic T-cells and some NK cells.

- CD10: Also known as CALLA (common acute lymphoblastic leukemia antigen); normally expressed by immature B-cells, follicular T helper cells, follicular germinal center cells, and neutrophils. Lack of CD10 expression in B lymphoblastic leukemia/lymphoma (B-ALL) may be associated with a worse prognosis.
- CD11b: Expressed only in maturing granulocytes and monocytes.
  - CD13 is expressed in myeloid cells and monocytes.
  - CD14 is only expressed in monocytes.
  - CD15 is expressed only in more mature neutrophils (but not on immature progenitors) and monocytes.
- CD16: Expressed in maturing granulocytes and monocytes but also in natural killer (NK) cells.
- CD19: Expression of this antigen is restricted to B-cells and is present throughout B-cell maturation, from B lymphoblasts to plasma cells. Plasma cells neoplasms are typically CD19 negative.
- CD20: This is expressed by mature B-cells. Normal plasma cells are typically negative for CD20. Because CD20 is expressed later on in the maturation process of B-cells, B-ALL cases should have more CD19 cells than CD20 cells. CD20 expression can vary in B-cell lymphomas; for example, CD20 is typically dim in CLL/SLL and bright in hairy cell leukemia (HCL).
- CD22: B-cells during early maturation state have cytoplasmic expression of CD22, and with maturation, CD22 is expressed as a surface antigen.
- CD23: Weakly expressed in resting B-cells and increased expression with activation. Aberrant expression is often seen in CLL/SLL.
- CD11c, CD25, and CD103: Expressed in HCL.
- CD26: Expressed in immature and activated T-cells and NK cells. Most CD4 positive cells are also CD26 positive. In cutaneous T-cell lymphoma/Sézary syndrome (SS), the tumor cells are CD4 positive and CD26 negative.
- CD30: Expressed in activated T and B-cells (immunoblasts) and monocytes; expressed in classic Hodgkin lymphoma, ALK positive and ALK negative anaplastic large cell lymphoma (ALCL), primary cutaneous CD30 positive T-cell lymphoproliferative disorders (lymphomatoid papulosis and cutaneous ALCL), a subset of peripheral T-cell lymphomas, primary effusion lymphoma, primary mediastinal large B-cell lymphoma, and frequently in plasmablastic lymphoma.
- CD33: Expressed in myeloid cells and monocytes.
- CD34: Expressed in immature B and T-cells and myeloblasts.
- CD36: Expressed in monocytes, early erythroids, and megakaryocytes.
- CD38: Expressed in myeloid, monocytic cells, erythroid precursors, immature B and T-cells, and plasma cells.
- CD41 and CD61 are both expressed in megakaryocytes.
- CD56 and CD57: Expressed in NK cells.
- CD117: Expressed in immature myeloid cells and mast cells.
- CD200: Useful in differentiating CD5 positive B-cell lymphomas CLL/SLL (CD200 positive) from mantle cell lymphoma (usually CD200 negative).
- CD235a: Expressed in erythroid precursors.
- HLA-DR: (human leukocyte antigen-DR): Observed in myeloblasts, monocytes, B-cells, and activated T-cells.
- FMC7: Also useful in differentiation CD5 positive B-cell lymphomas CLL/SLL (FMC7 negative) and mantle cell lymphoma (FMC7 positive).
- MPO (myeloperoxidase): Observed in myeloid cells and granulocytes.

Flow cytometry and mature B-cell lymphoid neoplasms

Mature B-cell lymphoid neoplasms are distinguished from non-neoplastic cells by immunoglobulin light chain class restriction and aberrant antigen expression. Mature B-lymphoid cell lineage can be established with B-cell markers such as CD19, CD20, or CD79a. B-cells also exhibit surface immunoglobulins. If the cells are neoplastic, then these cells may express either kappa or lambda class of light chain alone, known as light chain restriction. Light chain restriction can be seen in

- Mature B-cell lymphoid neoplasms
- Rare reactive B-cell populations, e.g., tonsils of children (which have been shown to exhibit lambda light chain restriction), multicentric Castleman disease
- Some cases of florid follicular hyperplasia

Therefore, light chain restriction is not synonymous with neoplasm, and findings of flow cytometry should be interpreted in conjunction with other findings.

Aberrant antigen expression may also be observed, for example, expression of CD5 in B-cells. However, CD5 expression in B-cells can also be a normal phenomenon and is seen in some B-cells in peripheral blood, mantle zone cells in lymph nodes, and subset of hematogones in bone marrow. CD5 positive B-cell lymphomas can be observed in following situations:

- CLL/SLL
- Mantle cell lymphoma (MCL)
- Miscellaneous: B-cell prolymphocytic leukemia (B-PLL: sometimes), diffuse large B-cell lymphoma (DLBCL: rarely), marginal zone lymphoma (MZL: rarely), and lymphoplasmacytic lymphoma (LPL: very rarely)

B-cell markers

Normal B-cells express pan cell markers (see also list below) including CD19, CD20, CD22, and CD79a and show a polytypic mixture of kappa and lambda surface light chains. They also express CD23 (partial) and CD200 (partial) and lack expression of CD5 and CD10 in discrete populations. It is important to note that CD10 is expressed in normal germinal center B cells, so a small polytypic CD10 positive population is normal to see in a setting in which there are reactive germinal centers, i.e., a lymph node. However, a large discrete population might raise the possibility of FL (see below).

There are various markers of B-cells that are described below:

- CD19: This expression is restricted to B-cells and is present throughout B-cell maturation, from B-lymphoblasts to plasma cells. However, neoplastic plasma cells are typically CD19 negative.
- CD20: This is expressed by mature B-cells. As CD20 is expressed later during the maturation process of B-cells, more CD19 positive cells should be observed in B-ALL compared with CD20 cells. Normal plasma cells are negative for CD20, and plasma cells of plasma cell neoplasms are also CD20 negative, thus CD20 should not be used to distinguish normal and abnormal plasma cells. Bright CD20 staining can be seen in normal follicular center cells, cells of FL, and in HCL. Weak intensity (dim) staining for CD20 can be seen in CLL/SLL. Patients receiving Rituximab, an anti-CD20 drug, will show lack of staining for CD20.

- CD22: Early B-cells have cytoplasmic expression of CD22 and with maturation (similar to CD20), surface expression of CD22 is observed. In addition, similar to CD20, weak (dim) staining of CD22 is observed in CLL/SLL and bright staining in cases with HCL.
- CD23: Weakly expressed in resting B-cells and increased expression with activation. CD23 is found to be positive in CLL/SLL cases.
- CD10: Normally expressed by immature B-cells/hematogones, follicular T helper cells, follicular germinal center cells, and neutrophils.

Markers useful in diagnosing B-cell lymphomas include the following:

- CD10: Expressed in FL, Burkitt lymphoma, germinal center phenotypes DLBCL, and B-ALL.
- CD5: Aberrantly expressed in CLL/SLL, mantle cell lymphoma, B-PLL (sometimes), DLBCL (rarely), MZL (rarely), and LPL (very rarely).
- CD200: Useful in differentiating CD5 positive B-cell lymphomas CLL/SLL (CD200 positive) from mantle cell lymphoma (usually CD200 negative).
- FMC7: Also useful in differentiation of CD5 positive B-cell lymphomas CLL/SLL (FMC7 negative) and mantle cell lymphoma (FMC7 positive).
- CD11c, CD25, and CD103: Expressed in HCL.

Flow cytometry is useful in diagnosis of various B-cell lymphomas.

Chronic lymphocytic leukemia/small lymphocytic lymphomas

CLL/SLL has a characteristic immunophenotype and very often involves the peripheral blood, making flow cytometry a very convenient tool for its diagnosis. CLL/SLL typically expresses CD5, CD19, CD20 (dim), CD22 (dim), CD23, CD200 (bright uniform), and with dim monoclonal expression of a surface light chain (kappa or lambda); it does not express FMC7 or CD10.

B-cell prolymphocytic leukemia

B-PLL shows bright expression of CD19, CD20, CD22, CD79a, and FMC7; it does not typically express CD5 or CD23.

Mantle cell lymphoma

MCL is another CD5 positive lymphoma but has a distinct immunophenotype from CLL/SLL. MCL expresses CD5, CD19, CD20 (bright), CD22, and FMC7 with monoclonal expression of a surface light chain, but it does not typically express CD23, CD200, or CD10.

Follicular lymphoma

FL is very rarely detected in the peripheral blood, but lymph node or bone marrow samples can be used for flow cytometry. Although the phenotype is not entirely specific, expression of CD10, CD19, CD20, and CD22 with monoclonal expression of a surface light chain and no expression of CD5 can be suggestive of FL.

Hairy cell leukemia

HCL has such a characteristic immunophenotype that it has essentially become a flow cytometry diagnosis. Furthermore, it can be commonly detected in the peripheral blood. HCL expresses CD19 (bright), CD20 (bright), CD22, CD79a, FMC7, CD123, and CD200 as well as bright expression of the characteristic markers CD11c, CD25, and CD103 [4]. HCL does not typically express CD5 or CD10.

Burkitt lymphoma

Burkitt lymphoma will express CD10 (bright), CD19, CD20, and CD22 but also require correlation with morphology (starry sky), immunohistochemistry (BCL2−, BCL6+, and ∼100% KI-67), and fluorescence in situ hybridization (FISH) for *MYC* rearrangement.

Diffuse large B-cell lymphoma

DLBCL does not have a highly specific immunophenotype, but certain features can serve as clues toward the diagnosis. There might be a monoclonal B-cell population with expression of pan B-cell markers (CD19, CD20, and CD22) and monoclonal surface light chain. High forward scatter might raise suspicion of larger cell size. DLBCL with a germinal center B-cell−like immunophenotype will express CD10. However, non-germinal center phenotype will be negative for CD10. Some DLBCLs can express CD5. Furthermore, because of the large cell size, false negatives can occur by flow cytometry and require correlation with the morphology and immunohistochemistry.

Other B-cell lymphomas, such as MZL and LPL, can also be detected by flow cytometry in the peripheral blood, bone marrow, or tissue, by identifying a discrete monoclonal B-cell population, but their immunophenotype is rather nonspecific by flow cytometry alone.

Flow cytometry and mature T and natural killer−cell lymphoid neoplasm

Mature T- and NK-cell lymphoid neoplasms can be identified by flow cytometry by detection of aberrant antigen expression. This can be achieved by observing loss of one or more pan T-cell markers (e.g., loss of CD5 or CD7) or the presence of antigens not normally expressed (e.g., expression of CD13, CD15, CD33 or NK cells expressing CD5).

Normal T-cells will express pan T-cell antigens such as CD2, CD3 (surface and cytoplasmic), CD5, and CD7 and show a mixture of CD4 and CD8 cells with a typical CD4:CD8 ratio of about 2:1. Most T-cells are TCRαβ with a small subset being TCRγδ.

Normal T-cell large granular lymphocytes typically express CD2, surface CD3, CD8, CD56 (subset), and CD57 but dim to absent CD16.

Normal NK cells typically express CD2, cCD3, CD7, CD8 (some), CD16, CD26, CD56, and CD57 (subset); they characteristically do not express surface CD3.

Characteristics T and NK cell markers include

- CD1a: Expressed in immature T-cells and therefore positive in T lymphoblastic leukemia (T-ALL).
- CD2: Expressed in T and NK cells; may be aberrantly expressed in AML.

- sCD3: Expressed in more mature T-cells; may be aberrantly lost and not observed in T-cell lymphomas.
  - cCD3: Expressed in all T-cells and also NK cells.
- CD4: Expressed in T helper cells and monocytes (in monocytic leukemias disproportionate increase in CD4 positive cells is observed).
- CD5: Expressed in T-cells; may be lost in T-cell lymphomas and may be aberrantly expressed in some B-cell lymphomas.
- CD7: Expressed in T and NK cells; may be lost and not observed in T-cell lymphomas.
- CD8: Expressed in cytotoxic T-cells and some NK cells.
- CD10: Normally expressed by immature B-cells, follicular T helper cells, follicular germinal center cells, and neutrophils. Expressed in AITL.
- CD16: Expressed in NK cells and maturing granulocytes.
- CD26: Expressed in immature and activated T-cells, and NK cells. Most CD4 positive cells are also CD26 positive. In cutaneous T-cell lymphoma/SS, the tumor cells are CD4 positive and CD26 negative.
- CD30: Expressed in activated T and B cells and monocytes; T-cell neoplasms with CD30 expression include ALK positive and negative ALCL, primary cutaneous CD30 positive T-cell lymphoproliferative disorders (lymphomatoid papulosis and cutaneous ALCL), and a subset of peripheral T-cell lymphomas.
- CD56 and CD57: Expressed in NK cells and T-cell large granular lymphocytics (T-LGLs).

Detection of clonal or restricted population of T and natural killer cells

The enormous diversity of T-cell receptor specificities is created by germline variable (V), diversity (D), junctional (J), and constant (C) region genes, and of many V-segments available in the germline configuration, only one is incorporated into each chain of rearranged receptor [5]. Therefore, normal T-cells usually show a mixture of cells with variable expression of V-beta family subtypes. In T-cell lymphomas, the neoplastic cells will demonstrate restricted V-beta expression that may be identified by flow cytometry. Clonal T-cell receptor gene rearrangement studies by polymerase chain reaction (PCR) can also be used. However, false positive and false negative results may be observed by

Table 8.1 T-cell lymphomas and CD4 and CD8 positivity.

CD4 positive T-cell lymphomas	CD8 positive T-cell lymphomas:	CD4 and CD8 negative T-cell lymphomas:
• Peripheral T-cell lymphoma, NOS (can be CD4+ and CD8+) • Anaplastic large cell lymphoma • Angioimmunoblastic T-cell lymphoma • Adult T-cell leukemia/lymphoma (can be CD4+ and CD8+) • T-cell polymorphocytic leukemia (can be CD4+ and CD8+) • Cutaneous T-cell lymphoma/Sézary syndrome	• T-cell large granular lymphocyte leukemia • Subcutaneous panniculitis—like T-cell lymphoma • Hepatosplenic T-cell lymphoma (can be CD4− and CD8−)	• Enteropathy-associated T-cell lymphoma • Hepatosplenic T-cell lymphoma

both methods. NK cells lack T-cell receptor gene expression. Therefore, either of the abovementioned tests is not feasible. However, flow cytometric analysis of NK-receptor expression has been developed to provide evidence of NK cell clonality. Various CD positive T-cell lymphomas are listed in Table 8.1.

Flow cytometry of selected T-cell lymphomas is discussed in the following section.

Sézary syndrome and mycosis fungoides

The diagnosis of both SS and mycosis fungoides (MF) can be facilitated by flow cytometry. They typically express CD2, CD3, CD4, and CD5; but they frequently lose expression of CD7 and CD26, which assist in identifying the aberrant population. There is also often a markedly increased CD4:CD8 ratio [4].

T-cell prolymphocytic leukemia

T-PLL usually expresses CD4, but a significant proportion of cases (approximately 25%) have double expression of CD4 and CD8. Unlike MF/SS, T-PLL often retains expression of CD7 and CD26. However, a characteristic feature of T-PLL by flow is super bright (increased intensity) expression of several T-cell antigens, including CD2, CD5, CD7, and CD26 [4].

T-cell large granular lymphocytic leukemia

T-LGL leukemia very often can be detected in the peripheral blood, making flow cytometry a convenient method to identify the abnormal population. T-LGL typically expresses CD3, CD8, CD16 (abnormally increased), CD57 (bright, uniform), and TCRαβ; CD5 expression is often markedly decreased; aberrant complete loss of CD56 expression can also be seen in T-LGL [4].

Chronic lymphoproliferative disorder of natural killer cells

This is essentially the NK cell equivalent of T-LGL leukemia and is also readily identified by flow cytometric immunophenotypic aberrancies. CLPD NK has expression of CD2, CD7, CD8 with abnormal expression of CD56 (decreased to absent), and CD57 (increased) [4].

Plasma cell dyscrasias

Plasma cell dyscrasias are a heterogenous group of disorders caused by the monoclonal proliferation of plasma cells in the bone marrow. Multiple myeloma is the most serious and prevalent plasma cell dyscrasias with a median age of onset of 60 years where symptoms result from lytic bone disease, anemia, renal failure, and immunodeficiency. Monoclonal gammopathy of undetermined significance (MGUS) affecting up to 3.2% of patients over the age of 50 years is related to plasma cell dyscrasia and progresses to myeloma in some patients. Most patients with multiple myeloma show evidence of bone marrow plasmacytosis, a monoclonal gammopathy in serum or urine, and lytic bone lesions may be present in up to 60% of patients [6].

Flow cytometry is also useful in diagnosis of plasma cell dyscrasias. Plasma cells are CD38 positive (bright) and CD138 positive (although less sensitive). Plasma cells in the blood and tonsils are CD45 positive, whereas in the bone marrow, plasma cells may be either CD45 positive or negative. Normal plasma cells are CD19 positive and CD20 negative. Abnormal plasma cells are typically CD19 negative and CD56 positive. Up to 20% of cases of neoplastic plasma cell neoplasms are CD117

Table 8.2 Phenotypic aberrancies that can be seen in abnormal plasma cells.	
Positive	**Negative**
CD28	CD19
CD56	CD27
CD117	CD45
	CD81

positive. Typical phenotypic aberrancies seen in abnormal plasma cells are expression of CD28, CD56, and CD117, and lack of expression of CD19, CD27, and CD81 (Table 8.2).

Flow cytometry and acute leukemia

Flow cytometry has proven an invaluable tool in acute leukemia; it can greatly assist in lineage determination, quantitation of blasts, subclassification, and the detection of minimal residual disease (MRD). Blasts are identified as having low side scatter and weak CD45, which are present at left of the lymphocyte population in scattergram, CD45 versus side scatter. Basophils, having lost granules during processing may also be found in the same area, but can be differentiated because basophils are CD123 positive. Myeloblasts in addition should demonstrate markers of immaturity, CD34 (immature myeloid and lymphoid marker) and CD117 (immature myeloid marker), and pan myeloid markers (MPO, CD13, and CD33) but lack of mature myeloid markers such as CD11b, CD15, and CD16. Lymphoblasts in addition should demonstrate markers of immaturity, CD34, and TdT (terminal deoxynucleotidyl transferase). Lymphoblasts, if B-lymphoblasts, will also lack surface immunoglobulin and show lack of CD20 (more mature B-cell marker, more cells will be positive for CD19 than CD20). These lymphoblasts are also positive for cCD22. Lymphoblasts, if T-lymphoblasts, will show CD1a (immature T-cell marker) and lack surface CD3. These cells are however, positive for cCD3. In acute leukemia, blast populations are usually discrete, dense populations, and typical maturation patterns are absent.

In AML, there may be aberrant expression of lymphoid antigens, and those that are frequently expressed include CD2, CD7, CD19, and CD56. Similarly, in ALL, there may be aberrant expression of one or two myeloid antigens.

Flow cytometry and subtypes of acute myeloid leukemia

Markers that help to establish a myeloid lineage include CD13, CD15, CD33, and MPO. MPO is considered to be a lineage-defining antigen; however, a subset of AML can be negative for MPO, notably AML with minimal differentiation (AML M0). Therefore, MPO expression on an expanded blast population is highly specific but not entirely sensitive for the diagnosis of AML. CD4, CD11b, CD38, CD64, and CD123 are expressed at varying levels during normal myeloid development; increased or decreased expression for that stage of development is an abnormal feature on blasts. Aberrant expression of nonmyeloid lineage antigens (such as CD2, CD5, CD7, CD19, and CD56) can also be a feature of abnormal blasts.

Certain genetic subtypes of AML may have features that can be identified by flow cytometry and are listed below:

- AML with t(8;21) frequently shows co-expression of B-cell antigens, particularly CD19. CD56 is also commonly aberrantly co-expressed. CD34 and HLA-DR tend to be bright in these cases. CD117 and MPO are typically positive. The pan myeloid markers CD13 and CD33 are also positive but often show decreased expression [2]. Morphologically, these cases usually correlate with AML with maturation (FAB, M2).
- Acute promyelocytic leukemia (APL): The abnormal promyelocytes are typically CD34 negative and completely or partially negative for HLA-DR. They are also negative or have low expression for the more mature markers CD11b and CD15 (normal promyelocytes should have high levels of CD15). The abnormal promyelocytes lack CD10 and CD16, which are mature myeloid markers. APL is positive for CD13 (heterogenous) and CD33 (homogenous and bright), CD117, and MPO (bright) [2]. It has been reported that combination of absence of CD11b, CD11c, and HLA-DR identifies 100% of APLs [7]. CD2 positivity is also a feature of APL. Although the morphology and flow cytometry immunophenotypic findings can strongly suggest APL, rapid confirmatory testing with FISH or PCR for t(15;17) is still required for definitive diagnosis.

Of note, CD34 and HLA-DR negative and CD117 positive AML are not exclusive to APL and can also be seen in AML with *NPM1* mutation.

- AML with minimal differentiation (FAB, M0): Blasts are positive for CD34 and HLA-DR. Myeloid-specific antigens such as CD13 and CD33 are positive, but MPO is often negative or present in only a small fraction of the blasts.
- AML without maturation (FAB, M1): The flow appearance is similar to AML, M0 except MPO is expressed to a greater degree (>3%).
- AML with maturation (FAB, M2): Here, there is a reduced percentage of blasts with evidence of maturation. CD45 side scatter shows a continuum of cells from the myeloblast region to the maturing myeloid cell regions. Expression of CD19 and less often CD56 in cases of M2 is associated with presence of t(8;21).
- Acute myelomonocytic and monocytic leukemia (FAB M4 and M5): Positive for monocytic markers, CD4, CD11b, CD14, CD36, and CD64. Significant populations of cells positive for CD4, when the same cells are negative for CD3, indicate that the cells are monocytic. Reactive monocytes in the peripheral blood may be CD16 positive. Monoblasts are positive for CD4, CD15, CD33 (bright), CD36, CD64 (bright), and HLA-DR (bright); monoblasts are typically negative for CD13, CD14, and CD16 [2, 4]. Presence of CD2 correlates with M4Eo.
- Pure erythroid leukemia (FAB, M6b): Positive for erythroid differentiation markers CD71, CD235a (glycophorin), and CD36 [2]. CD36 is positive in monocytes, erythroid cells, and megakaryocytes. Erythroid leukemia may be positive for CD117 but is negative for CD34 and HLA-DR as well as CD64. If the case is not megakaryocytic (i.e., CD41 and CD61 negative) and CD64 is negative (i.e., not monocytic), then it favors diagnosis of erythroid leukemia.
- Acute megakaryocytic leukemia (FAB M7): Megakaryoblasts are positive for CD41 (glycoprotein IIb) and CD61 (glycoprotein IIIa); CD36 is also usually positive but is not specific (it can also be seen in monocytes and erythroid cells). However, other myeloid and monocytic

markers, such as CD14, CD16, CD64, and MPO, are absent. Megakaryocytic leukemia is associated with t(1;22) or inv(3).

Flow cytometry of B-lymphoblastic leukemia/lymphoma

The blasts of B-lymphoblastic leukemia/lymphoma express CD10, CD19, cytoplasmic CD22, cytoplasmic CD79a, CD45 (dim), and TdT; they usually express CD34 and can variably express CD20. They do not typically express surface light chains. Nonlineage-specific myeloid markers, such as CD13 and CD33, can sometimes be seen; however, MPO expression should be absent and its presence would raise the possibility of mixed phenotype acute leukemia (see below).

Some cases of *BCR-ABL*—like B-ALL, a high-risk subtype of B-ALL, show strong expression of *CRLF2*, which can be useful for its diagnosis [8].

Flow cytometry of T-lymphoblastic leukemia/lymphoma

The blasts of T-lymphoblastic leukemia/lymphoma express cytoplasmic CD3, CD7 (uniform, bright), and usually TdT. Other markers often expressed are CD1a, CD2, and CD5. CD10, CD34, and CD117 can be variably expressed. Surface CD3 is often absent. Nonlineage-specific myeloid markers, such as CD13 and CD33, can also sometimes also be seen in T-ALL; however, MPO expression should be absent and its presence would raise the possibility of mixed phenotype acute leukemia (see below). Although partial CD19 can sometimes be seen, these cases should still be classified as T-ALL and not MPAL.

Early T-cell precursor (ETP) ALL is a recently recognized subtype of T-ALL composed of ETPs that have very limited T-cell differentiation and also still have the ability to differentiate to myeloid precursors. As such, the diagnosis of ETP ALL is facilitated by flow cytometry with expression of CD3 (cytoplasmic) and CD7; they do not express or only weakly express CD5 (<75% of the blasts), CD1a, or CD8, and they express at least one myeloid/stem cell marker (CD11b, CD13, CD33, CD34, CD65, CD117, or HLA-DR) [4].

Flow cytometry of mixed phenotype acute leukemia

Mixed phenotype acute leukemia (MPAL) has essentially become a flow cytometry diagnosis with the lineages/populations identified and defined by flow antigens. An MPAL can be a combination of myeloid and B cell, myeloid and T cell, B-cell and T-cell lineages, or (rarely) all three. MPAL can manifest either as a single population with different lineages (biphenotypic) or discrete separate populations (bilineage). Lineage-specific markers can be, and most commonly and accurately are, identified by flow cytometry. Assigning a lineage to the blast population requires the following:

Myeloid lineage: MPO (by flow cytometry or cytochemistry) or two or more monocytic markers (CD11c, CD14, CD64).

T-cell lineage: cytoplasmic CD3.

B-cell lineage: strong CD19 with at least one of CD79a, cytoplasmic CD22, C10, or weak CD19 with two more of CD79a, cytoplasmic CD22, CD10.

Flow cytometry of blastic plasmacytoid dendritic cell neoplasm

Blastic plasmacytoid dendritic cell neoplasm (BPDCN) is a malignancy of precursor plasmacytoid dendritic cells that most often involves the skin and bone marrow. Flow cytometry can be useful in its diagnosis with the cells expressing CD4, CD45 (dim), CD56, CD123 (bright), CD303, and TCL1. MPO is not expressed, nor are B-cell or T-cell lineage—specific markers [4].

Flow cytometry and myelodysplastic syndrome

There are several features in of myelodysplastic syndrome (MDS) that can be identified by flow cytometry [9]. Granulocytes may be hypogranular, and the normal population of granulocytes as seen in the side scatter versus CD45 is distorted. The hypogranular cells occupy a lower position in the scattergram compared with a normal population. Abnormal antigen intensity (i.e. increased or decreased expression) or homogeneous expression of an antigen that should display varying intensity is both indications of an abnormally maturing population. Aberrant antigen expression from other lineages (CD7 or CD56 for example) can also be seen [2, 4]. Normal myeloid precursors should show high CD34 expression with low to absent CD38 expression followed by increased acquisition of CD38 and loss of CD34 as they mature. They then also acquire CD13, CD15, CD33, and CD117. Two other plots that are also helpful are CD11b versus CD16 and CD13 versus CD16. Immature granulocytes start with low expression of CD16 and subsequently express CD11b, expression of which becomes brighter and brighter with maturation. Subsequently, CD16 expression becomes very high. Thus, mature neutrophils are positive for both CD11b and CD16. This is the normal pattern of expression. When the CD13 versus CD16 plot is considered, immature cells have a bright expression of CD13 but with very little CD16 expression. With maturation, CD13 becomes dim (at the stage of metamyelocyte), still with low levels of CD16. Subsequently both CD13 and CD16 become bright. Thus, mature neutrophils are positive for both CD13 and CD16 [10]. In MDS, these normal patterns may be lost. Another feature of abnormal myeloid differentiation is gain of CD56 in the myeloid cells with lack of CD16. Normal distribution of CD11b versus CD16 is shown in Fig. 8.1. Fig. 8.2 shows normal distribution of CD13 versus CD16.

Flow cytometry and hematogones

Hematogones are maturing B-cell precursors representing a normal component of the bone marrow. Typically, these cells decrease with age and usually constitute 1% or fewer bone marrow cells.

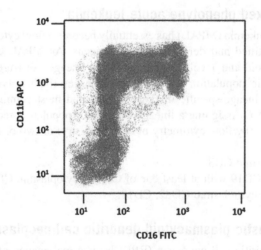

FIGURE 8.1

Normal distribution of CD11b versus CD16.

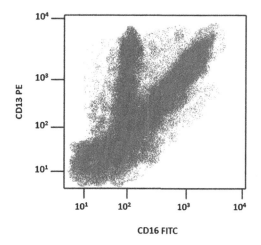

FIGURE 8.2

Normal distribution of CD13 versus CD16.

However, there may be hyperplasia of hematogones especially in the setting of regeneration after chemotherapy or stem cell transplantation or in patients with congenital or immune cytopenias. Hematogones may be confused with neoplastic B-cells as there is morphologic and immunophenotypic overlap. The earliest hematogones are positive for CD10, CD19, CD34, CD79a, and TdT but negative for and CD20. Subsequently CD10 and CD34 are down regulated with progressive gain of CD20. CD34 expression in hematogones is biphasic (either positive or negative), and expression of CD20 and CD34 is mutually exclusive. There are three stages (early, intermediate and late) of hematogones based on antigen expression, which can be seen by flow cytometry as a normal pattern of B-cell development [2].

Key points

- Mature B-cell lymphoid neoplasms are distinguished from nonneoplastic cells by immunoglobulin light chain class restriction and aberrant antigen expression.
- Light chain restriction can be seen in mature B-cell lymphoid neoplasms, rare reactive B-cell populations, such as tonsils of children (which have been shown to exhibit lambda light chain restriction), multicentric Castleman disease, and some cases of florid follicular hyperplasia.
- CD5 positive B cell lymphomas are CLL/SLL, MCL, and B-PLL (sometimes), DLBCL (rarely), MZL (rarely) and LPL (very rarely).
- CD10 positive B-cell neoplasms include B-ALL, FL, DLBCL (GCB type), Burkitt lymphoma, and HCL (uncommon, 10% of cases).
- CD30 positive neoplasms include classic Hodgkin lymphoma, ALK positive and ALK negative anaplastic large cell lymphoma (ALCL), primary cutaneous CD30 positive T-cell lymphoproliferative disorders (lymphomatoid papulosis and cutaneous ALCL), a subset of peripheral T-cell lymphomas, primary effusion lymphoma, primary mediastinal large B-cell lymphoma, and frequently in plasmablastic lymphoma.

- Mature T and NK cell lymphoid neoplasms are identified by flow cytometry by detection of aberrant antigen expression. For example, aberrant loss of one or more pan T-cell markers (such as loss of CD5 or CD7) or the presence of antigens not normally expressed (such as CD15, CD13, CD33 or NK cells expressing CD5).
- NK cells by flow cytometry are negative for surface CD3 and positive for CD16, CD56, and CD57. NK cells are also positive for CD2, cCD3, CD7, CD8 (subset), and CD26.
- Plasma cells are CD38 positive (bright) and CD138 positive (although less sensitive). Normal plasma cells are CD19 positive and CD20 negative. Typical phenotypic aberrancies seen in abnormal plasma cells are expression of CD28, CD56, and CD117, and lack of expression of CD19, CD27, and CD81.
- Myeloblasts will demonstrate markers of immaturity, CD34 (immature myeloid and lymphoid marker), and CD117 (immature myeloid marker); pan myeloid markers (MPO, CD13, and CD33); and lack of mature myeloid markers such as CD11b, CD15, and CD16.
- The B-lymphoblasts of B-ALL typically express CD10, CD45 (dim) and TdT as well as B-cell lineage markers such as CD19, cCD22, cCD79a. They usually also express CD34. However, they do not express surface light chains.
- The lymphoblasts of T-ALL express T-cell lineage markers such as cCD3, CD7 (uniform, bright), CD1a (immature T-cell marker), CD2, and CD5. TdT is usually expressed; CD10, CD34 and CD117 can be variably expressed. However, sCD3 is often absent.
- ETP ALL is a recently recognized subtype of T-ALL composed of ETPs with expression of CD3 (cytoplasmic) and CD7; they do not express (or only weakly) express CD5 (<75% of the blasts), CD1a, or CD8, and they express as least one myeloid/stem cell marker (CD11b, CD13, CD33, CD34, CD65, CD117, or HLA-DR).
- AML can sometimes show aberrant expression of non-myeloid antigens (including T-cell markers, such as CD2 or CD7). Notably, AML with t(8;21) can show co-expression of B-cell antigens (particularly CD19) and CD56.
- APL: Abnormal promyelocytes are positive for CD117, CD13 (heterogenous) and CD33 (homogenous and bright). They are negative for CD34 and completely or partially negative for HLA-DR. They are also negative for CD11b and CD15. The abnormal promyelocytes lack CD10 and CD16, which are mature myeloid markers.
- Acute monocytic leukemia: Positive for monocytic markers, CD4, CD11b, CD14, CD36, and CD64.
- Acute erythroid leukemia: Positive for CD235a (glycophorin) and CD36 in absence of CD64. CD36 is positive for monocytes, erythroid cells, and megakaryocytes. If the case is not megakaryocytic (i.e., CD41 and CD61 negative) and CD64 is negative (i.e., not monocytic) then it favors diagnosis of erythroid leukemia.
- Acute megakaryocytic leukemia: Positive for CD41 and CD61.
- Several features of MDS can be identified by flow cytometry. Granulocytes may be hypogranular as evidenced by decreased granulocyte side scatter. Abnormal antigen intensity or homogeneous expression of an antigen that should display varying intensity are both indications of an abnormally maturing population. Aberrant antigen expression from other lineages (such as CD7 or CD56) can also be seen.
- Hematogones are maturing B-cell precursors representing a normal component of bone marrow. The earliest hematogones are positive for CD10, CD19, CD34, CD79a, and TdT but negative for CD20. Subsequently CD10 is downregulated with progressive gain of CD20. There are three stages of hematogones based on antigen expression, which can be seen by flow cytometry as a normal pattern of B-cell development.

References

[1] Brown M, Wittwer C. Flow cytometry: principles and clinical applications in hematology. Clin Chem 2000; 46:1221−9.

[2] Cherian S, Wood B. Flow cytometry in evaluation of hematopoietic neoplasms a case based approach. CAP press; 2012.

[3] Faure GC, Amsellem S, Arnoulet C, Bardwet V, et al. Mutual benefits of B-ALL and HLDA/HCDM HLDA 9th Barcelona 2010. Immunol Lett 2011;134:145−9.

[4] Wang SA, Hasserjian RP, editors. Diagnosis of blood and bone marrow disorders. Springer International Publishing; 2018.

[5] Spencer J, Choy MY, MacDonald TT. T cell receptor Vβ expression by mucosal T cells. J Clin Pathol 1991; 44:915−8.

[6] Barlogie B, Alexanian R, Jagannath S. Plasma cell dyscrasias. JAMA 1992;268:2946−51.

[7] Dong HY, Kung JX, Bhardwaj V, McGill J. Flow cytometry rapidly identifies all acute promyelocytic leukemias with high specificity independent of underlying cytogenetic abnormalities. Am J Clin Pathol 2011;135:76−84.

[8] Konoplev S, Lu X, Konopleva M, et al. CRLF2-positive B-cell acute lymphoblastic leukemia in adult patients: a single-institution experience. Am J Clin Pathol 2017;147(4):357−63.

[9] Bellos F, Kern W. Flow cytometry in the diagnosis of myelodysplastic syndromes and the value of myeloid nuclear differentiation antigen. Cytometry B Clin Cytom 2017;92:200−6.

[10] Monaghan SA, Surti U, Doty K, Craig FE. Altered neutrophil maturation patterns that limit identification of myelodysplastic syndromes. Cytometry B Clin Cytom 2012;82:217−28.

Cytogenetic and genetic abnormalities in hematologic neoplasms

Introduction

Cancer is the second most common cause of death in the developed countries and the leading cause of untimely deaths (>35% of deaths before the age of 65 years). Cancer is a genetic disorder, and prognosis sometimes depends on the abnormality. For example, leukemia with inv(3)(q21.3q26.2) *GATA2-MECOM* has a poor prognosis, whereas other defects causing leukemia, such as Philadelphia chromosome positive chronic myeloid leukemia (CML), may have good prognosis [1]. Cytogenetics is a subdiscipline of genetics dealing with cytological and molecular analysis of chromosomes during cell division, including their movement as well as the location of genes on the chromosome. Cytogenetics plays an important role in diagnosis and determining prognosis of various hematological disorders. An in-depth discussion on various cytogenetics techniques is beyond the scope of this book, but a brief description is provided in this chapter.

Various different specimens can be used for cytogenetic analysis, including peripheral blood (such as blood lymphocytes), bone marrow, amniotic fluid, cord blood, tissue specimens, and even urine (for fluorescence in situ hybridization [FISH] analysis in a suspected case of bladder cancer). However, blood is the most commonly utilized specimen. The development of chromosome banding technique was the first major advancement in cytogenetic analysis. The first banding technique Q-banding (with quinacrine dihydrochloride and further examination with fluorescence microscopy) was later replaced mostly by G-banding (staining of chromosomes with Giemsa solution). The polymeric region of the chromosome can be visualized by C-banding. NOR-banding stains the nuclear organizing region. In 1986, FISH was developed and was another breakthrough discovery in molecular cytogenetics. This technique allows detection of specific nucleic acid sequences in morphologically preserved chromosomes, thus visualizing small segments of DNA using specific probes. Later, development of spectral karyotyping and multicolor FISH techniques allowed visualization of all 24 different chromosomes using unique color combination for each chromosome using various combinations and concentrations of fluorescent dyes. More recently, microarray-based methods have been developed using large insert genomic clones, c-DNA, or oligonucleotides that can replace metaphase chromosomes as DNA targets for analysis [2]. Common chromosomal abnormalities are numerical abnormalities (aneuploidy; deviation of normal 46 chromosomes) and structural rearrangements of chromosome (terminal and interstitial deletion abbreviated as "del," inversion abbreviated as "inv," and translocation abbreviated as "t"). In general, translocation means exchange between two or more chromosomes; deletion means loss of part of a chromosome; and inversion indicates rearrangement within an individual chromosome. Other abnormalities, such as duplications, ring chromosome, and isochromosomes, are also

Hematology and Coagulation. https://doi.org/10.1016/B978-0-12-814964-5.00009-7

considered as structural rearrangements of chromosomes. The short arm of chromosome is denoted as "p" while the long arm as "q". For reporting purpose, numerical abnormalities precede structural abnormality of chromosomes. For example, Philadelphia chromosome is denoted as t(9;22)(q34.1; q11.2), indicating this chromosome is formed due to reciprocal translocation of chromosome 9 and 22 involving region q34 (long arm) in chromosome 9 and region q11 in chromosome 22.

Cytogenetic abnormalities in chronic myeloid leukemia

CML is associated with an abnormal chromosome 22 known as the Philadelphia chromosome, which was the first discovered genetic abnormality associated with human cancer. The Philadelphia chromosome t(9;22)(q34.1;q11.2) results in the formation of a unique fusion gene product, *BCR-ABL1* (*ABL1* gene from chromosome 9 fuses with *BCR* gene on chromosome 22). The *BCR-ABL* fusion gene encodes and enhances tyrosine kinase activity that is implicated in the development of CML, and it is the primary target for the treatment of this disorder using tyrosine kinase inhibitors, such as imatinib mesylate. The Philadelphia chromosome is present in hematopoietic cells from patients with CML but not in nonhematopoietic tissues, including bone marrow fibroblasts.

The *ABL1* gene (also known as Abelson murine leukemia viral oncogene) had been previously identified as the cellular homolog of the transforming gene of the Abelson murine leukemia virus. The *ABL1* gene is located on chromosome 9q34 and encodes a nonreceptor protein-tyrosine kinase, c-ABL. The *ABL1* gene has 11 exons with 2 alternative 5′ first exons and a very large first intron of over 250 kilobases (kb). The *BCR* gene, which stands for breakpoint cluster region, is located on chromosome 22 and has 25 exons. The BCR gene product is a 160-kDa cytoplasmic phosphoprotein denoted BCR.

Formation of Philadelphia chromosome is due to a break in the first intron of the *ABL1* gene. Depending on the location of the breakpoint within the major *BCR* region, the consequence of the t(9;22) translocation in CML is to fuse the first 13 or 14 exons of the *BCR* gene upstream of the second exon of the *ABL1* gene. The two alternative fusion genes are traditionally described according to the original *BCR* exon nomenclature as b2a2 and b3a2 fusions or by the subsequent nomenclature as e13a2 or e14a2, respectively.

A third minor *BCR* region (mu-*BCR*) on chromosome 22 resulting in fusion of *BCR* exon 19 to *ABL1* exon 2 (e19a2) has been described in several patients, leading to generation of a p230 form of BCR-ABL1. There are three common variants of the BCR-ABL1 fusion protein:

- $p210^{BCR-ABL1}$: Created by the fusion of the *ABL1* gene at exon 2 (a2) with a breakpoint in the major *BCR* region at either exon 13 (e13/b2) or exon 14 (e14/b3) to produce an e13a2 (b2a2) or e14a2 (b3a2) transcript. Transcription of the e13a2 fusion gene followed by RNA splicing results in the generation of a novel 8.5-kb (230 base pairs) fusion *BCR-ABL1* mRNA that is translated into a 210 kDa protein. The protein product of the e14a2 fusion is 25 amino acids longer (305 base pairs) than the e13a2. The p210 protein is present in most patients with CML and one-third of those with Philadelphia positive B-cell acute lymphoblastic leukemia (Philadelphia chromosome positive B-cell ALL).
- $p190^{BCR-ABL1}$: If the breakpoint on chromosome 22 is at a different location (minor or m-bcr), which leads to the first exon of *BCR* (e1) fused upstream of *ABL1* exon 2 (a2), then this is e1a2. This results in a protein 190 kD in size, $p190^{BCR-ABL1}$. That is, it is created by the fusion of the *ABL1* gene at a2 with a breakpoint in the minor *BCR* region at e1 to produce an e1a2 transcript

that is translated into a 190 kDa protein. This variant is present in two-thirds of those with Philadelphia chromosome positive B-cell ALL and a minority of patients with CML. The presence of p190 in chronic phase CML is correlated with monocytosis and a low neutrophil/monocyte ratio in the peripheral blood. When compared with other patients with CML, patients with the p190 fusion protein may have an inferior outcome when treated with tyrosine kinase inhibitors.

- p230$^{BCR-ABL1}$: Created by the fusion of the *ABL1* gene at a2 with a breakpoint in the mu BCR region at exon 19 (e19) to produce an e19a2 transcript that is translated into a 230 kDa protein. This variant is seen in some patients with chronic neutrophilic leukemia.

The World Health Organization (WHO) diagnostic criteria for CML require the detection of the Philadelphia chromosome or its products, the *BCR-ABL1* fusion mRNA, and the BCR-ABL1 protein. This can be accomplished through conventional cytogenetic analysis (karyotyping), FISH analysis, or by reverse transcription polymerase chain reaction (RT-PCR).

Conventional cytogenetics technique requires in vitro culture and is time- and labor-intensive. In addition, this method can detect about 5% Philadelphia positive cells in a population of normal cells, and it can give false negative results in cells with complex chromosomal rearrangements. However, FISH employs large DNA probes linked to fluorophores and permits direct detection of the chromosomal position of the *BCR* and *ABL1* genes when employed with metaphase chromosome preparations. It can also be utilized on interphase cells from bone marrow or peripheral blood, in which physical colocalization of *BCR* and *ABL* probes is indicative of the presence of the *BCR-ABL* fusion gene [3]. The specificity of metaphase FISH is somewhat higher than MGG (May−Grunwald−Giemsa) banding for detection of the Philadelphia chromosome and allows easy identification of complex chromosomal rearrangements that may mask the t(9;22) translocation. The specificity of interphase FISH is lower, by approximately 10%, because of false positive results from coincidental colocalization of nonfused *BCR* and *ABL1* genes in interphase nuclei.

RT-PCR is a highly sensitive technique that employs specific primers to amplify a DNA fragment from *BCR-ABL1* mRNA transcripts. Depending on the combination of primers used, the method can detect the e1a2, e13a2 (b2a2), e14a2 (b3a2), and e19a2 fusion genes. The use of nested primers and sequential PCR reactions renders the technique extremely sensitive, capable of routine detection of one Philadelphia positive cell in 10^5-10^6 normal cells. Several features must be considered in the interpretation of RT-PCR data:

- There are many variables in the RT-PCR assay, including which internal standard to use and how to compare results obtained in different laboratories. Although initial efforts to create an international standardized scale for quantitative PCR for *BCR-ABL* measurements look promising, it is preferable to obtain serial samples from individual patients at the same laboratory, if possible.
- Patients with rare fusions, such as e6a2 or b2a3, may not be detected with standard primer sets.
- Patients with 5′ m-bcr breakpoints can sometimes exhibit both e13a2 (b2a2) and e14a2 (b3a2) transcripts, whereas M-bcr (major breakpoint-bcr) can also produce e1a2 transcripts at lower levels, probably due to alternative splicing.
- *BCR-ABL1* fusion transcripts (M-bcr or m-bcr) can be detected at very low level (one cell in 10^8-10^9) in hematopoietic cells from some normal individuals; this defines the limits of useful sensitivity of RT-PCR for diagnosis of leukemia.

Cytogenetic abnormalities in myelodysplastic syndrome

Patients with myelodysplastic syndrome (MDS) may have chromosomal abnormalities during diagnosis, or abnormal clones may appear later during the course of the disease. These abnormalities include a numerical change in chromosome numbers (i.e., monosomy or trisomy), a structural abnormality involving only one chromosome (e.g., inversion and interstitial deletion), or less commonly, a balanced translocation involving two chromosomes. Approximately 10%–15% of MDS patients exhibit complex karyotypes with multiple abnormalities. Additional chromosomal aberrations may evolve during the course of MDS, or an abnormal clone may emerge in a patient with a previously normal karyotype; these changes appear to accelerate progression of disease to acute leukemia. All of these common chromosome abnormalities observed in MDS are also commonly seen in other myeloid diseases (i.e., acute myeloid leukemia (AML) and myeloproliferative neoplasms).

Clonal chromosomal abnormalities can be detected in bone marrow cells in 40%–70% of patients with primary MDS, as opposed to 70%–80% detected in patients with AML de novo. The likelihood of chromosomal abnormalities is increased in patients with advanced MDS. Although rates vary depending on the technique used and the population studied, the most common chromosomal abnormalities seen in MDS is del(5q) or (−5), which is deletion of the long arm of chromosome 5 (5q) occurring in approximately 15% of cases overall. The exact size of the deletion of chromosome 5 can vary but always involves deletion of bands q31–q33.

Approximately 10% of patients with de novo MDS and up to 50% of patients with therapy-related MDS demonstrate −7 or del(7q), either alone or as part of a complex karyotype. Approximately 90% of cases have loss of a whole chromosome 7 (−7) and 10% are actual deletions, del(7q). Trisomy 8 is seen in less than 10% of patients with MDS and is considered an intermediate risk finding.

Deletions of the long arm of chromosome 20, del(20q), occur in less than 5% of cases of MDS and are also seen in patients with AML and myeloproliferative disorders. When found as the sole chromosomal abnormality, del(20q) is associated with a favorable prognosis. However, loss of the Y chromosome is common in older men without hematologic disorders and is not thought to play a role in the pathogenesis of MDS.

In addition, the presence of one of the following chromosomal abnormalities is presumptive evidence of MDS in patients with otherwise unexplained refractory cytopenia and no morphologic evidence of dysplasia such as unbalanced abnormalities: del(7q), del(5q), del(13q), del(11q), del(12p) or t(12p), del(9q), idic(X)(q13), t(17p) or isochromosome (17q) (i.e., loss of 17p), and balanced abnormalities: t(11;16)(q23;p13.3), t(3;21)(q26.2;q22.1), t(1;3)(p36.3;q21), t(2;11)(p21;q23), inv(3)(q21q26.2), and t(6;9)(p23;q34) (Table 9.1).

The ability of cytogenetic analysis to predict the outcome of individual patient with MDS is difficult because many patients die from persistent and profound pancytopenia, regardless of whether or not progression to AML occurs. Patients with loss of Y or del(11q) are considered to have a very good prognosis by the Comprehensive Cytogenetic Scoring System for MDS [4]. Patients with normal karyotype, isolated del (5q), isolated del(12p), isolated del (20q), or double (including del(5q)) have low risk of developing AML and are considered to have a good prognosis. Patients with del(7q), trisomy 8, trisomy 19, isochromosome 17q, unspecified single or double abnormalities, or two or more noncomplex clones are considered to be an intermediate prognostic group. Patients with loss of chromosome 7, inv(3), t(3q) or del(3q), double (including loss of chromosome 7 or del (7q)) or

Table 9.1 Chromosomal abnormalities in patients with myelodysplastic syndrome (MDS).	
Common chromosomal abnormalities in MDS	**Chromosomal abnormalities that may be presumptive evidence of MDS in patients with unexplained refractory cytopenia but without morphologic evidence of dysplasia**
• del(5q): Deletion of the long arm of chromosome 5 (5q) (most common). The exact size of the deletion of chromosome 5 can vary but always involves deletion of bands q31−q33. • del(7q): Either alone or as part of a complex karyotype. Approximately 90% of cases have loss of a whole chromosome 7 (−7), and 10% are actual deletions, del(7q). • Trisomy 8∗: Seen in about 10% of patients with MDS and is considered an intermediate risk finding. • del(20q)∗: Occur in 5%−8% of cases of MDS and are also seen in patients with acute myeloid leukemia and myeloproliferative disorders. If only abnormality, del(20q) is associated with a favorable prognosis. • Loss of the Y chromosome∗: May not play a role in pathogenesis of MDS	• Loss of chromosome 7 or del(7q) • del(5q) • Loss of chromosome 13 or del(13q) • del(11q) • del(12p) or t(12p) • del(9q) • idic(X)(q13) (isodicentric x chromosome) • t(17p) (unbalanced translocations) or isochromosome (17q) (i.e., loss of 17p) • t(11;16)(q23.3;p13.3) • t(3;21)(q26.2;q22.1) • t(1;3)(p36.3;q21.2) • t(2;11)(p21;q23.3) • inv(3)(q21.3q26.2) • t(6;9)(p23;q34.1)

∗Gain of chromosome 8, del(20q), and loss of Y chromosome are not considered to be sufficient to serve as presumptive evidence of myelodysplastic syndrome if they are the sole cytogenetic abnormality in the absence of morphologic criteria.

complex with 3 abnormalities are in a poor prognostic subgroup. Patients with more than three abnormalities (in these situations typically chromosomes 5 and 7 are affected) have high risk of progression to AML and are considered to be in a very poor prognostic subgroup.

Abnormalities in certain genes have been identified in patients with MDS and AML with or without the presence of chromosomal abnormalities. Although these gene mutations have been found to affect DNA methylation, tumor suppressor genes, and oncogenes, the prognostic significance of such mutations is not well characterized. Although many of these gene mutations are even more frequent in AML, particularly in AML with a normal karyotype, these gene mutations are thought to provide insight into the pathobiology of MDS and its progression to AML.

- Splicing factor mutations (*SF3B1, SRSF2, U2AF1, ZRZR2*) are seen in about half of all cases of MDS and result in altered RNA splicing.
- *SF3B1*: Splicing factor 3B1 is the most common splicing factor mutation. It is seen in high frequency (80%) in MDS with ring sideroblasts (MDS-RS). It has been incorporated into the definition of MDS-RS; the presence of an *SF3B1* mutation makes 5% ring sideroblasts sufficient to render the diagnosis (as opposed to a cutoff of 15% ring sideroblasts without the mutation). *SF3B1* mutations are associated with a better prognosis and less pronounced cytopenias in MDS.
- *SRSF2*: Second most common splicing factor mutation (10%−15%) of MDS. Confers worse prognosis in MDS.
- *U2AF1*: Also associated with a worse prognosis and higher risk of progression to leukemia.
- *ZRSR2*: Least common of the splicing factor mutations. Thought to confer neutral prognostic impact (but data is conflicting).

- *ASXL1*: Common in MDS, de novo AML, CMML, PMF, and as a clonal hematopoiesis of indeterminate potential (CHIP). *ASXL1* is involved in histone modification. *ASXL1* mutations appear to confer a reduced overall survival in AML with intermediate risk cytogenetics and higher risk of progression of MDS to AML. They also have been associated with a more proliferative phenotype of CMML, with higher WBC, higher monocyte count, and circulating immature monocytic precursors.
- *TET2* mutations: Somatic mutations in *TET2* (Ten-Eleven Translocation family of genes) occur in approximately 20%−30% of patients with MDS. Loss of function mutations of *TET2* results in increased DNA methylation and silencing of genes that are normally expressed [5].
- *RUNX1* transcriptional core-binding factor gene: *RUNX1* (runt-related transcription factor 1 family of gene also known as AML 1 gene) gene mutations are seen in about 10% of cases of de novo MDS and impart a poorer prognosis.
- *TP53* tumor suppressor gene: This tumor suppressor gene is located on 17p and mediates cell cycle arrest in response to a variety of cellular stressors. In patients with MDS, approximately 5%−10% of cases have known *TP53* mutations at the time of diagnosis. Abnormalities in p53 are more common in patients with MDS associated with prior exposure to alkylating agents of radiation (i.e., therapy-related MDS). Loss of wild-type *TP53* is associated with resistance to treatment and is a marker of poor prognosis independent of the IPSS risk score.
- *RAS* oncogenes: Mutations of *RAS* (these genes encode RAS proteins that were first isolated in rat sarcoma and now abbreviated as RAS) have been identified in about 5% of cases of MDS and in a subset of patients with AML. The majority of the mutations in MDS are in the *NRAS* gene, while *KRAS* and *HRAS* mutations occur less frequently. In both AML and MDS, RAS mutations have been reported more frequently in cases with a monocytic morphology (e.g., chronic myelomonocytic leukemia). *NRAS* mutations are considered to impart a poorer prognosis; in addition, *RAS* mutations are associated with MDS characterized by −7/del(7q) and impart a poorer prognosis.
- *IDH* mutations: Mutations in the isocitrate dehydrogenase oncogenes (i.e., *IDH1* and *IDH2*) have been reported in about 5% of cases of MDS resulting in DNA hypermethylation and alteration of gene expression. Presence of *IDH* mutation is considered a neutral prognostic factor (but data is also conflicting).
- *FLT3* gene mutations: Although mutations in the *FLT3* gene (human analogue of the murine fetal liver tyrosine kinase gene) are uncommon in MDS, they have been associated with a worse prognosis.

Prognosis of various mutation associated with MDS are listed in Table 9.2.

Cytogenetic abnormalities in patients with myeloid malignancies

Approximately 50%−60% patients with de novo AML (acute myeloid leukemias or acute myelo-blastic leukemia) demonstrate abnormal karyotypes. These are categorized by the WHO as AML with recurrent genetic abnormalities. Commonly observed abnormalities are listed in Table 9.3.

Table 9.2 Prognosis of various mutations associated with MDS.
• *SF3B1*(splicing factor 3B1) RNA splicing factor: Favorable prognosis; strongly associated with MDS with ring sideroblasts.
• *TET2* (ten-eleven translocation family of gene) mutation: neutral prognosis (but data is conflicting)
• *ASXL1*: involved in histone modification and imparts a poorer prognosis in MDS.
• *RUNX1* (runt-related transcription factor 1 family of gene) mutation: Poorer prognosis than *TET2* mutations
• *TP53* Tumor suppressor gene mutations: Poor prognosis
• *RAS* Oncogenes mutations of *RAS* (these genes encode RAS proteins that were first isolated in rat sarcoma and abbreviated as RAS): NRAS mutations impart a poorer prognosis in MDS
• *IDH* mutations: Mutations in the isocitrate dehydrogenase oncogenes (i.e., *IDH1* and *IDH2*) are considered neutral prognostic factors (but data is also conflicting).

Table 9.3 Common karyotypic abnormalities seen in patients with acute myeloid leukemia (AML).
• t(15;17)(q24.1;q21.2)
• t(8;21)(q22;q22.1)
• inv(16)(p13.1;q22)/t(16;16)(p13.1;q22)
• 11q23 rearrangements
• Trisomy 8

The following cytogenetic abnormalities, if found, result in the diagnosis of AML regardless of blast count:

- AML with the t(8;21)(q22;q22.1); *RUNX1-RUNX1T1* (previously *AML1-ETO*): This is seen in approximately 5% of adults with newly diagnosed AML and is one of the the most frequent abnormalitiest in children with AML. The *RUNX1* (previously *AML1* or core-binding factor alpha-2) gene on chromosome 21 and the *RUNX1T1* (previously *ETO* or *MTG8*) gene on chromosome 8 form a chimeric product that regulates the transcription of a number of genes that are critical to hematopoietic stem and progenitor cell growth, differentiation, and function. Cases of AML with the t(8;21) have a morphologically distinct phenotype, such as perinuclear clearing (hof), azurophilic or "salmon-colored" granules, and Auer rods. It imparts a favorable prognosis in adults due to good response to intensive chemotherapy.
- AML with the inv(16)(p13.1q22) or t(16;16)(p13.1;q22); *CBFB-MYH11*(inversion of chromosome 16 produces fusion transcript involving *CBFB*, the core binding factor gene and *MYH11*, the smooth muscle myosin heavy chain gene): Patients with inv(16) represent approximately 5-8% of newly diagnosed AML and typically demonstrate abnormal eosinophils at all stages of maturation in the bone marrow. Patients with inv(16) generally have a good response to intensive chemotherapy and a favorable prognosis.
- Acute promyelocytic leukemia (APL) with t(15;17)(q24.1;q21.2); *PML-RARA*: This translocation is highly specific for APL, where the gene coding for the retinoic acid receptor alpha (*RARA*) normally located on chromosome 17 (17q21.2) is disrupted by t(15;17) and is fused with

promyelocytic leukemia gene (PML) on chromosome 15 [6]. This disruption results in *PML/RARA* (promyelocytic leukemia/retinoic acid receptor fusion protein) fusion protein that is thought to lead to persistent transcriptional repression, thereby preventing differentiation of promyelocytes. Either a short or long transcript isoform can be produced depending on where PML breakpoint occurs. There are also variant *RARA* translocations that can occur in acute leukemia. These variant fusion partners include *ZBTB16* (previously *PLZF*) on 11q23.2, *NUMA1* on 11q13.4, *NPM1* at 5q35.1, and *STAT5B* at 17q21.2. In addition to cytogenetic abnormalities described above, the following four AML subtypes, which require at least 20% myeloid blasts for diagnosis, have been identified as subcategories of AML with genetic abnormalities of prognostic significance:

- AML with t(9;11)(p21.3;q23.3); *KMT2A-MLLT3*: *KMT2A* (previously the mixed lineage leukemia gene or *MLL*) located at chromosome 11q23.3 encodes a transcription factor with pleiotropic effects on hematopoietic gene expression and chromatin remodeling. *MLL* gene can fuse with one of more than 50 different known chromosomal patterns, but most common translocation is *MLLT3* (chromosome 9p21.3), which is seen more often in children, and is associated with monocytic features [7]. These patients tend to have an intermediate response to standard therapy.
- AML with t(6;9)(p23;q34.1); *DEK-NUP214*: This abnormality is seen in approximately 1% of patients with newly diagnosed AML and has increased incidence in pediatric populations. It is characterized by basophilia, pancytopenia, and multilineage dysplasia. Patients with this abnormality have a poor response to standard therapy, which may be a result of the lesion itself or its association with *FLT3*-ITD, which is also known to convey a poor prognosis.
- AML with inv(3)(q21.3;q26.2) or t(3;3)(q21.3;q26.2); *GATA2/MECOM* (previously *EVI1*): The t(3;3) and inv(3) account for approximately 1% of AML cases and are associated with a poor response to therapy. These cytogenetic changes lead to the activation of the *MECOM* gene, which is thought to result in the peripheral blood thrombocytosis and increased atypical megakaryocytes in the bone marrow characteristic of cases with this abnormality.
- AML (megakaryoblastic) with t(1;22)(p13.3;q13.1); *RBM15-MKL1*: This type of leukemia is also rare but is typically a megakaryoblastic process occurring in infants and children, although it is not seen in patients with Down syndrome.
- AML with *BCR-ABL1*: A newly recognized provisional entity. It is a de novo AML that carries the characteristic t(9;22)(q34.1;q11.2) usually resulting in the p210 fusion; but importantly, there is no evidence of CML either before or after therapy. There are also usually additional cytogenetic abnormalities (including complex karyotypes) present in addition to the t(9;22). Cases of mixed phenotype AML, therapy-related AML, and AML with other recurrent genetic abnormalities are excluded from this category. It is a rare form of AML, occurs in adults, and carries a poor prognosis.

In addition to chromosomal abnormalities described above, other specific gene mutations also occur in AML. These are categorized under the broader WHO heading of AML with gene mutations:

- Nucleophosmin (*NPM1*): Previously provisional, now considered a distinct entity. Mutations usually involve exon 12 and often display myelomonocytic or monocytic features. It is most common in de novo AML with a normal karyotype. In the absence of *FLT3*-ITD, it imparts a favorable prognosis.

- *CEBPA* (biallelic): Previously provisional, now a distinct entity, but only if the mutation is biallelic (i.e., one mutation occurs on one allele and a second mutation occurs in the opposite allele). Biallelic *CEBPA* mutations are also frequently observed in de novo AML patients with a normal karyotype and, in absence of *FLT3*-ITD, impart favorable prognosis.

The following mutations are not considered distinct WHO entities but are also commonly encountered in myeloid neoplasms:

- *FLT3*; fms-related tyrosine kinase 3: Although not categorized as a distinct entity in the WHO classification, it is a frequently occurring mutation, seen most often in AML with normal karyotype, AML with t(6;9), and APL. The two primary types of *FLT3* mutations are internal tandem duplications (*FLT3*-ITD) and mutations affecting codons 385 or 386 of the second tyrosine kinase domain (*FLT3*-TKD). *FLT3*-ITD mutations are associated with an adverse outcome.
- *KIT*: Observed in AML with t(8;21) and inv (16)/t(16;16), and the presence of *KIT* mutations imparts poor prognosis.
- *NRAS* and *KRAS*: *RAS* mutations are seen in AML with monocytic differentiation. *RAS* mutations also provide clonal diagnostic support for CMML and are associated with a more aggressive clinical course. These mutations can also be seen in MDS and are associated with a worse prognosis in low-risk disease.

Characteristic chromosomal abnormalities have also been identified in patients who developed a MDS or AML after chemotherapy and/or radiation therapy for an earlier disorder. The most common type of therapy-related MDS/AML is due to damage from alkylating agents and is characterized by abnormalities in chromosomes 5 and/or 7. The second most common subtype of therapy-related AML occurs in patients who have been treated with chemotherapeutic drugs that inhibit DNA topoisomerase II. AML associated with these drugs is often characterized by balanced translocations involving the *KMT2A* gene at 11q23 or the *RUNX1* gene at 21q22.

Cytogenetic abnormalities in myeloproliferative neoplasms

JAK2: Seen in >95% of PV and about 50%–60% of ET and PMF. The most common mutation is the activating V617F mutation.

MPL: Codon 515 mutations are seen in approximately 3% of ET and 8% of PMF and confer an intermediate prognosis. *MPL* mutations are mutually exclusive with *JAK2* mutations.

CALR: Mutations are seen in about 30% of ET and PMF. *CALR*-mutated PMF has a better survival than *JAK2*-mutated PMF.

Cytogenetic abnormalities in acute lymphoblastic leukemia

Chromosomal abnormalities are seen in about 80% of B-cell ALL.

The t(4;11)(q21;q23.3) is seen in up to 60% of infants (up to 1 year of age) with ALL but is rarely observed in adult patients with ALL. This translocation results in a fusion gene involving *KMT2A* on 11q23.3 and *AFF1* on 4q21 and is associated with a poor prognosis. Individuals with ALL and t(4;11) have high leukocyte counts, an immature immunophenotype, B-cell lineage, and frequent coexpression of myeloid antigens.

The t(9;22)(q34.1;q11.2) that produces the Philadelphia chromosome is observed in about 2%–5% of children and about 25% of adults. Most childhood cases involve a p190 BCR-ABL fusion protein. In adults, about half of cases are p190 and the other half p210 (the most commonly seen in CML). This translocation is associated with a poor prognosis. However, therapy with tyrosine kinase inhibitors has greatly improved patient outcomes.

The t(1;19)(q23;p13.3), *TCF3-PBX1*, occurs in approximately 6% of children with ALL with a pre–B-cell phenotype (positive for CD10, CD19, and cytoplasmic mu heavy chain). It is less common in adults. *TCF3* is located on 1q23 and *PBX1* on 19p13.3. The resulting fusion protein is an oncogenic transcriptional activator.

The t(12;21)(p13.2;q22.1), resulting in the fusion *ETV6/RUNX1* gene, is present in about 25% of children with B-lineage ALL and 3% of adults with ALL. *ETV6* is located on 12p13.2 and *RUNX1* on 21q22.1. Patients with ALL and t(12;21) have a favorable prognosis with a >90% cure rate in children.

The t(5;14)(q31.1;q32.1), resulting in the fusion *IGH/IL3* gene, is rare and seen in <1% of ALL. The *IGH* gene is on 14q32.1, and the *IL3* gene is on 5q31.1. B-ALL with this fusion has a characteristic morphologic finding of increased circulating eosinophils; however, these are reactive and not part of the neoplastic clone.

B-ALL, *BCR-ABL*–like is a new provisional entity in the 2017 WHO, which lacks the *BCR-ABL1* translocation but has a gene expression profile similar to that of B-ALL with *BCR-ABL*. It is seen in 10%–25% of cases, and these patients have a poor prognosis especially if also coharboring a *CRLF2* translocation. However, some patients have shown good responses to tyrosine kinase inhibitors.

B-ALL with iAMP21 is another new provisional entity in the 2017 WHO. It has amplification of chromosome 21 ($\geq$5) often detected by FISH studies for *ETV-RUNX1*. Risk for this type of B-ALL is greatly increased in patients with the Robertsonian translocation rob(15;21). It is associated with a poor prognosis in patients treated with standard therapy, but the risk can be remedied with a more intensive regimen.

Patients who have hyperdiploidy with more than 50 chromosomes often have an excellent prognosis (>90% cure rate in children). It is more common in childhood (25%) than in adult (7%–8%) ALL. The individual structural abnormalities do not appear to influence outcome in patients with hyperdiploidy except for the t(9;22), which is associated with a poor prognosis.

Approximately 5%–6% of ALL patients, independent of age, have clonal loss of various chromosomes, resulting in a hypodiploid clone with fewer than 45 chromosomes. ALL patients with hypodiploid clones generally have a poor prognosis, especially patients with near-haploid and low-hypodiploid clones. Deletion of 9p is an unfavorable risk factor associated with a high rate of relapse in precursor B-ALL in children.

Among children with T-cell ALL, approximately 60% have an abnormal karyotype. These patients have a distinct pattern of recurring karyotypic abnormalities involving both T-cell receptor (TCR) and non-TCR gene loci, including activating mutations of *NOTCH1* (in more than 50%). Relatively common translocations can involve the *TLX1* or *TLX3*. Other translocations include t(10;11)(p13;q14); *PICALM-MLLT10*, and translocations involving *KMT2A*. Additional translocated genes include *MYC*, *TAL1*, *LMO1*, *LMO2*, and *LYL1*.

The new provisional entity, early T-cell precursor ALL, is unique in that the mutational landscape (such as mutations in *FLT3*, *DNMT3A*, *KRAS/NRAS*, or *IDH1/IDH2*) is more similar to that of myeloid neoplasms than other T-ALLs.

Cytogenetic abnormalities in multiple myeloma

Conventional karyotyping detects cytogenetic abnormalities in 20%–30% of patients with myeloma. In contrast, FISH detects abnormalities in almost all cases. The most common chromosomal translocations in multiple myeloma involve 14q32, the site of the immunoglobulin heavy chain (IgH) locus, and are known as "primary IgH translocations". These include t(11;14)(q13;q32), t(6;14)(p25;q32), t(12;14)(p13;q32), t(4;14)(p16;q32), t(8;14)(q24;q32), t(14;16)(q32;q23), and t(14;20)(q32;q11).

Abnormalities that impart a standard risk for developing multiple myeloma include t(11;14) and t(6;14) and hyperdiploidy. Abnormalities that impart an intermediate risk include t(4;14), del13, and hypodiploid.

Abnormalities that impart poor prognosis include t(14;16), t(14;20), and/or del17p13. The poor prognosis of these high-risk factors may be abrogated by the presence of at least one trisomy.

Deletions of 17p, including 17p13, the p53 locus, are found in 10% of multiple myeloma patients and are associated with a shorter survival after both conventional chemotherapy and hematopoietic cell transplantation. There is a low rate of complete response, rapid disease progression, advanced disease stages, plasma cell leukemia, and central nervous system multiple myeloma.

Trisomy (hyperdiploidy) is associated with a favorable outcome in multiple myeloma. Hyperdiploidy occurs in myeloma typically due to trisomies of odd-numbered chromosomes. Trisomies occurring in high-risk patients (by FISH) can reduce the risk to that of standard risk.

Cytogenetic and genetic abnormalities in B- and T-cell lymphomas

CLL/SLL: Trisomy 12 is seen in approximately 20% of cases of chronic lymphocytic leukemia (CLL) and is associated with poor prognosis, whereas del(13q14) is seen in about 50% of cases and is associated with a favorable prognosis. Other deletions seen in CLL include deletions of 11q22-23 (*ATM* and *BIRC3*) and 17p13 (*TP53*). The 11q deletions are the most common type of karyotypic evolution over time. *SF3B1* mutations are associated with adverse prognostic features (such as unmutated IGHV and increased CD38 expression) and adverse prognosis. Mutations in *NOTCH1* and *TP53* are associated with adverse prognosis and a higher risk to progression to diffuse large B-cell lymphoma (DLBCL) (Richter syndrome). Other mutations associated with adverse prognosis in CLL include *ATM* and *BIRC3*. Mutations in *BTK* and *PLCG2* are important because they are associated with resistance to ibrutinib therapy.

Follicular lymphoma (FL): The classic cytogenetic abnormality observed is t(14;18)(q32;q21).

At the molecular level, this translocation juxtaposes the *bcl-2* proto-oncogene (band 18q21) with the Ig heavy chain gene (band 14q32) resulting in deregulation of *bcl-2* gene expression and elevation of *bcl-2* mRNA and protein. The classic abnormality is seen in 90% of cases of FL, grades I and II. In addition to the classic abnormality, other alterations are seen in 90% of FL. The number of additional chromosomal alterations increases with histologic grade and transformation. Abnormalities of 3q27 and/or *BCL6* rearrangements are seen in 5%–15% of FL, mostly grade 3B.

Mantle cell lymphoma (MCL): This is characterized by the presence of a balanced chromosomal translocation, the t(11;14)(q13;q32), which is seen in the vast majority (>95%) of mantle cell lymphomas. This abnormality juxtaposes the *CCND1* gene (11q13) with the *IgH* (14q32) gene, resulting in cyclin D1 overexpression. This abnormality promotes cell cycle progression from the G1 phase to the S phase.

Marginal zone lymphoma: The most common anomalies in extranodal marginal zone B-cell lymphoma (MZBCL) of MALT type (MALT stands for mucosa-associated lymphoid tissue) include

- The translocation t(11;18)(q21;q21)/*BIRC3-MALT1* fusion: 20%−50% incidence; associated with low-grade MALT lymphoma of the stomach and of the lungs.
- The translocation t(14;18)(q32;q21)/*IGH-MALT1* fusion, leading to enhanced *MLT1* expression, may occur in 10%−20% of all MALT lymphomas. It is associated with MALT lymphoma of the ocular adnexa, orbit, salivary gland, liver, skin, and lungs.
- The translocation t(1;14)(p22;q32)/*BCL10-IGH* and/or the corresponding deregulation or rearrangement of *BCL10* at 1p22 is another recurrent chromosome aberration in a minority of cases. It appears to be more frequent in high-grade MALT than in low-grade MALT lymphoma and is associated with MALT of the lungs and small intestine.
- The translocation t(3;14)(p14.1;q32)/*FOXP1-IGH* fusion may occur in 10% of all MALT lymphomas. It is associated with MALT lymphoma of the orbit, ocular adnexa, thyroid, and skin. It is not found in MALT lymphoma of the stomach, of the salivary gland, and in other forms of MZBCL.
- Trisomy 3 and trisomy 18 were reported in low-grade and high-grade MALT lymphoma.

In splenic MZBCL, the 7q deletions are the most common (about 40% of cases) abnormality observed. Other abnormalities include total or partial trisomy 3; however, recurrent chromosomal translocations have been identified. Nodal MZBCL does not have a distinct cytogenetic profile.

Diffuse large B-cell lymphoma: Different cytogenetic abnormalities are observed in DLBCLs. These abnormalities include

- The translocation t(3;v)(q27;v)/*Bcl6* rearrangement is seen in 30% of cases of DLBCLs.
- The translocation t(14;18)(q32;q21)/*Bcl2* rearrangement, feature of FL is seen in 15%−25% of cases.
- *MYC* rearrangement is seen in about 8−14% of cases and is associated with a worse overall survival. The *MYC* partner is *IGH* gene, which is observed in 60% of cases and non-Ig gene in the remainder of cases. DLBCL cases with only a *MYC* rearrangement are still classified as DLBCL, NOS; however, due to the worse prognosis conferred by *MYC* rearrangements, they are often specified as DLBCL, NOS with *MYC* rearrangement.
- High grade B-cell lymphoma with *MYC* and *BCL2* and/or *BCL6* rearrangements: DLBCL cases that harbor a *MYC* rearrangement along with either a *BCL2* and/or *BCL6* rearrangement are now classified under this new category. These cases have also been referred to as "double hit" (if *MYC* rearrangement together with either *BCL2* or *BCL6* rearrangements are present) or "triple hit lymphoma" (if all three rearrangements are present) and carry a worse prognosis.
- Large B-cell lymphoma with *IRF4* rearrangement is a newly recognized subtype in the 2017 WHO classification. It is characterized by strong expression of IRF4/MUM1 by immunohistochemistry and is most often also accompanied rearrangement of *IRF4*, usually with *IGH*. It is most often seen in younger age groups (children and young adults) and often involves the head and neck region, particularly the Waldeyer's ring.

- DLBCL of the testis: *MYD88* L265P mutation is seen in about 80% and *CD79B* mutations in about 50%.
- Primary DLBCL of the central nervous system: *MYD88* L265P mutation is seen in about 80% and *CD79B* mutations in about 60%. Biallelic loss of *CDKN2A/p16* is seen in 60%−90% of cases.

Burkitt lymphoma: Various cytogenetic abnormalities are observed in Burkitt lymphoma. These abnormalities include

- The translocation t(8;14)(q24;q32), which is seen in the vast majority (80%) of cases: *MYC* gene is on chromosome 8 and *IgH* gene is on chromosome 14. However, 5%−10% of classic Burkitt lymphoma cases appear to lack a *MYC* translocation [8].
- The translocation t(8;22)(q24;q11): Gene for kappa light chain is on chromosome 22.
- The translocation t(2;8)(p12;q24): Gene for lambda light chain is on chromosome 2.
- Burkitt-like lymphoma with 11q aberration; provisional entity in the 2017 WHO classification. They resemble classical Burkitt lymphoma by morphology and immunohistochemistry, but they lack *MYC* rearrangement. However, they harbor either gains or losses at chromosome 11q and often have a more complex karyotype.

Hairy cell leukemia

- *BRAF* V600E is seen in almost 100% of cases.
- More than 85% of cases have hypermutated *IGHV* genes.
- However, there are no specific cytogenetic abnormalities associated with hairy cell leukemia.

Anaplastic large cell lymphoma: In all cases of ALK, positive anaplastic large cell lymphoma (ALCL) and ALCL rearrangement involving the anaplastic lymphoma kinase (*ALK*) gene on chromosome 2p23 is observed. There are several translocations and inversions involving ALK, the most common one being t(2;5)(p23;q35), encoding a nuclear phosphoprotein *NPM/ALK* fusion protein (70%−75% of cases).

ALK negative ALCL lacks *ALK* rearrangement but has been shown recently to harbor other recurring genetic abnormalities. Rearrangements of *DUSP22-IRF4* on chromosome 6p25.3 occur in about 30% of cases and are associated with an excellent prognosis. In contrast, rearrangements of *TP63* occur in about 8% of cases and are associated with a dismal prognosis. These rearrangements have not been reported in ALK positive ALCL [9,10].

Key points

- The Philadelphia chromosome t(9;22)(q34.1;q11.2) results in the formation of a unique gene product (*BCR-ABL1*), tyrosine kinase. There are three common variants of *BCR-ABL1*: p210$^{BCR-ABL1}$ is present in most patients with CML and one-third of all those with Philadelphia chromosome positive B-ALL; p190$^{BCR-ABL1}$ is present in two-thirds of those with Philadelphia chromosome positive B-cell ALL and a minority of patients with CML; p230$^{BCR-ABL1}$ seen in some patients with chronic neutrophilic leukemia.
- Patients with MDS with a normal karyotype, isolated del (5q), isolated del (20q), and loss of Y have low risk of developing AML, whereas patients with three or more abnormalities (in these situations

typically chromosomes 5 and 7 are affected) or abnormalities of chromosome 7 have high risk of progression to AML. All other abnormalities have intermediate risk of developing AML.

- The following cytogenetic abnormalities, if found, result in the diagnosis of AML regardless of blast count: AML with t(8;21)(q22;q22); *RUNX1-RUNX1T1,* AML with inv(16)(p13.1q22), or t(16;16)(p13.1;q22); *CBFB-MYH11* and APL with t(15;17)(q24.1;q21.1); *PML-RARA.* Patients with these abnormalities typically respond well to therapy.
- AML with inv(16) represents approximately 5%–8% of newly diagnosed AML and typically demonstrates abnormal eosinophils at all stages of maturation in the bone marrow.
- AML with t(9;11)(p22;q23); *MLLT3-MLL*: This is seen more often in children and is associated with monocytic features.
- AML with t(6;9)(p23;q34); *DEK-NUP214*: This is characterized by basophilia, pancytopenia, multilineage dysplasia, and poor prognosis.
- AML with t(3;3) and inv(3) is characteristically associated with peripheral blood thrombocytosis, increased atypical megakaryocytes in the bone marrow, and a poor prognosis.
- AML (megakaryoblastic) with t(1;22)(p13;q13); *RBM15-MKL1*: This is typically a megakaryoblastic process occurring in infants, although it is not seen in patients with Down syndrome.
- *FLT3* mutations are seen most often in AML with normal karyotype, AML with t(6;9), and APL. The two primary types of *FLT3* mutations are internal tandem duplications (*FLT3*-ITD) and mutations affecting codons 385 or 386 of the second tyrosine kinase domain (*FLT3*-TKD). *FLT3*-ITD mutations are associated with an adverse outcome.
- Nucleophosmin (*NPM1*) and biallelic *CEBPA* mutations are frequently observed in AML patients with a normal karyotype and in absence of *FLT3*-ITD impart favorable prognosis.
- Chromosomal abnormalities are seen in about 80% of B-cell ALL. The t(4;11) is seen in up to 60% of infants (up to 1 year of age) with ALL but is rarely observed in adult patients with ALL. The t(9;22) that produces the Philadelphia chromosome is observed in about 2–5% of children and about 30% of adults. This translocation is associated with a poor prognosis. B-ALL with t(12;21); *ETV6/RUNX1*: This is present in about 25% of children with B-ALL and 3% of adults and has a favorable prognosis. Patients who have hyperdiploidy with more than 50 chromosomes often have a good prognosis. B-ALL patients with hypodiploid clones (<45 chromosomes) generally have a poor prognosis, especially patients with near-haploid and low-hypodiploid clones. Deletion of 9p is an unfavorable risk factor associated with a high rate of relapse in precursor-B ALL in children.
- Among children with T-cell ALL, approximately 60% have an abnormal karyotype. These patients have a distinct pattern of recurring karyotypic abnormalities involving both TCR and non-TCR gene loci, including activating mutations of *NOTCH1* in more than 50%.
- Cytogenetic abnormalities are observed in multiple myeloma. The most common chromosomal translocations in MM involve 14q32, the site of the immunoglobulin heavy chain (IgH) locus, and are known as "primary IgH translocations." Abnormalities that impart a standard risk are t(11;14) and t(6;14) and hyperdiploidy. Abnormalities that impart poor prognosis are t(14; 16), t(14; 20), and/or del17p13. The poor prognosis of these high-risk factors may be abrogated by the presence of at least one trisomy.

- In CLL, trisomy 12 is associated with poor prognosis and del(13q14) with a favorable prognosis. 11q deletions are the most common type of karyotypic evolution over time. *SF3B1*, *NOTCH1*, *ATM*, *BIRC3*, and *TP53* mutations are associated with adverse prognosis. Mutations in *BTK* and *PLCG2* are associated with resistance to ibrutinib therapy.
- FL: The classical cytogenetic abnormality is t(14;18)(q32;q21).
- MCL is characterized by the balanced chromosomal translocation (11;14)(q13;q32). This abnormality juxtaposes the *CCND1* gene (11q13) with the *IgH* (14q32) gene, resulting in cyclin D1 overexpression.
- Marginal zone lymphoma: The most common anomalies in extranodal MZBCL of MALT type include t(11;18)(q21;q21)/*BIRC3-MALT1* fusion, most common (20%−50% incidence), and associated with low-grade MALT lymphoma of the stomach and of the lung; t(14;18)(q32;q21)/ *IGH-MALT1* is associated with MALT lymphoma of the ocular adnexa, orbit, salivary gland, liver, skin, and lung; t(3;14)(p14.1;q32)/FOXP1-*IGH* is associated with MALT lymphoma of the orbit, ocular adnexa, thyroid, and skin. t(1;14)(p22;q32)/*BCL10-IGH* is associated with MALT of the lung and small intestine.
- DLBCL: t(3;v)(q27;v)/Bcl6 rearrangement is seen in 30% of cases of DLBCL. t(14;18)(q32; q21)/Bcl2 rearrangement, feature of FL, is seen in 15%−25% of cases. *MYC* rearrangement is seen in about 8−14% of DLBCLs and is associated with a worse overall survival.
- High grade B-cell lymphoma with *MYC* and *BCL2* and/or *BCL6* rearrangements are DLBCL cases that harbor a *MYC* rearrangement along with either a *BCL2* and/or *BCL6* rearrangement. These cases have also been referred to as "double hit" or "triple hit lymphoma" and carry a worse prognosis.
- DLBCL with *IRF4* rearrangement occurs in younger age groups and often involves the head and neck region, particularly the Waldeyer's ring.
- DLBCL of the testis: *MYD88* L265P mutation is seen in about 80% and *CD79B* mutations in about 50%.
- Primary DLBCL of the central nervous system: *MYD88* L265P mutation is seen in about 80% and *CD79B* mutations in about 60%. Biallelic loss of *CDKN2A/p16* is seen in 60%−90% of cases.
- Burkitt lymphoma: t(8;14)(q24;q32) is seen in the vast majority of cases; *MYC* gene is on chromosome 8 and *IgH* gene is on chromosome 14. t(8;22)(q24;q11): gene for kappa light chain is on chromosome 22. t(2;8)(p12;q24): gene for lambda light chain is on chromosome 2.
- Burkitt-like lymphoma with 11q aberration resembles classical Burkitt lymphoma by morphology and immunohistochemistry but they lack *MYC* rearrangement. They harbor either gains or losses at chromosome 11q and often have a more complex karyotype.
- Hairy cell leukemia has *BRAF* V600E in almost 100% of cases, and more than 85% of cases have hypermutated *IGHV* genes.
- ALCL: All cases of ALK positive ALCL have a rearrangement involving the *ALK* gene on chromosome 2p23. The most common is t(2;5)(p23;q35) resulting in an *NPM1/ALK* fusion protein (70%−75% of cases).
- ALK negative ALCL lacks *ALK* rearrangement but has been shown recently to harbor other recurring genetic abnormalities. Rearrangements of *DUSP22-IRF4* on chromosome 6p25.3 are associated with an excellent prognosis and rearrangements of *TP63* with a dismal prognosis. These rearrangements have not been reported in ALK positive ALCL.

References

[1] Huret JL, Ahmad M, Arsaban M, Bernheim A, et al. Atlas of genetic and cytogenetics in oncology and hematology in 2013. Nucleic Acids Res 2013;41(Database issue):D920–4.

[2] Kannan TP, Zilfalil BA. Cytogenetics: past, present and future. Malays J Med Sci 2009;16:4–9.

[3] Primo D, Tabernero MD, Rasillo A, Sayagues JM, et al. Pattern of BCR/ABL gene rearrangement by interphase fluorescence in situ hybridization (FISH) in BCR/ABL leukemias: incidence and underlying genetic abnormality. Leukemia 2003;17:1124–9.

[4] Schanz J, Tüchler H, Solé F, Mallo M, et al. New comprehensive cytogenetic scoring system for primary myelodysplastic syndromes (MDS) and oligoblastic acute myeloid leukemia after MDS derived from an international database merge. J Clin Oncol 2012;30:820–9.

[5] Abdel-Wahab O, Mullally A, Hedvat C, Garcia-Manero G, et al. Genetic characteristics of TET1, TET2 and TET3 alterations in myeloid malignancies. Blood 2009;114:144–7.

[6] Chen Z, Chen SJ. RARA and PML genes in acute promyelocytic leukemia. Leuk Lymphoma 1992;8: 253–60.

[7] Chandra P, Luthra R, Zuo Z, Yao H, et al. Common properties of dysregulated Ras pathway signaling and genomic progression characterized de novo and therapy related cases. Am J Clin Pathol 2010;133:686–93.

[8] Said J, Lones M, Yea S. Burkitt lymphoma and MYCL what else is new? Adv Anat Pathol 2014;21:160–5.

[9] Parrilla Castellar ER, Jaffe ES, Said JW, et al. ALK-negative anaplastic large cell lymphoma is a genetically heterogeneous disease with widely disparate clinical outcomes. Blood 2014;124(9):1473–80.

[10] Chihara D, Fanale MA. Management of anaplastic large cell lymphoma. Hematol Oncol Clin N Am 2017;31: 209–22.

Benign lymph node

Introduction

The lymph system is an important part of body's immune system, and lymph nodes, which are small bean-shaped glands and an integral part of lymph system, are located in clusters in various parts of the body including neck, armpit, and groin as well as inside the center of the chest and abdomen. A lymph node consists of cortex and medulla, encased in a fibrous capsule. Subcapsular sinus is located just underneath the capsule. Outer cortex consists mainly of B cells arranged as follicles, and T cells are found in deeper cortex. Paracortex is the portion of the lymph node immediately surrounding the medulla and contains a mixture of mature and immature T cells. The medulla contains lymphocytes, plasmacytoid lymphocytes, plasmablasts, and plasma cells.

There are two types of follicles seen in the cortex:

- Primary and
- Secondary

The unstimulated follicles are primary follicles and consist of small lymphocytes and follicular dendritic cells. The small lymphocytes are predominantly B lymphocytes, which exhibit surface IgD and surface IgM, and are positive for CD5. Follicular dendritic cells are stained by CD21, CD23, and CD35. Because the primary follicles are B-cell areas, they will stain positively with B-cell markers such as CD20, CD79a, and PAX-5. The primary follicle stains with B-cell lymphoma 2 protein (Bcl-2) but is negative for CD10 and Bcl-6.

Secondary follicles are derived from primary follicles following antigenic stimulation. Secondary follicles have an inner germinal center and an outer mantle zone. Within the germinal center in a reactive follicle, a light zone and a dark zone are present. This is referred to as polarity. The light zone consists of centrocytes and follicular dendritic cells. Mitotic rate is lower in this zone. Tingible body macrophages are less seen in this area. In contrast, the dark zone has centroblasts with a high mitotic rate and increased number of tingible body macrophages. The germinal center of a reactive follicle stains with CD10 and Bcl-6 (B-cell lymphoma 6 protein) and is negative for Bcl-2. As the secondary follicle is a B-cell area, B-cell markers such as CD20 will highlight both the germinal center and the mantle zone. A few T cells are found in the germinal center. The Ki-67 staining (Ki-67 is a nuclear protein and the name is derived from Kiel, Germany; 67 denotes the clone number) of the germinal centers demonstrates a high proliferation index.

The outer mantle zone of the secondary follicle is composed of B lymphocytes. These B lymphocytes are positive for CD5 and Bcl-2. Outer to the mantle zone is the marginal zone, which is still a

Hematology and Coagulation. https://doi.org/10.1016/B978-0-12-814964-5.00010-3

Table 10.1 Immunohistochemical pattern of benign lymph nodes.

Immunohistochemical marker	Comment
Pan B-cell markers (e.g., CD20, CD79a)	Stains B cells of mantle zone, marginal zone, germinal center. and scattered B cells in paracortex
CD10	Stains B cells in the germinal center
Bcl-6	Stains B cells in the germinal center
Bcl-2	Does not stain cells in the germinal center, stains B cells in the mantle zone and marginal zone
CD3	Stains T cells in the paracortex. Stains the T cells in the paracortex; also stains the few T cells in the germinal center
CD21, CD23, CD35	Stains the follicular dendritic cells
S100	Stains the interdigitating cells in the paracortex.

B-cell area, but the cells are more loosely packed. The area beyond the marginal zone is the paracortex and is a predominantly T-cell area. The paracortex is thus easily stained by CD3, a pan T-cell marker. Bcl-2 also stains T cells. Thus, Bcl-2 staining pattern of a reactive lymph node includes staining of the paracortex, outer parts of the secondary follicle, but not the germinal center.

The primary follicle consists of naïve B cells, which are as yet not stimulated. When stimulated by an antigen, the naïve B cells move into the germinal center and transform into centroblasts, which in turn form centrocytes. From the centrocytes are derived plasma cells and memory cells. The plasma cells move into the medulla, an ideal location to release immunoglobulins into the circulation. Memory cells typically reside in the marginal zone, whereas naïve B cells are found in the mantle zone layer.

The paracortex consists predominantly of T cells, interdigitating dendritic cells and high endothelial venules. A few B cells are also present in the paracortex. Cells once activated may transform into immunoblasts. Immunohistochemical pattern of benign lymph nodes is summarized in Table 10.1.

Reactive lymphoid states

Reactive lymphoid hyperplasia may be due to numerous causes, and the typical patterns of lymphoid hyperplasia are follicular hyperplasia, paracortical hyperplasia, sinusoidal expansion, granulomatous lymphadenitis, and mixed pattern (Table 10.2). Infections are a common cause of both follicular and paracortical hyperplasia. Viral infections can result in both follicular and paracortical hyperplasia. However, it is important to differentiate follicular hyperplasia from follicular lymphoma. Follicular lymphoma is the most common indolent type of non-Hodgkin lymphoma and second most common form of non-Hodgkin lymphoma involving B cells (please see Chapter 11). In Table 10.3, features of follicular hyperplasia and follicular lymphoma are listed.

Viral lymphadenopathy

Most often viral infections lead to paracortical hyperplasia where expansion of the paracortex with presence of transformed lymphocytes is observed. There is also hyperplasia of the interdigitating dendritic cells, and these cells have pale cytoplasm, which causes the paracortex to have a

Table 10.2 Patterns of lymphoid hyperplasia.
Follicular hyperplasia (bacterial and viral infections, rheumatoid arthritis, Castleman disease, IgG4-related sclerosing disease)
Paracortical hyperplasia (viral infections), drug (e.g., phenytoin)-related lymphadenopathy
Nodular paracortical T-cell hyperplasia
Sinusoidal expansion
Granulomatous
With necrosis
Mixed pattern

Table 10.3 Characteristics features of follicular hyperplasia and follicular lymphoma.

Follicular hyperplasia	Follicular lymphoma
Follicles are of variable size	Back-to-back follicles; follicles are uniform in size and may occupy the medulla
Distinct mantle zone present	Mantle zone indistinct
Germinal center with light and dark zone	Distinction of light and dark zone lost
Germinal center has TBM[a]	Germinal center lacks TBM
Lymphocytes do not extend beyond capsule	Expansion beyond capsule
high Ki67. However in high grade FL, Ki67 may also be high	Ki67 usually low
Bcl-2 negative in germinal center	Bcl-2 positive in germinal center

[a]*TBM: Tingible body macrophage.*

mottled appearance. Follicular hyperplasia may also be present. The large transformed cells are immunoblasts and may mimic large cell lymphoma. On occasion, these cells may appear like Reed−Sternberg cells. The immunoblasts may be B cells (more often) or T cells. Sometimes, with the presence of B immunoblasts, CD20 expression is downregulated. These cells frequently express CD30, thus mimicking Reed−Sternberg cells. However, these are negative for CD15. In cases of CMV (*Cytomegalovirus*) infections, intranuclear viral inclusions with a halo may be observed. In HSV (herpes simplex virus) infections, multinucleated giant cells with ground glass nuclei may be seen. Similar to CMV, intranuclear inclusions may also be present. In measles, Warthin−Finkeldey giant cells may be seen.

In patients with HIV (human immunodeficiency virus) infections, lymph nodes typically exhibit follicular hyperplasia. Three patterns have been described:

- In pattern A (early pattern), there are reactive follicles with reduced mantle zone. There are folliculosis and interfollicular hemorrhage. Aggregates of monocytoid B cells are evident, and scattered Warthin−Finkeldey cells may also be present.
- In pattern C (late stage), the follicles are atrophic with vascular proliferation, which is highly similar to the "lollipop" appearance of follicles often seen in hyaline vascular Castleman disease (HVCD).

- Pattern B is a transition from pattern A to C and has features in between.

Lymph node CD4+ T cells are selectively depleted during HIV infection. However, except for end-stage disease, lymph node changes do not correlate with those of peripheral blood, and therefore, peripheral blood is a more sensitive parameter to monitor immunologic changes in various stages of HIV infection [1].

Bacterial infections and lymphadenopathy

In most instances, bacterial infections involve the lymph node or the spleen as a secondary site of dissemination. A notable exception includes cat-scratch disease, which primarily affects the lymph nodes. Most of the bacterial infections produce nonspecific changes, although some infections do produce morphologic changes characteristic of a specific organism. Cat-scratch disease, the most common cause of chronic lymphadenopathy among children and adolescents, features subacute regional lymphadenitis with an associated inoculation site due to a cat scratch or bite and the causative agent is *Bartonella henselae* [2,3]. An important differential diagnosis for bacterial lymphadenitis is neutrophil-rich variant of anaplastic large cell lymphoma. Various bacterial infections and associated morphologic changes are described in Table 10.4.

Table 10.4 Bacterial infections and lymphadenopathy.

Bacterial infection	Comments
Staphylococcus and *Streptococcus*	Acute inflammatory cells with necrosis (central); abscess formation may be seen. Inflammation of the capsule is evident. Increased number of macrophages will be seen and subsequently granulation tissue and fibrosis.
Actinomyces	Similar to above; in addition, sulfur granules, which are large bacterial colonies, may be seen
Bartonella Henselae (cat-scratch disease)	Follicular hyperplasia with necrosis and acute inflammation. Areas of necrosis are surrounded by palisading histiocytes. Giant cells may be seen. Warthin–Starry stain highlights the bacteria similar to immunostains.
B. Henselae (bacillary angiomatosis)	Bacillary angiomatosis is typically seen in immunocompromised individuals especially HIV/AIDs patients. The lymph node shows nodules, composed of vessels. The endothelial cells of the vessels have vacuolated cytoplasm. In between the blood vessels are amphophilic materials, which are clumps of bacteria. Also, present are foamy macrophages between the blood vessels.
Tropheryma whipplei	Dilatation of the sinuses by large lipid vacuoles and multinucleated giant cells. Also found are macrophages with vacuolated cytoplasm. These macrophages contain the bacteria, which are diastase resistant Periodic Acid Schiff (PAS) positive. Overall, the appearance resembles to that of "Swiss cheese."
Treponema pallidum	Follicular hyperplasia with significant interfollicular hyperplasia is seen. There is also vascular proliferation with vasculitis. The capsule may become thickened. Granulomas may be seen.
Tuberculosis	Tuberculosis by *Mycobacterium tuberculosis* forms granulomatous lymphadenitis. Granulomas are focal collection of epithelioid macrophages. Some may be converted to giant cells. There is a peripheral collar of fibroblasts, plasma cells, and lymphocytes. There is central coagulative necrosis, referred to as caseous necrosis.

Cat-scratch lymphadenopathy

Cat-scratch lymphadenopathy is characterized by necrotizing granulomatous inflammation. Initially, there is deposition of extracellular amorphous eosinophilic proteinaceous material within germinal centers. Small abscesses with areas of necrosis develop. Eventually, macrophages surround these abscesses, leading to formation of stellate granulomas.

The causative organism is *B. henselae* and can be identified by silver stains such as Warthin–Starry stain.

Toxoplasma gondii and lymphadenopathy

Toxoplasma gondii is an intracellular protozoan capable of infecting many species including human. The sexual form of the parasite is found in intestinal epithelium of domestic cats (only known host), and oocytes are shed in the feces of cats. Contact with cat's feces or eating undercooked meat of an animal infected with *T. gondii* may cause human infection. Toxoplasmosis is a very common human infection of the world, but importance of this infection to humans refers mainly to primary infection during pregnancy, resulting in congenital toxoplasmosis, stillbirth, or even spontaneous abortion [4]. Toxoplasma lymphadenopathy is characterized by a triad of histologic findings:

* Follicular hyperplasia
* Clusters of epithelioid macrophages, within or in close proximity to germinal centers and
* Monocytoid B cells within sinuses.

Granulomatous lymphadenopathy

Granulomatous lymphadenopathy is an important category of lymphadenopathy, and there is a wide variety of causes. Granuloma formation is a chronic inflammatory reaction where macrophages and other inflammatory cells are involved. In granulomatous lymphadenopathy caused by cat's-scratch disease and tularemia, monocytoid B lymphocytes with T cells and macrophages contribute to formation of granuloma. However, granulomatous lymphadenopathy may be due to noninfective causes, for example, sarcoidosis and sarcoid-like reaction [5]. Sarcoidosis classically presents with hilar lymphadenopathy and rounded, nodular nonnecrotizing granulomas. Common causes for granulomatous lymphadenopathy are listed in Table 10.5.

Table 10.5 Common causes of granulomatous lymphadenopathy.
• Infections (viral, fungal, mycobacterial, bacterial)
• Systemic (sarcoidosis, rheumatoid arthritis, vasculitis)
• Associated with tumors (lymphomas, carcinomas, melanoma)
• Immunodeficiency states
• Foreign body

Necrotizing lymphadenopathy

In necrotizing lymphadenopathy, focal or geographic areas of necrosis occur, which is associated with a cellular response. Both bacterial (e.g., cat-scratch disease, lymphogranuloma venereum, tularemia) and viral (e.g., Epstein–Barr virus, herpes simplex virus) infections are associated with necrotizing lymphadenopathy. An important differential diagnosis of necrotizing lymphadenopathy is Kikuchi-Fujimoto disease.

Progressive transformation of germinal centers

In typical cases of progressive transformation of germinal centers (PTGC), large lymphoid nodules in a background of follicular hyperplasia are observed. These large nodules are progressively transformed germinal centers. It is thought that the cells of the mantle zone infiltrate and disrupt the germinal center, and in this process, germinal center becomes ill defined. The follicular center cells are reduced. Tingible body macrophages are rare or absent. In the majority of cases, PTGC are seen to affect the lymph nodes focally. PTGC is considered as a benign reactive change. However, it is associated with lymphomas, especially nodular lymphocyte predominant Hodgkin lymphoma (NLPHL).

Regressive changes in germinal center

This is characterized by small compact B-cell nodules where the germinal center lymphocytes are lost. The stromal components are prominent, including the follicular dendritic cells. These features are commonly associated with HVCD. Other conditions associated with similar findings are Epstein–Barr virus infection, autoimmune diseases, postradiation or postchemotherapy, HIV/AIDS, and syphilis.

Specific clinical entities with lymphadenopathy

Lymphadenopathy may be related to many reasons, and in this section, association of lymphadenopathy with certain specific diseases is addressed. Lymphadenitis, which refers to inflammation of lymph nodes, is often associated with lymphadenopathy, and it is often difficult to distinguish between two terms because these words may be used interchangeably.

Kikuchi-Fujimoto disease

The etiology of Kikuchi's disease remains elusive, and various infective organisms have been implicated including viral infection. This disease was first reported in Japan in 1972 and observed predominantly in Japan and other Asian countries and is a rare disease, which is typically self-limiting within one to 4 months, but a possible recurrence rate of 3%–4% has been reported. The disease presents as painless cervical lymphadenopathy in young adults affecting both males and females, although females are four times more likely to be affected by this disease than males. Diagnosis is based on histology of the affected lymph node specimen obtained by excisional biopsy [6]. Morphologically

- Areas of necrosis in the paracortex are observed along with the presence of abundant karyorrhectic debris. Interestingly, neutrophils are absent.

- The necrotic areas contain bright eosinophilic fibrinoid deposits and are surrounded by activated large T lymphocytes and numerous histiocytes. The histiocytes have abundant cytoplasm.
- Plasma cells and plasmacytoid monocytes are also present.
- The areas of necrosis and the surrounding numerous histiocytes are the reason for the synonym of Kikuchi's disease as necrotizing histiocytic lymphadenitis.

Kimura disease

The etiology of Kimura disease is also unknown, and like Kikuchi's disease, it is seen more often in Asians. However, it may be also observed in other ethnic groups [7]. It is a chronic inflammatory disorder of the subcutaneous tissue and affected regional lymph nodes. The cervical area is the most common site of involvement. Affected lymph nodes exhibit

- Follicular hyperplasia
- Intense eosinophilia with eosinophilic microabscess with infiltration of the germinal centers. Warthin–Finkeldey polykaryocytes may be seen in the germinal centers.
- Hyperplasia of capillaries is also a feature.
- This disease is associated with peripheral blood eosinophilia and elevated levels of IgE.

Kawasaki disease

This is also known as mucocutaneous lymph node syndrome. It was first reported in Japan, and although seen more often in the Japanese, it is also seen in Western countries affecting mostly children. The etiology, however, remains unclear. Morphologically, expansion of interfollicular areas with loss of normal architecture is observed. There are areas of necrosis with presence of neutrophils and nuclear debris. Vessels may be increased.

Dermatopathic lymphadenitis

This is seen in the lymph node draining areas of skin involved with irritation, infection, or inflammation. Langerhans cells present in skin migrate to regional lymph nodes on antigenic stimulation. Subsequently, T-cell activation takes place and as a result T-cell hyperplasia in the lymph node is observed. Therefore, the paracortical area is expanded. Macrophages with brown pigments are found within the paracortex. The pigment is typically melanin. There are also interspersed interdigitating dendritic cells and Langerhans cells.

Lymphadenopathy in autoimmune diseases

Autoimmune diseases such as systemic lupus erythematosus (SLE) and rheumatoid arthritis are important causes of lymphadenopathy. In SLE, necrotic areas within the paracortex are seen with eosinophilic debris within the necrotic areas. With time, histiocytes and lymphocytes surround the necrotic areas. Multiple hematoxylin bodies, which are ill-defined structures representing degenerated nuclei, are seen. In cases of lymphadenopathy due to rheumatoid arthritis, the lymph nodes exhibit florid follicular hyperplasia. Plasmacytosis is seen in the interfollicular areas. There also occurs deposition of hyaline material within the lymph node.

Rosai—Dorfman disease

This disorder is also known as sinus histiocytosis with massive lymphadenopathy [8]. The disease presents as bilateral painless cervical lymphadenopathy. Typically, it is a benign disorder, which resolves spontaneously. Morphologically, the lymph nodes demonstrate dilatation of the sinuses by large macrophages with round vesicular nuclei, nucleoli, and foamy cytoplasm. Emperipolesis (which is the presence of intact lymphocytes within histiocytes) is present and is the hallmark of the disease. Eosinophils are not prominent, which is a feature of Langerhans cell histiocytosis (LCH). The macrophages are positive for S100 and typical macrophage/histiocytic markers (e.g., CD68). However, macrophages are negative for CD1a. In contrast, CD1a is positive in LCH.

Langerhans cell histiocytosis

LCH is a group of disorders, ranging from focal lesions (typically of bone) to systemic disease. The lymph nodes and spleen may be sites of primary disease or secondary involvement. Affected lymph nodes show sinusoidal dilation by Langerhans cells. These cells have elongated irregular nuclei with a linear groove and inconspicuous cytoplasm. Multinucleated cells and eosinophils are also seen. Langerhans cells are positive for S100 and histiocytic markers, and they are also CD1a positive.

Castleman disease

Castleman disease, also known as angiofollicular hyperplasia, can be broadly classified under two categories, localized and systemic (also known as multicentric). The types of localized disease include HVCD type and plasma cell Castleman disease (PCCD) type. The etiology of Castleman's disease is still not fully elucidated, but one known etiologic agent is HHV-8 (human herpesvirus-8). HHV-8 encodes for a homolog of IL-6, which is a B-cell growth factor. High levels of IL-6 may also be from endogenous sources. The HVCD variant is seen in 80%—90% of all cases of unicentric Castleman disease. The mediastinum is a common site of involvement and typically forms a mass. There occurs follicular hyperplasia and the follicles may show two or more germinal centers. The germinal centers are depleted of small lymphocytes. They are occupied by numerous follicular dendritic cells. Hyaline deposits are found within the germinal canters. The mantle zone of the follicles is expanded, and in some, the mantle zone is composed of concentric rings (onion skin appearance). Sclerotic blood vessels are seen to radially penetrate the follicles (lollipop appearance). In the interfollicular areas, there are numerous high endothelial venules, with plasma cells, small lymphocytes, immunoblasts, eosinophils, and plasmacytoid monocytes.

In the PCCD, the interfollicular area has a large number of plasma cells. Vascular proliferation in the interfollicular areas may also be present. The follicles in PCCD are usually hyperplastic. The histologic findings in multicentric Castleman disease (MCD) are similar to that of PCCD.

Castleman disease may predispose to a variety of neoplasms. Patients with HVCD are at risk of developing follicular dendritic cell sarcoma. Malignant lymphoma may develop in patients with PCCD.

IgG4-related sclerosing disease—associated lymphadenopathy

This disease is probably autoimmune in nature. It is characterized by enlargement of one or more exocrine glands or extranodal sites with raised serum IgG4 levels and a lymphoplasmacytic infiltration associated with sclerosis. Examples of sites being involved are

- Pancreas
- Biliary tract
- Liver
- Lungs
- Submandibular gland
- Lacrimal gland
- Breast
- Kidneys
- Lymph nodes (mediastinal, intraabdominal, and axillary sites most frequently involved)

  Morphology of affected lymph nodes:

- Follicular hyperplasia with interfollicular plasmacytosis
- Intrafollicular plasma cells may also be seen
- PTGC
- In some cases, paracortical expansion with plasma cells
- Focal areas of fibrosis may be present
- Increased IgG4 positive plasma cells by immunostains (IgG4+ cells >50/HPF and IgG4/IgG + cell ratio >40%

Key points

- Primary follicles consist of small lymphocytes and follicular dendritic cells. The small lymphocytes are predominantly B lymphocytes, which exhibit surface IgD and surface IgM and are positive for CD5. Follicular dendritic cells are stained by CD21, CD23, and CD35.
- Secondary follicles are derived from primary follicles following antigenic stimulation. Secondary follicles have an inner germinal center and an outer mantle zone. Within the germinal center in a reactive follicle, there is a light zone and a dark zone. This is referred to as polarity. The light zone consists of centrocytes and follicular dendritic cells. Mitotic rate is lower in this zone. Tingible-body macrophages are less seen in this area. In contrast, the dark zone has centroblasts with a high mitotic rate and increased number of tangible body macrophages. The germinal center of a reactive follicle stains with CD10 and Bcl-6, and it is negative for Bcl-2.
- The outer mantle zone of the secondary follicle is composed of B lymphocytes. These B lymphocytes are positive for CD5 and Bcl-2. Outer to the mantle zone is the marginal zone, which is still a B-cell area, but the cells are more loosely packed. The area beyond the marginal zone is the paracortex and is a predominantly T-cell area. The paracortex is thus easily stained by CD3, a pan T-cell marker. Bcl-2 also stains T cells. Thus, Bcl-2 staining pattern of a reactive lymph node include staining of the paracortex, outer parts of the secondary follicle, but not the germinal center.
- Most often viral infections lead to paracortical hyperplasia, which results in expansion of the paracortex with presence of transformed lymphocytes. There is also hyperplasia of the

interdigitating dendritic cells, and these cells have pale cytoplasm. This causes the paracortex to have a mottled appearance. Follicular hyperplasia may also be present. The large transformed cells are immunoblasts and may mimic large cell lymphoma. On occasion, these cells may appear like Reed—Sternberg cells.

- In cases of CMV infections, intranuclear viral inclusions with a halo may be evident. In HSV infections, multinucleated giant cells with ground glass nuclei may be seen. Similar to CMV, intranuclear inclusions may also be present. In measles, Warthin—Finkeldey giant cells may be seen.

- In patients with HIV infections, lymph nodes typically exhibit follicular hyperplasia. Three patterns are described. In pattern A (early pattern), there are reactive follicles with reduced mantle zone. There is folliculosis, interfollicular hemorrhage. Aggregates of monocytoid B cells are evident, and scattered Warthin—Finkeldey cells may also be present. In pattern C (late stage), the follicles are atrophic with vascular proliferation. Pattern B is a transition from pattern A to C and has features in between.

- Toxoplasma lymphadenopathy is characterized by a triad of histologic findings. These are follicular hyperplasia, clusters of epithelioid macrophages, within or in close proximity to germinal centers, and monocytoid B cells within sinuses.

- In typical cases of PTGC, large lymphoid nodules in a background of follicular hyperplasia are observed. These large nodules are progressively transformed germinal centers. It is thought that the cells of the mantle zone infiltrate and disrupt the germinal center making germinal center ill defined. The follicular center cells are reduced. Tingible body macrophages are rare or absent. In the majority of cases, PTGC are seen to affect the lymph nodes focally. PTGC is considered as a benign reactive change. However, it is associated with lymphomas, especially NLPHL.

- Kikuchi-Fujimoto disease is seen predominantly in Asians. The disease presents as painless cervical lymphadenopathy in young adults. Morphologically, areas of necrosis in the paracortex are observed along with the presence of abundant karyorrhectic debris. Interestingly, neutrophils are absent. The necrotic areas contain bright eosinophilic fibrinoid deposits. The necrotic areas are surrounded by activated large T-lymphocytes and numerous histiocytes, which have abundant cytoplasm. Plasma cells and plasmacytoid monocytes are also present. The areas of necrosis and the surrounding numerous histiocytes are the reason for the synonym of Kikuchi-Fujimoto disease as necrotizing histiocytic lymphadenitis.

- Kimura's disease: Affected lymph nodes exhibit follicular hyperplasia and intense eosinophilia with eosinophilic microabscess with infiltration of the germinal centers. Warthin—Finkeldey polykaryocytes may be seen in the germinal centers. Hyperplasia of capillaries is also a feature. This disease is associated with peripheral blood eosinophilia and elevated levels of IgE.

- Kawasaki disease: This disorder is also known as mucocutaneous lymph node syndrome. Morphologically, there occurs expansion of interfollicular areas with loss of normal architecture. There are areas of necrosis with presence of neutrophils and nuclear debris. Vessels may be increased.

- Dermatopathic lymphadenitis is seen in the lymph node draining areas of skin involved with irritation, infection, or inflammation. Langerhans cells present in skin migrate to regional lymph nodes on antigenic stimulation. Subsequently, there is T-cell activation and T-cell hyperplasia in the lymph node. Thus, the paracortical area is expanded. Macrophages with brown pigments are found within the paracortex. The pigment is typically melanin. There are also interspersed interdigitating dendritic cells and Langerhans cells.

- Rosai—Dorfman disease: This disorder is also known as sinus histiocytosis with massive lymphadenopathy. The disease presents as bilateral painless cervical lymphadenopathy. Typically, it is a benign disorder that spontaneously resolves. Morphologically, the lymph nodes demonstrate dilatation of the sinuses by large macrophages with round vesicular nuclei, nucleoli, and foamy cytoplasm. Emperipolesis is present and is the hallmark of the disease. Eosinophils are not prominent, which is a feature of LCH. The macrophages are positive for S100 and typical macrophage/histiocytic markers (e.g., CD68). They are negative for CD1a. In contrast, CD1a is positive in LCH.
- LCH: It is a group of disorder, ranging from focal lesions (typically of bone) to systemic disease. The lymph nodes and spleen may be sites of primary disease or secondary involvement. Affected lymph nodes show sinusoidal dilation by Langerhans cells. These cells have elongated irregular nuclei with a linear groove and inconspicuous cytoplasm. Multinucleated cells and eosinophils are also seen. Langerhans cells are positive for S100, histiocytic markers, and also CD1a.
- Castleman disease (also known as angiofollicular hyperplasia): There are essentially two forms of the disease, localized and systemic (also known as multicentric). The types of localized disease include hyaline vascular (HVCD) type and plasma cell (PCCD) type. The HVCD variant is seen in 80%—90% of all cases of unicentric CD. There occurs follicular hyperplasia, and the follicles mat show two or more germinal centers. The germinal centers are depleted of small lymphocytes. They are occupied by numerous follicular dendritic cells. Hyaline deposits are found within the germinal center. The mantle zone of the follicles is expanded, and in some, the mantle zone is composed on concentric rings (onion skin appearance). Sclerotic blood vessels are seen to radially penetrate the follicles (lollipop appearance). In the interfollicular areas, there are numerous high endothelial venules, with plasma cells, small lymphocytes, immunoblasts, eosinophils, and plasmacytoid monocytes. In the PCCD, the interfollicular area has a large number of plasma cells. Vascular proliferation in the interfollicular areas may also be present. The follicles in PCCD are usually hyperplastic. The histologic findings in MCD are similar to that of PCCD. CD may predispose to a variety of neoplasms. Patients with HVCD are at risk of developing follicular dendritic cell sarcoma. Malignant lymphoma may develop in patients with PCCD.

Lymph nodes affected by IgG4-related sclerosing disease exhibit follicular hyperplasia with interfollicular plasmacytosis; PTGC; and focal areas of fibrosis may be present. There is increased IgG4-positive plasma cells by immunostains (IgG4 + cells >50/HPF and IgG4/IgG + cell ratio >40%).

References

[1] Wood GS. The immunohistology of lymph nodes in HIV infection. Prog AIDS Pathol 1990;2:25—32.
[2] Lamps LW, Scott MA. Cat-scratch disease: historical, clinical and pathological perspectives. Am J Clin Pathol 2004;121(Suppl.):S71—80.
[3] Nelson CA, Saha S, Mead PS. Cat-scratch disease in the United States, 2005-2013. Emerg Infect Dis 2016;22: 1741—6.

[4] Pappas G, Roussos N, Falagas ME. Toxoplasmosis, snapshots: global status of *Toxoplasma gondii* seroprevalence and implications for pregnancy and congenital toxoplasmosis. Int J Parasitol 2009;39:1385–94.

[5] Asano S. Granulomatous lymphadenitis. J Clin Exp Hematop 2012;52:1–16.

[6] Dalton J, Shaw R, Democratis J. Kikuchi-Fujimoto disease. Lancet 2014;383(9922):1098.

[7] AlGhamdi FE, Al-Khatib TA, Marzouki HZ, AlGarni MA. Kimura disease: No age or ethnicity limit. Saudi Med J 2016;37:315–9.

[8] Averitt AW, Heym K, Akers L, Castro-Silva F, Ray A. Sinus histiocytosis with massive lymphadenopathy (Rosai Dorfman Disease): diagnostic and treatment modalities for this rare entity revisited. J Pediatr Hematol Oncol 2018;40:e198–202.

Precursor lymphoid neoplasms, blastic plasmacytoid dendritic cell neoplasm, and acute leukemias of ambiguous lineage

11

Introduction

Collectively, lymphoid neoplasms are the fourth most common cancer and the sixth leading cause of cancer death in the United States. In one report published in 2016, the authors speculated that 136,960 new lymphoid neoplasm cases were expected in 2016. Despite, decline in overall lymphoma incidences in recent years, precursor lymphoid neoplasm incidence rates increased from 2001 to 2012, particularly for B-cell neoplasms. Among the mature lymphoid neoplasms, the fastest increase was for plasma cell neoplasms. Rates also increased for mantle cell lymphoma (males), marginal zone lymphoma, hairy cell leukemia, and mycosis fungoides [1].

B-lymphoblastic leukemia/lymphoma, NOS

B-lymphoblastic leukemia (B-ALL) and its rare lymphoma counterpart, B-lymphoblastic lymphoma (B-LBL), mostly occur in children and adolescents and are regarded as one of top leading causes of death related to malignancies in this population. B-lymphoblastic leukemia/lymphoma is a neoplasm of precursor B-lymphoid cell, involving the bone marrow and blood (B-ALL) or occasionally presenting with primary involvement of nodal or extranodal sites (B-LBL). Nathwani and colleagues established the current morphologic description of "lymphoblastic lymphoma (LBL)" presenting with extramedullary mass lesions comprising cells indistinguishable from acute lymphoblastic leukemia (ALL) [2]. Adult B-ALL exhibits similar clinical course as pediatric population but has a worse prognosis in comparison with younger individuals. Ample evidence have shown that the clinical behavior, response rate, and clinical outcome of B-ALL rely largely on its genetic and molecular profiles, such as the presence of BCR-ABL1 fusion gene which is an independent negative prognostic predictor [3]. ALL is a disease of children, with majority being less than 6 years of age. Relative incidences of B-lymphoblastic leukemia/lymphoma and related disorders are summarized in Table 11.1.

B-ALL patients present with features of bone marrow failure. Lymphadenopathy and hepatosplenomegaly are observed frequently. Patients may have bone pain and joint pain. Patients with B-LBL are usually asymptomatic and most have limited-stage disease.

Hematology and Coagulation. https://doi.org/10.1016/B978-0-12-814964-5.00011-5

Table 11.1 Relative incidences of B-lymphoblastic leukemia/lymphoma and related disorders.

Disease	Relative incidence
B-ALL	80%–85% of cases
B-LBL	10% of cases
T-ALL and T-LBL	5%–10% of cases (T-LBL is more common than T-ALL)

B-ALL, *B-lymphoblastic leukemia;* B-LBL, *B-lymphoblastic lymphoma;* T-ALL, *T-Acute lymphoblastic leukemia;* T-LBL, *lymphoblastic lymphoma.*

Morphology and immunophenotype

Morphology includes presence of small- to medium-sized blasts, moderately condensed to dispersed chromatin and indistinct nucleoli. The nuclei may show convolutions. In 10% of cases the blasts may have azurophilic granules. Sometimes, the blasts may have cytoplasmic pseudopods (hand mirror cells). In LBL, lymph nodes are expected to have a diffuse growth pattern. Numerous mitotic figures are seen. Starry sky pattern may be present.

The blasts are dimly positive for CD45 and positive for B-cell markers (e.g., CD19, CD79a, CD22, Pax5). CD10 and TdT are positive in most cases. The expression of CD20 and CD34 is variable. Clonal surface immunoglobulin (IG) light chain is absent.

CD13 and CD33 may be expressed.

Pro–B-ALL (early precursor) includes TdT, CD19, cCD79a, and cCD22, which are positive.

Common B-ALL (intermediate stage): CD10 is positive.

Pre–B-ALL (most mature precursor): cytoplasmic mu chain.

Remission and cure rates

In children, the remission rate is >95%, and the cure rate is about 80%. In adults, the remission rate is 60%–80%, and the cure rate is <50%.

B-lymphoblastic leukemia/lymphoma with recurrent genetic abnormalities

These have distinct clinical or phenotypic properties, have important prognostic implications, and are biologically distinct. B-lymphoblastic leukemia/lymphoma with t(9;22): BCR-ABL1 is seen in 25% of adults with ALL, but less frequently in children. There is frequent expression of CD13 and CD33. CD25 is highly expressed, at least in adults. The prognosis is very poor.

B lymphoblastic leukemia/lymphoma with t(v;11q23.3): Rearrangement of KMT2A (also known as MLL) is most common leukemia in infants (up to 1 year of age). Patients have very high white count (>100K) and high frequency of CNS involvement. The tumor cells typically have pro–B-ALL immunophenotype with negative CD10. CD15 may be positive. The prognosis is poor.

B-lymphoblastic leukemia/lymphoma with t(12;21): ETV6-RUNX1 is not seen in infants. CD19, CD10 and CD34 are typically positive, and the prognosis is very favorable.

B-lymphoblastic leukemia/lymphoma with hyperdiploidy: Here blasts have >55 chromosomes (usually less than 66). Not seen in infants. CD19, CD10, and CD34 are typically positive, and prognosis is favorable. However, B-lymphoblastic leukemia/lymphoma with hypodiploidy has poor prognosis. B-lymphoblastic leukemia/lymphoma with t(5;14): IGH/IL3 is rarely observed. This disease is often present with eosinophilia. B-lymphoblastic leukemia/lymphoma with t(1;19): TCF3-PBX1 has a pre—B phenotype. The prognosis is poor.

B-lymphoblastic leukemia/lymphoma: BCR-ABL1-like is a provisional entity with a poor prognosis. B-lymphoblastic leukemia/lymphoma with iAMP21 is observed in older children who are often associated with low white blood cell (WBC) counts. The prognosis is poor [4].

T-lymphoblastic leukemia/lymphoma

T-lymphoblastic leukemia/lymphoma is a neoplasm precursor T lymphoid cell involving bone marrow and blood (T-ALL) or occasionally presenting with primary involvement of thymus or of nodal or extranodal sites (T-LBL). T-ALL typically presents with a high WBC count, mediastinal mass, and lymphadenopathy as well as hepatosplenomegaly. T-ALL often shows relative sparing of hematopoiesis compared with B-ALL. T-LBL often presents as a mediastinal mass. Pleural effusion is also common.

Morphology and immunophenotype

The blasts of T-ALL/LBL are indistinguishable from B-ALL/LBL.

The blasts are usually positive for CD1a, CD99, TdT, and CD34. T-cell markers are positive (CD2, CD3, CD4, CD5, CD7, and CD8) Only CD3 is considered to be lineage specific.

Prognosis

T-ALL is generally considered to be a higher risk disease than B-ALL.

NK-lymphoblastic leukemia/lymphoma

This is a provisional entity where there are immature cells expressing CD56, along with immature T-associated markers such as CD7, Cd2, and even including cCD3. The cells need to lack B cell and myeloid markers. T-cell receptor and (IG) genes are germline and blastic plasmacytoid dendritic cell neoplasm (BPDCN) should be excluded.

Blastic plasmacytoid dendrite cell neoplasm

This is an aggressive tumor derived from precursors of plasmacytoid dendritic cells. There is a high frequency of skin and bone marrow involvement. Most of the patients are in their sixties. Skin involvement is the most frequent initial presentation. Isolated nodule or isolated plaque or disseminated nodules, papules, or macules may be seen. Involvement of blood and bone marrow follows.

By morphology the infiltrate resembles blasts, either lymphoblasts or myeloblasts. The tumor cells are positive for CD4, CD43, CD56 and the plasma dendritic cell—associated markers, such as CD123, CD303, TCL1A, CD2AP, and SPIB.

Acute leukemias of ambiguous lineage

In acute undifferentiated leukemia the tumor cells express no markers considered to be specific for either myeloid or lymphoid lineage. There will be no more than one membrane marker for any given lineage. Thus, they lack cCD3, MPO and CD22, cCD79a, and strong CD19.

In mixed phenotype acute leukemia, the cells fulfill the criteria for B or T cells and myeloid cells. They can have one of the following features:

- Mixed phenotype acute leukemia with t(9;22); BCR-ABL1
- Mixed phenotype acute leukemia with t(v;11q23.3); KMT2A rearranged
- Mixed phenotype acute leukemia, B/myeloid, NOS
- Mixed phenotype acute leukemia, T/myeloid, NOS
- Mixed phenotype acute leukemia, NOS

Requirement for assigning lineage to a single blast population

B-cell lineage:

- Strong CD19 with one or more of CD10, cCD22, and CD79a or weak CD19 with two or more of CD10, cCD22, and CD79a

  T-cell lineage:

- cCD3 or surface CD3

  Myeloid lineage:

- MPO positivity or
- Monocytic differentiation with positivity for two or more with CD11c, CD14, CD64, and lysozyme

Key points

- ALL is a disease of children, with majority being less than 6 years of age. Blasts >25% in the bone marrow are used to define leukemia.
- In ALL, small- to medium-sized blasts are observed, which are moderately condensed to dispersed chromatin and indistinct nucleoli. The nuclei may show convolutions. The blasts are dimly positive for CD45 and positive for B-cell markers (e.g., CD19, CD79a, CD22, Pax5). CD10 and TdT are positive in most cases. The expression of CD20 and CD34 is variable.
- B-lymphoblastic leukemia/lymphoma with t(9;22): BCR-ABL1 is seen in 25% of adults with ALL, less in children. There is frequent expression of CD13 and CD33. CD25 is highly expressed, at least in adults. The prognosis is very poor.
- B lymphoblastic leukemia/lymphoma with t(v;11q23.3): Rearrangement of KMT2A (aka MLL) is most common leukemia in infants (up to 1 year of age). Patients have very high white count (>100K) and high frequency of CNS involvement.

- T-ALL is typically present with a high WBC count, mediastinal mass, lymphadenopathy and hepatosplenomegaly. T-ALL often shows relative sparing of hematopoiesis compared with B-ALL.
- T-LBL is often present as a mediastinal mass. Pleural effusion is also common.
- The blasts in T-ALL are usually positive for CD1a, CD99, TdT, and CD34. T-cell markers are positive (CD2, CD3, CD4, CD5, CD7, and CD8). Only CD3 is considered to be lineage specific.
- BPDCN is an aggressive tumor derived from precursors of plasmacytoid dendritic cells. There is a high frequency of skin and bone marrow involvement.
- Acute undifferentiated leukemia: Here the tumor cells express no markers, which are considered to be specific for either myeloid or lymphoid lineage. There will be no more than one membrane marker for any given lineage. Thus, they lack cCD3, MPO and CD22, cCD79a, and strong CD19
- Mixed phenotype acute leukemia: Here the cells fulfill the criteria for B or T cells and myeloid cells.

References

[1] Teras LR, DeSantis CE, Cerhan JR, Morton LM, et al. 2016 US lymphoid malignancy statistics by World Health Organization subtypes. CA Cancer J Clin 2016;66:443−59.
[2] Sheibani K, Nathwani BN, Winberg CD, Burke JS, et al. Antigenically defined subgroups of lymphoblastic lymphoma. Relationship to clinical presentation and biologic behavior. Cancer 1987;60:183−90.
[3] Zhang X, Rastogi P, Shah B, Zhang L. B lymphoblastic leukemia/lymphoma: new insights into genetics, molecular aberrations, subclassification and targeted therapy. Oncotarget 2017;8(39):66728−41.
[4] Boer JM, den Boer ML. BCR-ABL1-like acute lymphoblastic leukemia: from bench to bedside. Eur J Cancer 2017;82:203−18.

B-cell lymphomas

Introduction

Cancer constitutes a major cause of death worldwide. The occurrence of cancer is increasing because of the growth and aging of the population, as well as an increasing prevalence of established risk factors such as smoking, overweight, physical inactivity, and changing reproductive patterns associated with urbanization and economic development. Based on GLOBOCAN estimates, about 14.1 million new cancer cases and 8.2 million deaths occurred in 2012 worldwide. Over the years, the burden has shifted to less developed countries, which currently account for about 57% of cases and 65% of cancer deaths worldwide [1].

Lymphomas are cancer that arises from clonal proliferations of lymphoid cells at various stages of differentiation. Broadly lymphomas can be classified as Hodgkin lymphoma and non-Hodgkin lymphoma. American Cancer Society estimated that in 2014, there will be approximately 70,800 new cases of non-Hodgkin lymphoma and an estimated 18,990 people may die from these lymphomas, but there will be 9190 new cases of Hodgkin lymphoma in 2014 and an estimated 1180 deaths may occur from Hodgkin lymphoma. Currently, non-Hodgkin lymphoma is the fifth most common cancer in the United States.

Currently WHO (World Health Organization) classification of lymphomas are commonly used in clinic practice, and WHO recognizes about 70 different types of lymphomas and divided them into four broad groups including mature B-cell neoplasm, mature T cell and natural killer cell neoplasm, Hodgkin lymphoma, and immunodeficiency-associated lymphoproliferative disorders. Although Hodgkin lymphoma is a B-cell malignancy, it is characterized by Reed—Sternberg cells that are not found in any other B-cell lymphomas. This chapter focuses on B-cell lymphomas.

Of all the B-cell lymphomas the most frequent are diffuse large B-cell lymphomas (DLBCL) and Follicular lymphomas (FLs). The relative frequencies of the most common B-cell lymphomas are listed in Table 12.1.

B-cell lymphomas may be clinically low grade or high grade.

Low-grade B-cell lymphomas include the following:

- Follicular lymphomas (FL)
- Chronic lymphocytic leukemia (CLL)/small lymphocytic lymphoma (SLL)
- Marginal zone B-cell lymphoma
- Lymphoplasmacytic lymphoma (LPL)

Table 12.1 The relative frequency of the most common B-cell lymphomas.

B-cell lymphoma	Frequency
Diffuse large B-cell lymphoma	37%
Follicular lymphoma	29%
Chronic lymphocytic leukemia/small lymphocytic lymphoma	12%
Extranodal marginal zone lymphoma	9%
Mantle cell lymphoma	7%

High-grade B-cell lymphomas the following:

- DLBCL
- Burkitt lymphoma
- Lymphoblastic lymphomas
- Blastoid or pleomorphic mantle cell lymphomas
- High-grade B-cell lymphomas

Classical mantle cell lymphoma may be included in the low-grade category, but it tends to be more aggressive than the others in this group. Thus, it may be considered as intermediate grade.

Diffuse large B-cell lymphoma

DLBCL is a neoplasm of large B-lymphoid cells with the nuclear size equal to or exceeding normal macrophage nuclei that have a diffuse pattern. It is the most common lymphoma (second: FL) and accounts for about 25%–35% of adult non-Hodgkin lymphomas. The incidence is higher in developing countries. It presents with a rapidly enlarging, often symptomatic mass at a single nodal/extranodal site. In 40% of cases, it is initially confined to an extranodal site. Bone marrow involvement is low (10% cases) at diagnosis. Bone marrow may demonstrate a lower grade of lymphoma (referred to as discordant lymphoma). Li et al. recently reviewed DLBCL [2].

In DLBCL, the cells have nuclei that are larger than those of normal macrophages or more than twice the size of those of normal lymphocytes.

R-CHOP (rituximab, cyclophosphamide, hydroxydaunorubicin, Oncovin, prednisone) is still the standard treatment. With this treatment, 60% are cured, 30% relapse, and 10% are refractory to treatment.

DLBCL may occur as primary or de novo, but it may be also secondary to progression/transformation from other types of lymphoma including SLL (Richter transformation: 3.5% cases), FL (25%–33% cases), marginal zone lymphoma, nodular lymphocyte predominant Hodgkin lymphoma (NLPHL) and LPL.

Morphologic variants include

- Centroblastic (most common variant, approximately 80%): round, vesicular nuclei with two to four nucleoli
- Immunoblastic (second most common variant, approximately 10%): central nucleolus, basophilic cytoplasm
- Anaplastic: very large, polygonal cells with bizarre pleomorphic nuclei; may grow in cohesive pattern, mimicking carcinoma; tend to be CD30+

    Centroblastic variant: more likely to be germinal center B cell (GCB) subtype.
    Immunoblastic variant: increased incidence of myelocytomatosis (MYC) translocations, do worse.
    Anaplastic variant: these are associated with p53 mutations.

Molecular subtypes

- GCB subtype: more likely to have centroblastic morphology; generally does better. About 90% of double or triple hits are of this subtype.
- Activated B-cell subtype: more likely to be immunoblastic variant and generally does worse. Most double expressors are of this type (double expressors are those DLBCLs that express MYC and Bcl-2 by immunohistochemistry). Double expressor tends to do worse prognostically.

    Most of the primary extranodal large B-cell lymphomas tend to be of ABC subtype. MYD88 L265P mutations are found in ABC-like DLBCLs and may be seen in primary CNS lymphoma, primary cutaneous DLBCL, and leg type and primary testicular DLBCL. These tumors are sensitive to BTK inhibition by Ibrutinib.

    The Hans algorithm uses three markers to distinguish GCB from non-GCB subtype, and these are CD10, Bcl6 and MUM1. In cases of GCB subtype, CD10 is positive (over 30% of cells), or if CD10 is negative, Bcl6 is positive and MUM1 is negative. Here activation of NF-kB (nuclear transcription factor-kappa B) is absent. All other staining combinations are features of non-GCB subtype. Here activation of NF-kB is positive.

    Immunophenotype usually shows the presence of positive B-cell markers such as CD20, CD22, and CD79a. Surface immunoglobulins (IgM > IgG > IgA) are present in 50%−75%. Bcl-2 marker is positive in approximately 30% cases, but positive Bcl-6 is observed in much higher percentage of cases. Approximately 20% of DLBCL cases show positive CD30. However, CD10 could be positive if DLBCL was developed from FL.

    DLBCL and certain percentages that might be useful to remember:

- 10%: DLBCL may be positive for CD5 in 5%−10% of cases
- 20%: about 20% of cases of DLBCL are positive for CD30. This number is higher with anaplastic variant and Epstein−Barr virus (EBV)−positive DLBCL
- 30%: CD30, Bcl6, and MUM1 immunostains are considered to be positive if >30% of cells are positive
- 40%: immunostains for MYC are considered to be positive if >40% of tumor nuclei are positive
- 50%: Bcl2 is considered to be positive if >50% of tumor cells are positive

Immunohistochemistry for suspected diffuse large B-cell lymphoma

- Two B-cell markers (CD20 and another one such as CD79a or PAX5; CD20 should always be done because when this is positive, Rituximab may be used)
- CD3 and CD5: tumor cells should be negative. There is CD5-positive DLBCL, but it does not impact treatment. If the case is pleomorphic or blastoid mantle, CD5 could be positive.
- Cyclin D1: to rule out mantle cell lymphoma, especially the high-grade variants
- CD30: some may be positive; allows use of brentuximab
- Bcl-2, Bcl-6
- MYC: if MYC and Bcl-2 are positive, these are double expressors
- CD10, MUM-1: with BCl-6 to be used for Han's algorithm
- Anaplastic lymphoma kinase (ALK): to rule out ALCL; in addition, there is ALK-positive large B-cell lymphoma

Testing for EBV is useful in patients suspected with DLBCL during review of tissue biopsy. In biopsy tissue, molecular detection of EBV-encoded RNA transcript (EBER) by in situ hybridization is the gold standard for proving that a histopathological lesion is indeed EBV-related. Latent infection is characterized by EBER. In addition, LMP1 (latent membrane protein 1) immunostaining is also routinely performed to detect latent EBV in affected tissue [3]. DLBCL with EBV positivity is diagnosed as EBV-positive DLBCL, NOS. This entity was previously known as EBV-positive DLBCL of the elderly. The cases are mostly non-GCB subtype.

FISH studies for MYC, Bcl-2, and Bcl-6 should be done in all cases of DLBCL. Individuals with MYC rearrangement and double (MYC with Bcl-2 or Bcl-6) or triple (MYC and Bcl-2 and BCl-6) hit do worse.

Human herpesvirus-8 (HHV-8) is considered as the causative agent of Kaposi's sarcoma and is involved with primary effusion lymphoma and multicentric Castleman disease. HHV-8 is a gamma-herpesvirus that shows sequence hemology to EBV and herpesvirus saimiri. Testing for latency-associated nuclear antigen LANof HHV-8 in tissue sample is also useful in patients suspected with DLBCL because LNA is one of the few HHV-8—encoded proteins that is highly expressed in all latently infected tumor cells.

Other lymphomas of large B cells

- T cell—/histiocyte-rich DLBCL: here there are a limited number of scattered large B cells (<10%) embedded in a background of abundant T cells and histiocytes. They may arise de novo or from a background NLPHL.
- Primary central nervous system (CNS) lymphoma of brain: diffuse growth pattern with tumor cells in the perivascular area. Most are non-GCB (i.e., ABC subtype). CD10 positivity is seen <10% of cases. CD10 positivity should lead to search for systemic DLBCL that has spread to the CNS.
- Primary cutaneous DLBCL, leg type: like the primary DLBCL of the CNS and intravascular large-cell lymphoma, these are typically ABC subtype. Most cases are positive for Bcl2, Bcl6, and MUM 1. CD10 is usually negative.
- EBV-positive DLBCL, NOS: this was formerly known as EBV-positive DLBCL of the elderly. The cases are mostly non-GCB subtype.

- DLBCL associated with chronic inflammation: prototype is pyothorax-associated lymphoma that may also be associated with EBV.
- Lymphomatoid granulomatosis: Angiocentric and angiodestructive lymphoproliferative disease involving extranodal sites (most often lungs) where Epstein—Barr-induced immunodeficiency is associated with increased risk. Lungs are the commonest site, and there are three grades according to distribution of inflammatory cells and large cells. In grade 1, large cells are minimal, but in grade 3, large cells are most present.
- Large B-cell lymphoma with IRF4 rearrangement: it occurs mainly in children and young adults with predominant involvement of Waldeyer ring or head and neck region. Lesions can be diffuse, follicular, and diffuse or entirely follicular. There is strong expression of IRF4/MUM1. It is different from pediatric FL even when entirely follicular. Systemic therapy is required.
- Primary mediastinal large B-cell lymphoma: most likely thymus in origin with interstitial fibrosis and clear cell appearance of tumor cells. CD30 is positive in 80% of cases, but cells lack surface immunoglobulins. CD23 is positive in about 70% of cases.
- Intravascular large B-cell lymphoma: two major forms are described. One is the classic form, seen mainly in western countries. Features are mainly neurological or cutaneous. The other form is a hemophagocytic syndrome—associated form. This form is originally documented in Asians. The tumor cells can be CD5 positive.
- ALK-positive DLBCL: the tumor cells are immunoblastic or plasmablastic. ALK positivity is cytoplasmic and granular. This indicates expression of CLTC-ALK fusion protein. CD45 is weak/negative. CD30 is also negative. They are positive for EMA, plasma cell markers (CD138), and MUM 1. Tumors cells express cytoplasmic immunoglobulins (most often IgA) with light chain restriction.
- Plasmablastic lymphoma: diffuse proliferation of neoplastic cells most of which resemble immunoblasts. The tumor cells have the immunophenotype of plasma cells: CD138+, MUM1+, CD20−, Pax5−, EBER, and MYC are usually positive. Ki-67 is high. Involvement of oral cavity occurs commonly, and patients are positive for HIV.
- HHV-8—positive DLBCL: Large-cell lymphoma arising in HHV-8—associated Castleman disease. Patients are usually HIV positive, and large cells are plasmablastic.
- Primary effusion lymphoma: patients are usually positive for HIV, EBV, and HHV-8 virus (also known as Kaposi's sarcoma—associated herpesvirus). This disorder may show plasmablastic morphology. The tumor cells express CD45 but lack B-cell markers. They express CD30, CD138, and EMA.

Plasmablastic neoplasms

Plasmablastic neoplasms include the following:

- Plasmablastic myeloma
- DLBCL (typically immunoblastic variant)
- Plasmablastic lymphoma
- ALK+ large B-cell lymphoma
- Primary effusion lymphoma
- Large B-cell lymphoma in HHV8+ Castleman disease

CD5 positive aggressive B-cell lymphomas

- CD5+ DLBCL
- Intravascular large B-cell lymphoma
- Blastoid mantle cell lymphoma

High-grade B-cell lymphomas

This replaces the previous term B-cell lymphoma, unclassifiable with features intermediate between DLBCL and Burkitt lymphoma. There are two subtypes, HGBL with MYC and Bcl2 and/or Bcl6 rearrangement and HGBL, NOS. The first subtype is also known as double- or triple-hit lymphomas. In the NOS subtype, MYC rearrangement may be seen but without Bcl2 and Bcl6 gene rearrangement.

Double or triple hit lymphomas by morphology resemble typical DLBCL, NOS or mimic Burkitt lymphoma or have features intermediate between DLBCL and Burkitt lymphoma. Ki-67 tends to be high, but cases with low Ki-67 are also seen. As it is not possible to predict by morphology, all DLBCL, NOS should undergo testing by cytogenetic or molecular studies for MYC, Bcl2, and Bcl6 gene rearrangement. In general 7%−8% of DLBCLs are double-hit lymphomas.

The relative frequencies of gene rearrangement in double-hit lymphomas are as follow:

- MYC/Bcl2: 65%
- MYC/Bcl2/Bcl6: 21%
- MYC/Bcl6: 14%

Individuals with the first two types typically are GCB subtype. The third type may be either GCB or non-GCB.

Immunohistochemistry for MYC and Bcl2 should be done on all DLBCLs. If both are positive (positive MYC is when >40% of tumor cells are positive and positive Bcl2 is when >50% of tumor cells are positive), then these cases are referred to as double expressors. Approximately 30% of DLBCL cases are double expressors. Double expressors tend to non-GCB subtype.

Double expressors are not treated differently. However, R-CHOP fails in double- or triple-hit lymphomas.

B-cell lymphomas unclassifiable

B-cell lymphomas unclassifiable are B-cell lymphomas with features intermediate between DLBCL and classic Hodgkin lymphoma.

Follicular lymphoma

FL is defined as a neoplasm composed of follicular center B cells (centrocytes and centroblast), which usually have at least a partial follicular pattern.

The neoplastic cells are centrocytes (cleaved follicle center cells) and centroblasts (noncleaved follicle center B cells), and the relative portion of centrocytes to centroblast determines the grading scheme of this lymphoma. FL is predominantly observed in adults (median age of onset is 60 years) and is the second most common lymphoma diagnosed in the United States and Western Europe, accounting for 35% of all non-Hodgkin lymphoma and 70% of indolent lymphomas. Most patients have widespread disease at diagnosis (bone marrow is involved in 40%—70% of cases), but most patients are usually asymptomatic at diagnosis, except for lymph node enlargement. The median survival of patients is over 10 years.

Variants of FL include the following:

- In situ follicular neoplasia (ISFN)
- Duodenal-type FL
- Testicular FL
- Diffuse variant FL

In addition there are distinct and separate types of FL:

- Pediatric-type FL
- Primary cutaneous follicle center cell lymphoma

Usual follicular lymphoma

Out of all cases of FL approximately 80% are the so-called usual type.

FL predominantly involves lymph nodes. It may also involve the spleen, bone marrow, and peripheral blood. Pure extranodal involvement is uncommon.

In the lymph nodes, we expect to see increased follicles. The follicles will be seen in both cortex and medulla. The follicles may appear monotonous and may be back to back. Perinodal adipose tissue may be involved. The mantle zone will be poorly defined and the germinal centers lack polarization (centroblasts and centrocytes occupy different zones in a nonneoplastic follicle). Absence of tingible body macrophages is another feature of these follicles.

In a normal germinal center of a follicle, the centroblasts are present on one side, and this area appears relatively dark. The other area is occupied by centrocytes and they appear lighter. This variation of dark and light areas in the same germinal center is referred to as polarization. In FL, there is a loss of polarization. If a KI-67 immunostain is done, the darker area (with centroblasts) has a higher proliferation index than the lighter area.

Reactive germinal centers have tingible body macrophages. In FL, these are absent. In high grade (grade III) of FL, however, we may see increased tingible body macrophages. Thus, high-grade FL may be difficult to distinguish from reactive follicles.

Patterns of follicular lymphoma

Patterns of this disease is subdivided into follicular (>75% follicular), follicular and diffuse (25%—75% follicular), minimally follicular (<25% follicular), and diffuse (a diffuse area is defined as an area of tissue completely lacking follicles defined by CD21+/CD23+ FDC).

Grading of follicular lymphoma

Grading is from 1 to 3 depending on number of centroblasts per high power filed (hpf) (grade 1: 0—5 centroblasts, grade 2: 6—15 centroblasts, grade 3: over 15 centroblasts/hpf). Grade 3 is further subdivided into 3a (some centrocytes present) and 3b (solid sheets of centroblasts). Follicular dendritic cells (FDCs) are large cells with central nucleolus should not be counted as centroblasts. Centroblasts

Table 12.2 Morphological features and grading of follicular lymphoma.

Morphological characteristics of neoplastic follicles	Patterns	Grading
Poorly defined and closely packed Devoid of mantle zone Devoid of polarization (centroblast and centrocytes occupy different zones in a nonneoplastic follicle) Absence of tingible body macrophages	Follicular: >75% follicular Follicular and diffuse: 25%—75% follicular Minimally follicular: <25% follicular Diffuse: a diffuse area is defined as an area of tissue completely lacking follicles defined by CD21+/CD23+ FDC (follicular dendritic cells)	Grade 1: 0—5 centroblasts/hpf Grade 2: 6—15 centroblasts/hpf Grade 3: >15 centroblasts/hpf 3a: some centrocytes present 3b: solid sheets of centroblast

have multiple nucleoli located adjacent to nuclear membrane. Grades I and II can be reported as grades I—II/III. This is because grades I and II are clinically indolent and represent a continuum. Morphological features and grading of FL are summarized in Table 12.2. Bone marrow if involved typically show paratrabecular involvement.

Immunophenotyping in follicular lymphoma

In diagnosis of FL, it is important to demonstrate that the tumor cells are B cells and not T cells. Commonly used B-cell markers include CD19 (cluster of differentiation 19), CD20, CD79a, and PAX 5 (paired box protein 5 encoded by PAX 5 gene). For workup of lymphomas, typically two B-cell markers are used. One of them should be CD20, as this informs the clinicians that Rituximab (anti-CD20 antibody) may be used, if positive. FL is a tumor of GCBs. Markers for germinal center include CD10 and Bcl 6. The neoplastic follicles are positive for Bcl 2 (B-cell lymphoma 2 protein encoded by Bcl gene), in contrast to nonneoplastic ones. Thus, reactive follicles will be negative. Therefore, markers in FL include the following:

- CD20 and another B-cell marker should be positive
- T-cell marker such as CD3 (should be negative)
- CD10 and Bcl 6 (markers for germinal center and both should be positive)
- Bcl-2 (should be positive)
- Ki-67 (should be <20% in grades I—II and >20% in grade III)

In addition the tumor, B cells should exhibit surface light chain restriction (monotypic surface immunoglobulins).

Occasionally FLs can be Bcl-2 negative. Such Bcl-2 negative FL is as follows:

- High-grade FL
- Diffuse variant FL
- Primary cutaneous follicle center cell lymphoma
- Pediatric-type FL
- Testicular FL
- In addition, certain older antibodies used in the laboratory may result in false negativity.

At times FL can be CD10 negative. Such situations are seen with high grade FLs.

In about 20% of cases, there occurs a discrepancy between the histologic grade and the Ki-67 index, i.e., the histologic grade is that of I—II, but the Ki-67 is >20%. These are thought to be more aggressive cases.

Genetics of follicular lymphoma

FL is a malignant counterpart of normal germinal B-cells, and majority of patients with this disorder show chromosomal translocation: t(14;18)(q32;q21), which creates a derivative of chromosome 14 on which BCL-2 gene (B-cell lymphoma 2) is juxtaposed to immunoglobulin heavy chain gene (IgH) sequence. This abnormality results in overproduction of Bcl-2 protein, a family of proteins that blocks apoptosis. Therefore, as expected overproduction of Bcl-2 protein in these patients prevents cells from undergoing apoptosis. Two points regarding the genetic aspects of FL:

- Rearrangement of BCL-2 gene is the molecular consequence of the t(14;18)(q32;q21) chromosomal translocation, which is present in 75%—85% cases, and cytogenetics is the best way to detect karyotypic changes in the tumor specimen [4]. However, this abnormality is not associated with prognosis
- Proper function of BCL2 gene confers a survival advantage on B cells, but at the same time failure to switch off BCL2 during blast transformation may contribute to development of lymphoma by preventing apoptosis.

Prognosis for FL is indolent if it is grade 1 or 2. However, grade 3 FL is aggressive and is usually treated as DLBCL. Usually 25%—33% cases of FL may progress to DLBCL.

Variants of follicular lymphoma

In this section, variants of FLs are addressed.

In situ follicular neoplasia

This is seen in about 2% of otherwise unremarkable lymph nodes. Here there is partial or total colonization of germinal centers by clonal B cells carrying the BCL2 translocation in an otherwise reactive lymph node. It can also be seen in other reactive lymphoid tissue. Cells may also be seen in the peripheral blood with BCL2 rearrangement by polymerase chain reaction (PCR). These circulating cells are referred to as FL-like B cells.

Key histologic features include the following:

- The follicles are normal in size
- Intact mantle zone
- The cells in affected germinal centers are centrocytes
- Involved follicles are widely scattered
- The affected germinal centers demonstrate strong staining for BCL2 and CD10. In cases of partial involvement of the follicle, the intensity of staining is variable. Chance of future development of FL is very low (<5%).

Duodenal-type follicular lymphoma

These lesions are typically found as polyps in the second part of the duodenum. Most patients have localized disease. Prognosis is excellent without treatment. The tumor cells have the classic immuno-phenotype (for FL). The FDCs are pushed to the periphery, which can be documented by CD21/CD23. There is low risk of progression to nodal FL (<10%).

Testicular follicular lymphoma

These are seen mainly in children. They lack evidence of BCL2 translocation (i.e., Bcl-2 immunostain negative and t14;18 negative). Cytologically they are of high grade, usually grade 3A. However, they have good prognosis, even without additional therapy beyond surgical excision.

Diffuse variant of follicular lymphoma

This variant mainly occurs in the inguinal region. Here there is a diffuse growth pattern. Small follicles described as microfollicle are seen. The tumor cells are CD10 and CD23 positive. BCL2 is negative as well as t14;18.

Distinct and separate types of follicular lymphoma

In pediatric-type FL, the typical ages of these patients are 15–18 years. Here there are large expansile follicles. The follicles show a starry sky pattern and thin or absent mantle zones. The follicles have a monotonous population of cells and are intermediate in size with a blastoid appearance. Thus, the cells appear high grade (Ki-67 >30%). Immunostains demonstrate positivity for CD10 and Bcl-6. Bcl2 is negative as well as t14;18. Excision alone is sufficient in most cases.

Primary cutaneous follicle center lymphoma is a tumor of neoplastic follicle center cells (centrocytes and centroblasts) with a follicular, follicular and diffuse, or diffuse growth pattern, arising in the skin. It typically affects head or trunk. It is the most common B-cell lymphoma of the skin. Morphology resembles a high-grade lymphoma. There is expression of CD10 and BCl6. Bcl2 is typically negative. t14;18 may be seen in 10%–40% of cases. The prognosis is very good (>95%, 5 year survival). However, it may be difficult to ascertain whether this is a primary lymphoma of the skin or secondary involvement from another site. It is not required to grade this lymphoma or provide the pattern in the report. Please note that lymphomas with a diffuse growth pattern and a monotonous proliferation of centroblast and immunoblasts are classified, irrespective of site as primary cutaneous DLBCL, leg type. Thus, it is important to identify centrocytes in primary cutaneous follicle center cell lymphoma.

Chronic lymphocytic leukemia/small lymphocytic lymphoma

CLL and SLL are neoplasm of monomorphic small, round B lymphocytes in blood, bone marrow, and lymph nodes. CLL usually originates in the bone marrow and spills over into the blood. SLL starts in the lymphoid tissue. In fact, CLL and SLL are considered as the same entity with SLL restricted to tissue cases featuring no leukemia phase.

CLL is considered as the most common leukemia in western countries and also has the highest genetic predisposition. Majority of patients diagnosed with CLL are over 50 years of age, and unlike other leukemias, radiation exposure does not increase the risk of CLL. Most patients are asymptomatic although some may have autoimmune hemolytic anemia. The absolute lymphocyte count is >5000/mm^3 of blood and persists for more than 3 months. Monoclonal B-cell lymphocytosis is a condition seen in 3.5% of individuals over 40 years of age, and it is uncertain whether this is a forerunner of CLL.

CLL/SLL and certain percentages

- 10%: CLL may be familial in about 10% of cases. In about 10% of cases, monoclonal gammopathy (most often IgM) may be present.
- 20%: SLL is diagnosed in 10%−20% of cases, and of these, 20% evolve into frank CLL.
- 30%: hypogammaglobulinemia may be seen in about 30% of cases.

Morphology of chronic lymphocytic leukemia in peripheral blood

There is lymphocytosis with presence of smudge cells (aka basket cells). Increased smudge cells is associated with mutated IgH chain gene (better prognosis). Smudge cells may be prevented by pretreatment with albumin.

The lymphocytes are small and have typical chromatin (known as cracked mud or cracked window pane or soccer ball appearance). Prolymphocytes are usually seen as well. They are larger than the usual CLL lymphocytes and with bluer cytoplasm, and the nucleus has nucleoli. CLL has <10% prolymphocytes. If the prolymphocyte count exceeds 55%, then these cases are referred to as prolymphocytic leukemia (PLL). Cases between 10% and 55% are referred to as CLL/PLL.

A bone marrow aspiration and biopsy is not required for the diagnosis of CLL.

If a bone marrow aspiration is performed, it shows increased lymphocytes, typically greater than 30% of all cells. Bone marrow involvement may be interstitial or diffuse or nodular in pattern. Diffuse pattern carries worse prognosis. A case of CLL may be diagnosed from the peripheral blood by microscopic review and flow cytometry.

Immunophenotyping for CLL/SLL

CLL is a clonal B-cell lymphoproliferative disorder, and flow cytometry is useful in phenotyping. As expected in CLL, B-cell markers such as CD19 and CD20 should be positive. To demonstrate clonality, these B cells will show light chain restriction. The markers mentioned so far show dim expression. CLL cells exhibit aberrant expression of CD5 and CD23. Therefore, be coexpression of CD5 and CD23 should be observed in CD19- or CD20-positive cells. FMC7 is typically negative in CLL/SLL. CLL cells are also positive for CD43 and CD200.

CD200 is a useful marker to differentiate CLL from mantle cell lymphoma. CD200 positivity is very uncommon in mantle cell lymphoma.

It has been proposed that CLL may be diagnosed by flow cytometry when an individual has three points (out of maximum of five). One point is allotted for each of the following:

1. Weak expression of surface immunoglobulins

2. Positivity for CD5 and CD23
3. Negativity for FMC7
4. Negativity for CD22 and
5. Negativity for CD79b

Two additional markers such as CD38 and ZAP70 (zeta chain associated protein kinase 70 kDa molecular weight) should also be considered because presence of these markers indicate poor prognosis. Immunostains may be considered for SLL, which include B-cell markers (should be positive), T-cell marker (e.g., CD3 which should be negative), and CD5 and CD23 (both should be positive).

Staging of chronic lymphocytic leukemia

Two methods exist. One is the Rai staging and the other is Binet.

Rai staging

Stage 0: Lymphocytosis ($>$5000/mm^3 in peripheral blood; bone marrow (BM) lymphocytosis with lymphocytes $>$30% of cells)
Stage I: As above with lymphadenopathy
Stage II: Lymphocytosis, $\pm$lymphadenopathy but has hepatomegaly or splenomegaly
Stage III: Anemia, Hb $<$11 g/dL
Stage IV: Thrombocytopenia, platelets $<$100,000 K

Binet staging

Stage A: 0 or 1 or 2 areas involved (areas are cervical lymph nodes, axillary lymph nodes, inguinal lymph nodes, liver, and spleen). Hb $>$10 g/dL and platelets are $>$100,000 K
Stage B: 3 or 4 or 5 areas involved. Hb $>$10 g/dL and platelets are $>$100,000 K
Stage C: Hb is $<$10 g/dL or platelets are $<$100,000 K

Rai stage III and IV and Binet stage C represents poor prognosis.

Morphology of lymph nodes in CLL/SLL

- Effacement of lymph node architecture; predominant cells are small lymphocytes
- Pseudofollicular pattern, also known as vague nodules
- Pseudofollicle (proliferation center) contains prolymphocytes (medium-sized cells with dispersed chromatin and small nucleoli) and paraimmunoblasts (large cells with dispersed chromatin and central nucleoli). These proliferation centers exhibit high Ki67 proliferation rate.
- Plasmacytoid differentiation may be observed

In 5%−10% of cases, the neoplasm surrounds reactive follicles, the so-called interfollicular pattern.
Please note with IHC, CD23 is brighter in the proliferation centers.
Cyclin D1 is positive in the proliferation centers in about 20%−25% of cases.

LEF1 is a useful marker to identify CLL/SLL in tissues. It is aberrantly expressed in CLL/SLL but not in normal small B lymphocytes. Sometimes cells of proliferation centers may be positive for cyclin D1. These cells are negative for SOX11.

CLL/SLL and cytogenetics

The karyotypic investigation revealed the association of CLL with del13q14, trisomy 12, del 11q22-q23, and del 17p13 [5]. However, two most common abnormalities associated with CLL include the following:

- Del 13q14.3 (seen in 50%−60% of cases) is the most frequently observed chromosomal abnormality associated with CLL, but individuals with this abnormality usually have long survival
- Trisomy 12 (seen in approximately 15% cases): atypical morphology and aggressive clinical course (worse prognosis).

Del 11q22-23 and del 17p13 imparts poorer prognosis.

In recent years, new molecular prognostic factors such as the mutation status of the immunoglobulin variable heavy chain gene (IgVH gene), CD38, and ZAP-70 have emerged with significant improved prediction of prognosis of CLL [6]. Mutated IgVH gene from a postgerminal center or memory type "B" cell is associated with stable disease and long survival because such cells do not express ZAP-70.

CLL/SLL and prognosis

Important aspects of prognosis of CLL include the following:

Binet stage C or Rai stage III and IV

Presence of ZAP-70 and CD38 by flow indicates poor prognosis. If no somatic mutation of IgVH gene (nonmutated IgVH) is present, then these cells express ZAP-70, and it indicates worse prognosis. Somatic mutations are consistent with derivation from post-GCBs, and these cells do not express the tyrosine kinase ZAP-70. This is a good prognosis.

Diffuse involvement of bone marrow indicates poor prognosis.

Presence of trisomy 12 and 11q and 17 p deletions carry poor prognosis.

If the absolute lymphocyte doubling time is less than 1 year, it is also considered as poor prognosis.

CLL may transform into DLBCL (Richter transformation: 3.5% cases) and may also transform into Hodgkin's lymphoma (0.5% cases). Richter syndrome refers to DLBCL developing in individuals who have CLL/SLL irrespective of whether the two entities are clonally related or not. In 80% of cases, both entities are clonally related. If unrelated, the prognosis is better.

Monoclonal B-cell lymphocytosis

These are individuals who have lymphocytosis ($<5000/mm^3$) without lymphadenopathy, organomegaly, other extramedullary involvement, or any other feature of a B-cell lymphoproliferative disorder.

There are three subtypes:

1. CLL type (most common, 75% of cases): CD5, CD23+, surface Ig are dim
2. Atypical CLL type: here surface immunoglobulin expression is bright and CD23 may be negative
3. Non-CLL type: here surface immunoglobulin is bright and CD5 may be negative.

If the monoclonal B cell lymphocytosis (MBL) count is high (i.e., $>0.5 \times 10^9/L$), chance of progression to CLL is about 1%−2% per year.

B-cell prolymphocytic leukemia

B-cell PLL is due to neoplasm of B prolymphocytes. Most patients during diagnosis are over 60 years of age, and these patients often have marked splenomegaly and rapidly rising lymphocyte count. Leukemic cells derived from mantle cell lymphoma are excluded from diagnosis.

The characteristic of this leukemia is the presence of prolymphocytes (over 55% but often more than 90%) lymphoid cells in blood. Important features of these lymphoid cells include the following:

- Medium-sized (lymphocyte $\times$ 2)
- Round nucleus with central nucleolus
- Condensed chromatin
- Small amount of faintly basophilic cytoplasm

Bone marrow shows diffuse infiltrate of prolymphocytes. The spleen shows extensive red and white pulp involvement. Pseudofollicles are not seen in lymph nodes. Immunophenotype characteristic of B-cell PLL includes the following:

- B-cell markers are positive with light chain restriction
- Surface immunoglobulins (usually IgM $\pm$ IgD)
- CD5: positive in approximately 20%−30% cases
- CD23: positive in 10%−20% of cases
- FMC 7: positive

Mantle cell lymphoma

Mantle cell lymphoma is a B-cell neoplasm of monomorphous small- to medium-sized cells (that resemble centrocytes) with irregular contours. In 95% of cases, there is a CCND1 translocation. Mantle cell lymphomas constitute approximately 6% of all non-Hodgkin lymphomas. Median age of patients during diagnosis is 60 years, and a male predominance is also observed. In mantle cell lymphoma, pseudofollicle (proliferation center) or transformed cells (centroblast/paraimmunoblasts) are not observed. In addition, there is no transformation to large-cell lymphoma.

Most patients with mantle cell lymphoma present with lymphadenopathy and hepatosplenomegaly (stage III or IV). Bone marrow involvement at the time of diagnosis is observed in 50%−60% patients. Involvement of gastrointestinal tract is usually 30% (large gut: multiple lymphomatous polyposis and Waldeyer's ring).

Important morphological features of mantle cell lymphoma include the following:

- Monomorphic proliferation of small- to medium-sized lymphoid cells that resemble centrocytes
- Vague nodular/diffuse/mantle zone growth pattern
- Increase in hyalinized small blood vessels

In the majority (75%) of cases, there are histiocytes with eosinophilic cytoplasm (pink histiocytes). In some cases, there are naked germinal centers.

Immunophenotyping characteristics of mantle cell lymphoma include the following:

- B-cell markers positive with light chain restriction
- Strong expression of immunoglobulins (IgM ± IgD)
- CD5 positive but CD23 negative
- Positive CD43, BCL-2, and cyclin D1
- Negative CD10 and BCL-6

Positive for SOX11; may be useful for cyclin D1 negative cases. Negative for CD200; useful to differentiate from CLL/SLL.

Other causes of positive cyclin D1 include the following:

Plasma cell myeloma
Hairy cell leukemia (HCL)
DLBCL
CLL/SLL (proliferation centers)
lymphocyte predominant (LP) cells in nodular LP Hodgkin lymphoma

In mantle cell lymphoma, cyclin D1 expression is bright.

Mantle cell lymphoma is a distinct subtype of malignant lymphoma characterized by chromosomal translocation t(11;14)(q13;q32) involving cyclin D1 gene present in chromosome 11 and Ig heavy chain gene present in chromosome 14. This chromosomal translocation results in overexpression of cyclin D1 and cell cycle dysregulation in almost all cases. Clinically this disease displays an aggressive course with a continuous relapse pattern and a median survival of only 3−7 years. However, emerging strategies of including proteasome inhibitors, immune modulatory drugs, and mammalian target of rapamycin inhibitors along with use of other chemotherapeutic agents may improve the survival [7].

Mantle cell lymphoma variants include the following:

- Blastoid variant: cells resemble lymphoblasts with dispersed chromatin; high mitotic rate (>10/10 high power field)
- Pleomorphic (resembles DLBCL)
- Small
- Marginal zone−like

Prognosis

Mantle cell lymphoma generally is more aggressive than the other small B-cell lymphomas (e.g., FL, CLL/SLL, marginal zone lymphoma, lymphoplasmacytic lymphoma [LPL]). If the ki67 is greater than

30%, then these mantle cell lymphomas are even worse (treatment is with hyper CVAD). The histologic variants, blastoid, and pleomorphic are also more aggressive.

Indolent mantle cell lymphoma

Some cases of mantle cell lymphoma present with lymphocytosis, and the cells resemble CLL cells. Patients have minimal lymphadenopathy, and these cases are known as leukemic nonnodal mantle cell lymphoma. Some of these patients go on to develop classical mantle cell lymphoma.

In situ mantle cell neoplasia

This is defined as presence of cyclin D1—positive lymphoid cells with CCND1 rearrangements restricted to the mantle zone of otherwise hyperplastic-appearing lymphoid tissue. They are seen in <1% of all lymph nodes. In situ mantle cell neoplasia can occur in extranodal sites as well.

Marginal zone B-cell lymphoma

Marginal zone lymphoma represents 5%—17% of all non-Hodgkin lymphoma in adults, and WHO classification categorized marginal zone lymphoma into three categories including splenic marginal zone lymphoma, nodal marginal zone lymphoma, and extranodal marginal zone B-cell lymphoma involving mucosal tissues (mucosa-associated lymphoid tissues [MALT]). Although occurring in diverse anatomic lesions, lymphoma of MALT is a distinct clinical pathological entity originally described by Isaacson and Wright in 1983.

Key features of marginal zone lymphoma include the following:

- Heterogenous population of cells that include centrocyte-like cells, small lymphocytes, centroblasts, and immunoblasts as well as monocytoid cells (cells with abundant cytoplasm)
- Plasmacytic differentiation
- Dutcher bodies (intranuclear inclusions) in lymphocytes may be a useful feature.
- Tumor infiltrate in the marginal zone, which extends into the interfollicular area
- Under low power, the tumor appears "pink," due to the monocytoid B cells
- In epithelial tissue, the tumor cells infiltrate the epithelium forming lymphoepithelial lesions (three or more tumor cells with distortion or destruction of the epithelium)
- Possible transformation to DLBCL especially with high-grade MALT
- Gastrointestinal tract is the most common site of MALT lymphomas, and stomach is the most common location within the gastrointestinal tract
- Bone involvement in observed in 10%—20% of cases (bone marrow involvement in nodal marginal is higher than MALT, whereas bone marrow involvement in splenic marginal is highest)

Immunophenotyping for marginal zone lymphoma includes use of B-cell markers. Positive CD43 (aberrant expression) is seen in up to 50% of cases. CD5, CD10, and cyclin D1 are negative. Bcl2 is positive with negative Bcl6. However, there is no specific marker for MALT lymphoma. There are recent reports about the possibility of IRTA1 being a specific marker for marginal zone lymphomas.

Extranodal marginal zone lymphoma of mucosa-associated lymphoid tissue

This is an extranodal lymphoma composed of a heterogenous population of small B cells and includes marginal zone cells (centrocyte like), monocytoid cells, small lymphocytes and scattered larger immunoblasts, and centroblast-like cells. There may be plasmacytic differentiation. The tumor cells are in the marginal zones of the reactive B-cell follicles and extend in to the interfollicular areas. In epithelial tissue, the tumor cells infiltrate the epithelium forming lymphoepithelial lesions.

MALT lymphomas can occur in a variety of organ including the orbit, conjunctiva, salivary glands, skin, thyroid glands, lungs, stomach, and intestine. These tumors are often localized and have indolent clinical behavior. Pathological evaluation of tumor including using immunohistochemical and cytogenetics is useful in diagnosis. The discovery of an association between *Helicobacter pylori* infection and MALT lymphoma and cure of this disease with eradication of *H. pylori* indicates that underlying antigenic stimulation due to *H. pylori* infection may be related to this disorder. In the first study, the association of gastric MALT lymphoma and *H. pylori* was >90%. More recent studies demonstrate this association to be in a smaller proportion of cases (32%). *Chlamydia psittaci* and *Borrelia burgdorferi* are associated with ocular MALT lymphoma. Hashimoto's thyroiditis is risk factor for thyroid MALT lymphoma and Sjögren's syndrome for salivary gland MALT lymphoma. For treatment of MALT lymphoma of nongastric locations, radiotherapy, chemotherapy, or a combination of both may be used depending on grading of tumor [8].

The four most common sites for MALT lymphoma are stomach (28%), skin (20%), salivary glands (17%), and ocular adnexa (15%). Approximately 20% of individuals have regional lymph node involvement at the time of diagnosis. This has little impact on prognosis.

Chromosome translocation t(11;18)(q21;q21) is frequently associated with gastric MALT lymphoma and may be found in 30%−50% of all cases, but this translocation is rare in MALT lymphoma involving other sites. The chromosome translocation t(11;18)(q21;q21) leads to a fusion of the apoptosis inhibitor gene API2 on chromosome 11 and MALT lymphoma translocation 1 (MLT/MALT1 gene) on chromosome 18. Presence of this mutation may cause gastric MALT lymphoma less responsive to *H. pylori* eradication therapy using multiple antibiotics. Trisomy 3 may be present in up to 60% cases of MALT lymphomas.

The common translocations observed in MALT lymphoma are as follow:

- t(11;18) 20%−30%
- t(14;18) 10%−15%
- t(3;14) 10%
- t(1;14) <10%

Splenic marginal zone lymphoma

Splenic marginal zone lymphoma is a rare tumor where tumor cells surround the white pulp germinal centers with effacement of the mantle zone. The tumor cells expand into the outer marginal zone where cells with abundant pale cytoplasm are seen. Transformed cells are also seen in the outer area. The red pulp is infiltrated with small and larger tumor cells. Splenic hilar nodes and bone marrow are often involved. Tumor cells may be seen in the peripheral blood where these cells appear as lymphocytes with polar villi (in HCL, the villi are circumferential, not polar). The bone marrow when involved shows a sinusoidal pattern.

In approximately 30% of cases, paraproteins are identified. There is also an association with hepatitis C virus infection.

Nodal marginal zone lymphoma

In nodal marginal zone lymphoma, there is no evidence of extranodal or splenic marginal zone lymphoma. This lymphoma is in general indolent and uncommon. Morphology is similar to MALT lymphoma with immunotyping characteristics including the following:

- Presence of B-cell markers
- CD43+ (aberrant expression, seen in up to 50% of cases)
- Bcl-2 positive, but Bcl-6 negative
- CD5, CD10, and cyclin D1 negative

Burkitt lymphoma

This is a high grade, highly aggressive B-cell tumor, which was first recognized by British surgeon Denis Burkitt among children when he was working in Africa. Burkitt lymphoma is an aggressive form of disease, which if treated on a timely manner may be fatal [9]. In Africa, Burkitt lymphoma is common among children who also may be infected with malaria and EBV. Outside Africa, this type of lymphoma is relatively rare. There are three types of Burkitt lymphoma:

- Endemic: This type is observed mainly in Africa among children between ages of 4–7, and this disease is more common in boys than girls. Most often jaws and abdomen are involved, and EBV could be detected in most cases.
- Sporadic: Observed in both children and young adults (median age, 30), and EBV may be detected in up to 30% cases, and both HIV and EBV infection may be detected in up to 25% cases. In this type, involvement of kidneys, ovaries, and breasts are observed.
- Immunodeficiency associated: this is more common in the HIV setting.

Key morphologic features of Burkitt lymphoma include diffuse growth pattern with starry sky appearance, medium-sized cells with squared-off, well-defined borders, finely clumped chromatin, prominent nucleoli, and high mitotic rate (Ki67 ~ 100%). The cytoplasm is basophilic with lipid vacuoles.

Immunophenotyping indicates positive B-cell markers. CD10 and BCL6 (B-cell lymphoma 6 protein) are usually positive, but Bcl2 and TdT (terminal deoxynucleotidyl transferase, a DNA polymerase) are negative.

The genetic hallmark of Burkitt lymphoma is the translocation of MYC with IGH or kappa or lambda gene loci. Three specific chromosomal translocation including t(8;14), t(2,8), and t (8;22) are commonly present in cases involving Burkitt lymphoma. These translocations have been shown in many cases to be related to DNA molecular rearrangements of the immunoglobulin genes and c-myc genes (cellular homolog of retroviral myelocytomatosis oncogene). In translocation t(8;14), c-myc gene on chromosome 8 and Ig heavy gene on chromosome 14 are involved in rearrangement. In translocation t(2,8), kappa gene on chromosome 2 is involved, whereas in t(8;22), lambda gene on chromosome 22 is involved.

In about 10% of cases, it is not possible to identify MYC rearrangement, and these are designated MYC-negative Burkitt lymphoma. A subset of these has 11q aberrations. These tumors exhibit more cytologic pleomorphism. HGBL should be considered as differentials.

Lymphoblastic leukemia/lymphoblastic lymphoma

Lymphoblastic leukemia arises in the bone marrow, whereas lymphoblastic lymphoma arises in the lymphoid tissue. Lymphoblastic leukemia is most often B cell, and lymphoblastic lymphoma is most often T cell. Lymphoblastic lymphoma is a high-grade neoplasm that produces a mass lesion with 25% or fewer lymphoblasts in the bone marrow. There is diffuse effacement of architecture with starry sky pattern. The lymphoblasts are medium-sized cells with nuclei having condensed to dispersed chromatin and inconspicuous nucleoli. B-cell lymphoblastic leukemia affects skin, bone, and lymph nodes, whereas T-cell lymphoblastic lymphoma usually arises as a mediastinal mass.

Immunophenotyping involves positive B-cell markers. In addition TdT, CD10 and BCl-2 are all positive.

Lymphoplasmacytic lymphoma/waldenstrom macroglobulinemia

LPL is a low-grade lymphoma with a familial predisposition (in 20% cases) and associated with hepatitis C infection. Key features include the following:

- Involvement of bone marrow and sometimes lymph node and spleen.
- Spectrum of cells is seen including small B lymphocytes, plasmacytoid lymphocytes, and plasma cells. There are increases in mast cells as well as increase in epithelioid histiocytes.
- Serum monoclonal protein with hyperviscosity or cryoglobulinemia may be present.

Thus, LPL is a small B-cell lymphoma that usually involves the bone marrow and sometimes lymph nodes and spleen, which does not fulfill the criteria for any other small B-cell neoplasm.

Waldenstrom macroglobulinemia (WM) is LPL with bone marrow involvement with IgM monoclonal gammopathy (of any concentration).

Therefore, LPL is not synonymous to WM. However, 90% of LPL cases are WM.

Variants of LPL include lymphoplasmacytoid (small lymphs with occasional plasma cells), lymphoplasmacytic (small lymphs, plasmacytoid lymphs, and plasma cells), and polymorphous (increased large cells). This disease can transform into DLBCL.

Immunophenotype involves observation of positive B-cell markers. CD38 is also positive, but CD5, CD10, and CD23 are usually negative. Surface and cytoplasmic immunoglobulin may be present in some cells.

Common cytogenetic abnormality is chromosomal translocation t(9;14). The great majority (>90%) of LPLs have MYD88 L265P mutation. This abnormality is however, not specific.

Rearrangement of PXX-5 gene is observed in 50% cases.

MYD88 L265P mutation may also be seen in the following states:

- LPL and WM
- IgM monoclonal gammopathy of undetermined significance (MGUS) (50% of cases)
- Splenic marginal zone lymphoma (10% of cases)
- CLL (5% of cases)

Hairy cell leukemia

Hairy cell leukemia is a rare neoplasm of peripheral B cells, which was first described by Bouroncle et al. as leukemia reticuloendotheliosis and later renamed as HCL because of typical cytoplasmic projection (like hair) in tumor cells. The incidence of HCL is less than 1 in 100,000 people and comprises less than 2% of lymphoid leukemia affecting mainly male gender (male to female ratio, 4:1). The median age of diagnosis is 55 years. Most patients present with enlarged but rarely symptomatic splenomegaly and nonspecific symptoms such as fatigue and weakness. HCL is an indolent lymphoma involving bone marrow and splenic red pulp. Hepatomegaly is observed in nearly half of the patients, and lymph nodes are rarely enlarged [10]. Splenectomy was considered as a treatment of choice till 1984, which was associated with good therapeutic response, but more recently HCL is treated with purine analogs, pentostatin, and cladribine, which can results in complete remission in 76 to up to 98% cases, and today HCL is considered as a treatable disease [11].

In the peripheral blood, atypical lymphocytes with hairy projections are also observed. Typical features include pancytopenia and monocytopenia. The tumor cells are small- to medium-sized cells with oval-indented nuclei (heavy chromatin, absent nucleoli) and abundant pale cytoplasm. In the bone marrow, biopsy increased reticulin, and cells with pale cytoplasm (fried egg appearance) are observed. Red pulp disease with white pulp atrophy is observed in the spleen.

Immunophenotype shows positive CD11c, CD25, CD103, CD123, and annexin A1. FMC7, CD200, and cyclin D1 may also be positive. In addition, TRAP (tartrate resistant acid phosphatase) test in blood cell or bone marrow is also positive, which a characteristic of HCL.

Annexin A1 is the most specific marker. It is not expressed in any other B-cell lymphoma.

Presence of BRAF V600E mutation is seen in virtually 100% of cases of HCL.

Hairy cell leukemia variant (HCL-V) is a rare B-cell lymphoproliferative disease that shares many clinical, morphological, and immunophenotyping features of HCL with classic HCL including splenomegaly, neoplastic lymphocytes with cytoplasmic projections, or hair and bone marrow involvement. However, in contrast to classic HCL, patients with HCL-V have significantly elevated white blood cell count, easy-to-aspirate bone marrow (sometimes it is difficult to aspirate bone marrow in HCL patients), and neoplastic cells infrequently show positive TRAP test. In addition, patients with HCL-V might have anemia or thrombocytopenia but usually no monocytopenia, neutropenia, or pancytopenia. Most patients with HCL-V show diminished response or even lack of response to conventional therapy for HCL including therapy with interferon-alpha and cladribine [12].

Approach for diagnosis of lymphoma

Laboratory tests play an important role in diagnosis of various lymphomas. Complete blood count with differential and careful examination of peripheral blood smear to assess bone marrow function and also to observe or rule out abnormal cells is the first important step. Standard chemistry tests to evaluate renal and hepatic function, determination of blood glucose, calcium, albumin, lactate dehydrogenase, and beta-2-microglobulin is useful. Serum protein electrophoresis is also frequently ordered by clinicians to investigate if monoclonal gammopathy is present. HIV serology is also important because HIV infection is a risk factor for certain types of lymphomas.

Useful points to consider in the histologic evaluation of lymphomas:

- Lymphomas with presence of follicles: presence of follicles may mean reactive follicles or FL. Sometimes there may be neoplastic follicles, which represent follicular growth pattern of mantle cell lymphoma. In marginal zone, lymphoma benign follicles may be present with tumor cells in the interfollicular area. In LPL, there may be retention of normal lymph node architecture, or sometimes there may be follicular growth pattern. Peripheral T-cell lymphoma NOS has a variant called follicular pattern. Angioimmunoblastic T-cell lymphoma (AITL) is characterized by partial effacement of lymph node architecture
- Lymphomas with a nodular pattern (even vague nodules) include SLL (vague nodules), NLPHL, nodular sclerosis and lymphocyte rich classical Hodgkin lymphoma, and mantle cell lymphoma
- Lymphomas with increased vessels: this category include AITL, peripheral T-cell lymphoma, and mantle cell lymphoma
- Lymphomas with a starry sky pattern include Burkitt lymphoma, blastoid mantle, lymphoblastic lymphoma, and any aggressive high-grade lymphoma
- Lymphoma with a monotonous population of cells: mantle cell lymphoma
- Lymphoma with presence of plasmacytoid and plasma cells include LPL, SLL, and marginal zone lymphoma.
- Lymphomas with presence of intrasinusoidal tumor cells include ALCL and LPL
- Lymphomas with pale or pink areas (implies presence of cells with abundant cytoplasm) include marginal zone lymphoma. T-cell lymphomas (especially AITL) also may have cells with pale cytoplasm
- Lymphomas with characteristic cell sizes: small cells; SLL, intermediate cells; LL, Burkitt's lymphoma and large cells; DLBCL, ALCL
- B-cell lymphomas with CD10 positive tumor cells include FL, DLBCL, lymphoblastic lymphoma, and Burkitt's lymphoma (typically Bcl-2 negative and Bcl-6 positive)
- B-cell lymphomas that are typically Bcl-2 positive include FL, DLBCL, marginal zone lymphoma (typically Bcl-6 negative), and mantle cell lymphoma (typically Bcl-6 negative)

Immunophenotype of common low-grade B-cell lymphomas are listed in Table 12.3.

Key points

- Neoplastic follicles in FL are poorly defined and closely packed, devoid of mantle zone, devoid of polarization (centroblasts and centrocytes occupy different zones in a nonneoplastic follicle), lack tingible body macrophages.
- Patterns of FL include follicular (>75% follicular), follicular and diffuse (25%−75% follicular), minimally follicular (<25% follicular), and diffuse.
- Grading of FL include Grade 1: (0−5 centroblasts/hpf), Grade 2: (6−15 centroblasts/hpf), and Grade 3: (>15 centroblasts/hpf:3a: some centrocytes present, 3b: solid sheets of centroblasts). Bone marrow, if involved in FL, typically shows paratrabecular involvement.

Table 12.3 Immunophenotype of small B-cell lymphomas.

	Chronic lymphocytic leukemia	MCL	Lymphoplasmacytic lymphoma	MZL	Follicular lymphoma
CD19	+	+	+	+	+
CD20	+(dim)	+	+	+	+
Surface immunoglobulins	+(dim)	+	+	+	+
CD10	−	−	−	−	+
CD5	+	+	−	−	−
CD23	+	∓	±	±	±
Bcl2	+	+	+	+	+
Bcl6	−	−	− (Rarely blastoid may be +)	−	+
CyclinD1	−	+	−	−	−
SOX11	−	+	−	−	−
CD200	+	−	−	−	−

MCL: mantle cell lymphoma; MZL: marginal zone lymphoma.

- FL is a tumor of GCBs. Markers for germinal center include CD10 and Bcl 6. Bcl 2 will stain the tumor cells in the germinal center of the neoplastic follicles of FL. Reactive follicles will be negative. Grade 3 may be negative for Bcl2. In general, 25%−33% cases of FL may progress to DLBCL
- CLL is the most common leukemia in western countries, with the highest genetic predisposition, and radiation exposure does not increase risk of development of CLL, unlike other leukemias.
- Lymph nodes with SLL/CLL will show the presence of pseudofollicle (proliferation canters), which contains prolymphocyte (medium-sized cells with dispersed chromatin and small nucleoli) and paraimmunoblasts (large cells with dispersed chromatin and central nucleoli). Plasmacytoid differentiation may also be observed.
- CLL cells exhibit aberrant expression of CD5 and CD23. Therefore, coexpression of CD5 and CD23 should also be observed in these CD19- or CD20-positive cells. FMC7 is typically negative in CLL/SLL.
- Features of poor prognosis in CLL include diffuse marrow involvement, trisomy 12, those with no somatic mutations (i.e., ZAP70 positive), CD38 positive, and if the absolute lymphocyte count doubling time is less than 1 year.
- SLL/CLL may transform to DLBCL; Richter transformation (3.5% cases). In addition, SLL/CLL may also transform to Hodgkin's lymphoma (0.5% cases).
- Mantle cell lymphoma is a B-cell neoplasm of monomorphous small- to medium-sized cells that resemble centrocytes. It is characterized by absence of pseudofollicle/proliferation centers. In addition, transformed cells (centroblasts/paraimmunoblasts) are also absent, and no transformation to large-cell lymphoma is observed.

- Mantle cell lymphoma aberrantly expresses CD5 but not CD23. Blastoid variant of mantle is an aggressive high-grade lymphoma.
- Extranodal marginal zone B-cell lymphoma of MALT demonstrates a heterogenous population of cells, which include centrocyte-like cells, small lymphocytes, centroblasts, immunoblasts, and monocytoid cells (cells with abundant cytoplasm). Plasmacytic differentiation is also present. Marginal zone lymphoma does not have any classical markers. It is a diagnosis of exclusion. In about 50% of cases, there may be aberrant expression of CD43.
- Burkitt lymphoma is a high-grade, highly aggressive B-cell tumor, and key morphologic features include diffuse growth pattern with starry sky appearance, medium-sized cells with squared-off, well-defined borders, clumped chromatin, and prominent nucleoli with a high mitotic rate (Ki67 ~ 100%).
- Lymphoblastic leukemia arises in the bone marrow, whereas lymphoblastic lymphoma arises in the lymphoid tissue. Lymphoblastic leukemia is most often B cell, and lymphoblastic lymphoma is most often T cell. Lymphoblastic lymphoma is a high-grade neoplasm, which produces a mass lesion with 25% or fewer lymphoblasts in the bone marrow. There is diffuse effacement of architecture with starry sky pattern. The lymphoblasts are medium-sized cells with nuclei having condensed to dispersed chromatin and inconspicuous nucleoli
- LPL is a low-grade lymphoma with a familial predisposition (in 20%) and associated with hepatitis C infection. Key features include involvement of bone marrow and sometimes lymph node and spleen. In LPL, a spectrum of cells is seen including small B lymphocytes, plasmacytoid lymphocytes, and plasma cells. There is an increase in mast cells and epithelioid histiocytes.
- LPL with bone marrow involvement and IgM monoclonal gammopathy of any concentration is WM.
- DLBCL is a high-grade lymphoma with diffuse proliferation of large neoplastic B cells. It is the most common lymphoma (second: FL) and accounts for 30%—40% of adult non-Hodgkin lymphomas. It presents with a rapidly enlarging, often symptomatic mass at a single nodal/extranodal site. In 40% of cases, it is initially confined to an extranodal site.
- DLBCL may occur as primary or de novo or as secondary to progression/transformation from SLL (Richter transformation; approximately 3.5% cases), FL (25%—33% cases), marginal zone lymphoma, NLPHL, and LPL.
- Morphologic variants of DLBCL include centroblastic (round, vesicular nuclei with two to four nucleoli), immunoblastic (central nucleolus, basophilic cytoplasm), and anaplastic (very large, polygonal cells with bizarre pleomorphic nuclei; may grow in cohesive pattern, mimicking carcinoma; tend to be CD30+).
- Immunohistochemical subgroups of DLBCL include GCB like, CD10 positive (>30% of cells), CD10 negative but bcl 6 positive (MUM1 negative and NF-kB activation also negative), and non-GCB like (NF-kB positive).
- Typical features of HCL include pancytopenia and monocytopenia. The tumor cells are small- to medium-sized cells with oval-indented nuclei (heavy chromatin, absent nucleoli) and abundant pale cytoplasm. In the bone marrow biopsy, there is increased reticulin and cells with pale cytoplasm (fried egg appearance). HCL cells are positive for CD11c, CD25, CD103, CD123, and annexin A1.

References

[1] Torre LA, Bray F, Siegel RL, Ferlay J, et al. Global cancer statistics, 2012. CA Cancer J Clin 2015;65: 87–108.

[2] Li S, Young KH, Medeiros LJ. Diffuse large B-cell lymphoma. Pathology 2018;50:74–87.

[3] Gulley M. Molecular diagnosis of Epstein Barr virus related disease. J Mol Diagn 2001;3:1–10.

[4] Freedman A. Follicular lymphoma: 2014 update on diagnosis and management. Am J Hematol 2014;89: 429–36.

[5] Gaidano G, Foa R, Dalla-Favera R. Molecular pathogenesis of chronic lymphocytic leukemia. J Clin Investig 2012;122:3432–8.

[6] Amin NA, Malek SN. Gene mutations in chronic lymphocytic leukemia. Semin Oncol 2016;43:215–21.

[7] Dreyling M, Kluin-Nelemans HC, Bea S, Klapper W, et al. Update on the molecular pathogenesis and clinical treatment of mantle cell lymphoma: report of 11th annual conference of the European mantle cell lymphoma network. Leuk Lymphoma 2013;54:699–707.

[8] Tsang RW, Gospodarowica MK, Pintile M, Bezjak A. Stage I and II Malt lymphoma: results of treatment with radiotherapy. Int J Radiat Oncol Biol Phys 2001;50:1258–64.

[9] Rochford R, Moormann AM. Burkitt's lymphoma. Curr Top Microbiol Immunol 2015;390(Pt 1):267–85.

[10] Forconi F. Hairy cell leukemia: biological and clinical overview from immunogenetic insight. Hematol Oncol 2011;29:55–6.

[11] Maevis V, Mey U, Schmidt-Wolf G, Schmidt-Wolf IGH. Hairy cell leukemia: short review, today's recommendations and outlook. Blood Cancer J 2014;4:e184.

[12] Cessna MH, Hartung L, Tripp S, Perkins SL. Hairy cell leukemia variant: fact or fiction. Am J Clin Pathol 2005;113:132–8.

T- and natural killer—cell lymphomas

Introduction

Peripheral T-cell lymphomas comprise a variety of rare malignancies derived from mature (post-thymic) T-cells and natural killer (NK) cells. These malignancies are less common than B-cell lymphomas and account for 5%—10% of all cases of non-Hodgkin lymphoma in North American and Europe; In Asia, this percentage may be as high as 24%. T-cell lymphomas in general carry a poorer prognosis than B-cell lymphomas [1]. These lymphomas represent a heterogenous group of diseases differing in histology, tumor site, and cell origin. In addition, many subtypes are present in WHO (World Health Organization) classification for which clinical, morphological, molecular, and phenotypic data are necessary. For example, human T-cell leukemia virus type 1 (HTLV-1) provirus is necessary for the diagnosis of the adult T-cell leukemia/lymphoma. However, proper diagnosis is hampered by several difficulties including a significant morphological and immunophenotypic overlap across different entities and lack of characteristic genetic alterations in most of them [2]. To assist in organization, these lymphomas can be broadly classified under nodal, extranodal, cutaneous, and leukemic or disseminated type.

Nodal T-cell lymphomas

Nodal T-cell lymphomas are relatively rare, and diagnostic difficulties stem from their wide range of histological patterns. Most mature T-cell lymphomas retain some functional characteristics of non-neoplastic T-cells, such as capacity to secrete cytokines and costimulate immune cell growth, thus obscuring non-neoplastic immune cells [3].

Angioimmunoblastic T-cell lymphoma

Angioimmunoblastic T-cell lymphoma (AITL) is a peripheral T-cell lymphoma, characterized by systemic disease where the primary site of disease is the lymph node. It accounts for 1%—2% of all non-Hodgkin lymphoma and 15%—30% of all non-cutaneous T-cell lymphomas. Epstein—Barr virus (EBV) has been proposed as a possible infective agent involved in the pathogenesis of AITL due to the presence of EBV positive B-cells in the vast majority of cases; however, EBV may be negative early on in the disease and has also been postulated to be a secondary event in AITL.

AITL often presents as a systemic disease with lymphadenopathy, hepatosplenomegaly, skin rash, bone marrow involvement, polyclonal hypergammaglobulinemia, bilateral pleural effusions, and other systemic symptoms. In addition, secondary immunodeficiency can lead to infectious complications. Secondary neoplasms, notably EBV positive B-cell lymphomas (most commonly diffuse large B-cell lymphoma), are also known to occur in the setting of AITL [4].

Morphology of nodes includes polymorphous infiltrate of small- to medium-sized lymphocytes. These cells have clear cytoplasm and clear cytoplasmic border. The tumor cells may be admixed with other cells such as small lymphocytes, plasma cells, eosinophils, and histiocytes. There may also be admixed larger B-cells with immunoblastic or Hodgkin Reed—Sternberg—like morphology.

There is also an increase in expanded or "arborizing" high endothelial venules (HEV). In addition, expanded follicular dendritic cell meshworks (CD21+, CD23+), typically surrounding the HEV, are observed. Three histologic patterns (I—III) are recognized representing increasing amounts of architectural effacement and characteristic AITL morphology. In pattern I (earliest and least frequently seen pattern), the lymph node architecture is partially preserved with many hyperplastic lymphoid follicles; pattern II shows decreased follicles and regressive changes within the follicles; in pattern III (most frequently seen pattern), the lymph node architecture is lost with very few follicles. Mutations in *IDH2*, *TET2*, *DNMT3A*, and *RHOA* have been frequently documented.

Immunophenotyping shows positive pan T-cell markers (e.g., CD2, CD3, and CD5). AITL is thought to arise from T follicular helper cells (TFH) within the germinal centers; therefore, CD4 and CD10 are positive. Other markers for TFH cells including PD1 (programmed cell death protein 1), Bcl-6 (B-cell lymphoma 6 protein), CXCL13 (chemokine: C-X-C-motif ligand 13; also known as B-lymphocyte chemoattractant), and ICOS (inducible T-cell costimulator) are also positive. CD21 and CD23 highlight the expanded follicular dendritic cell meshworks. EBV-encoded RNA (EBER) in situ hybridization stains can show the B-cells that are infected with EBV, but the neoplastic T-cells themselves are EBV negative. Mutations in *IDH2*, *TET2*, *DNMT3A* and *RHOA* have been frequently documented.

Other nodal T-cell lymphomas of T follicular helper cell origin

Follicular T-cell lymphoma (FTCL): Neoplasm of TFH cells with a predominantly follicular growth pattern but lacking the characteristic features of AITL (such as HEVs and proliferations of follicular dendritic cells). It usually involves lymph nodes but is overall rare, accounting for only <1% of T-cell lymphomas. The neoplastic cells have a TFH phenotype; they are positive for pan T-cell markers but exhibit frequent loss of CD7. They are also positive for CD10, BCL6, CXCL13, ICOS, and PD1. B-cells are positive for EBV in about half of cases (far less than AITL). Approximately 20% of cases harbor a t(5;9)(q33;q22) *ITK-SYK*, which is relatively specific for FTCL and has not been reported in other peripheral T-cell lymphomas.

Nodal peripheral T-cell lymphoma with a TFH cell phenotype is essentially just a CD4 positive peripheral T-cell lymphoma with a TFH phenotype. At least two, but preferably three, TFH makers (CD10, BCL6, CXCL13, ICOS, PD1) are required for the diagnosis. It has a diffuse growth pattern but lacks the polymorphous background seen in AITL.

Peripheral T-cell lymphoma, not otherwise specified

Peripheral T-cell lymphoma is a mature T-cell lymphoma that does not meet criteria for any other subtype of mature T-cell lymphoma. It has a broad cytological spectrum involving both nodal and extranodal distribution (bone marrow, peripheral blood, liver, spleen, and skin).

Cells are most often medium or large, but clear cells may also be seen. Reed—Sternberg-like cells may also be present, and HEV may be increased. Inflammatory, polymorphous background is often seen. Three variants are known including:

- Lennert lymphoma (small lymphocytes with cluster of epithelioid histiocytes)
- Follicular lymphoma, previously in the 2008 WHO, has now been moved to the category of AITL (2017 WHO update) and other T-cell lymphomas of TFH cell origin (see previous section)
- T-zone (perifollicular growth pattern)

Immunophenotyping shows positive T-cell markers with frequent down regulation of CD5 and CD7. Usually CD4 is positive and CD8 is negative. Unlike AITL, CD10, Bcl-6, and CXCL13 are typically negative in PTCL.

Anaplastic large cell lymphoma

In this section, the different types of aplastic large cell lymphomas are discussed. They are broadly divided into ALK positive, ALK negative and the recently described breast implant associated ALCL.

ALK positive ALCL

Anaplastic lymphoma kinase (ALK) positive anaplastic large cell lymphoma (ALCL) is a T-cell neoplasm characterized by large cells with abundant cytoplasm with expression of ALK protein and CD30. ALCL frequently involves nodes and extranodal sites. ALK positive ALCL is more often seen in young adults. Morphological features include diffuse involvement with presence of "hallmark" cells (typically large cells with horseshoe/reniform nucleus). Some of these cells appear to have pseudoinclusions (doughnut cells).

Sinusoidal involvement by tumor cells is typically seen in ALCL. Morphologic patterns of ALCL include:

- Common pattern (majority of cases, typical ALCL morphology as described above)
- Lymphohistiocytic pattern (tumor cells admixed with a large number of histiocytes, which may mask the lymphoma cells themselves)
- Small cell pattern (small- to intermediate-sized lymphoma cells predominate, and there may be peripheral blood involvement at presentation; ALK staining is nuclear)
- Hodgkin-like pattern (mimics nodular sclerosis subtype of classical Hodgkin lymphoma)
- Composite pattern (shows more than one pattern)

Immunophenotyping shows positive T-cell markers such as CD2 and CD5, but CD3 is often negative. By definition, CD30 is positive and ALK straining is also positive. CD30 staining in ALCL is strong, uniform, and expressed on virtually all (>75%) the neoplastic cells. This staining pattern can help to differentiate from CD30+ peripheral T-cell lymphoma in which the CD30 staining tends to be more weak and heterogeneous (not expressed on all the neoplastic cells). Majority of cases are also positive for EMA (epithelial membrane antigen) cytotoxic markers (such as TIA1 [T-cell restricted intracellular antigen], granzyme B, and perforin) and CD4. However, CD8 is negative. If multiple T-cell markers are negative, then this is "null" phenotype. In addition, ALCL is consistently negative for EBV.

The chromosomal translocation t(2;5)(p23;q35) is a recurrent abnormality present in ALK-positive ALCL, and about 80% patients bear this signature translocation. This translocation results in a fusion gene involving nucleophosmin (*NPM1*) gene on chromosome 5q35 and a receptor tyrosine kinase gene known as *ALK* gene on chromosome 2p23 that encodes ALK protein. The NPM-ALK chimeric gene encodes a constitutively activated tyrosine kinase. Staining for ALK protein is both cytoplasmic and nuclear in cases with this signature translocation. Several variant partners to *ALK* have been described [5]. Staining for ALK is cytoplasmic only for most variant translocations. Membranous ALK staining is seen with t(2;X), and the partner gene is *MSN*.

Although ALK-positive ALCL typically presents with advanced stage disease, the lymphoma is sensitive to chemotherapy and has an overall good prognosis with a 5 year survival rate of 70%–80%. Relapses occur in approximately 30% of cases but generally remain chemosensitive.

ALK-negative ALCL

This is a T-cell neoplasm with strong and uniform CD30 expression on virtually all (>75%) the neoplastic cells that is indistinguishable from ALCL on morphological grounds (including the presence of characteristic "hallmark" cells) but lacks ALCL kinase (ALK) protein expression. It was a provisional entity in the 2008 WHO classification but is considered a unique and full entity in the 2017 WHO revision.

This tumor is seen more in older adults (ALK-positive ALCL is seen more often in children and young adults). Similar to ALK-positive ALCL, most cases present with advanced disease stage (III–IV). The tumor cells in ALK-negative ALCL may be larger and more pleomorphic than ALK positive ALCL, but in general they cannot be accurately differentiated on morphologic grounds alone.

ALK-negative ALCL was previously considered to have a far worse prognosis than ALK-positive ALCL, but recent studies demonstrating certain recurrent translocations have stratified ALK-negative ALCL into at least three distinct subgroups. Rearrangements of *DUSP22* on chromosome 6p25.3 occur in about 30% of cases and are associated with a good prognosis similar to that of ALK-positive ALCL. In stark contrast, rearrangements of *TP63* occur in about 8% of cases and are associated with a dismal prognosis. Cases of ALCL that lack *ALK*, *DUSP22*, or *TP63* rearrangements or "triple-negative" ALCL have prognosis similar to the ALK-negative ALCL as a whole group [6]. These rearrangements have not been reported in ALK-positive ALCL.

Extranodal natural killer–/T-cell lymphomas

Extranodal NK-/T-cell lymphomas can arise in a wide variety of nonlymph node sites including the nasal cavity, oral cavity, gastrointestinal tract, liver, spleen, and surrounding breast implants. Although cutaneous T-cell lymphomas are also extranodal lymphomas, they will be reviewed in a separate section.

Breast implant–associated anaplastic large cell lymphoma

Breast implant–associated anaplastic large cell lymphoma (BI-ALCL) is a recently described provisional entity in the 2017 WHO update and is uniquely associated with breast implants. It is morphologically and immunophenotypically similar to ALK-negative ALCL, including the presence of occasional "hallmark" cells and strong, uniform CD30 expression. Patients usually present with a unilateral peri-implant effusion approximately 9–11 years after implant placement [7]. The neoplastic cells are typically confined to the effusion fluid and within the fibrous capsule surrounding the breast implant. In general, BI-ALCL is slow growing and follows an indolent clinical course. It has an extremely good prognosis following complete capsulectomy (surgical excision of the implant and entire surrounding fibrous capsule) [8]. Occasionally, it may form a mass lesion or involve regional lymph nodes; such cases appear to have a more aggressive clinical course [9].

Extranodal natural killer—/T-cell lymphoma, nasal type

Extranodal NK-/T-cell lymphoma, nasal type is an extranodal lymphoma where vascular damage with consequent necrosis commonly occurs, and there is a very strong association of this lymphoma with EBV. The nasal cavity and adjacent areas (nasopharynx, paranasal sinuses) are the most frequent sites of involvement. Other extranodal sites include the skin and soft tissue. This lymphoma is more prevalent in Asians and Native Americans of Central and South America. The prognosis is typically poor.

Morphological features include diffuse infiltrate with angiocentric, angioinvasive, and angiodestructive pattern, where cells may be small or medium or even large. Nonneoplastic inflammatory cells (plasma cells, small lymphocytes, eosinophils, histiocytes) may accompany neoplastic cells. Immunophenotyping shows positive CD2 and CD56. Although surface CD3 is negative, cytoplasmic CD3ε is positive. Cytotoxic granules (e.g., granzyme, perforin, TIA1: T-cell restricted intracellular antigen) are present. Because extranodal NK-/T-cell lymphoma, nasal type is closely associated with EBV, in situ hybridization staining for EBV encoded RNA (EBER) is strongly positive in the neoplastic cells.

Intestinal T-cell lymphoma

The 2017 WHO update introduces this new category that includes the two lymphomas formerly known as enteropathy-associated T-cell lymphoma (EATL) and two new entities.

EATL: The previous category of EATL type 1 is now its own diagnosis, designated simply EATL. EATL is associated with celiac disease and the *HLA-DQA1*0501* and *HLA-DQB1*0201* genotypes, and it is more common in Northern Europeans. It often involves the jejunum or ileum and commonly presents as small, multifocal ulcerating mucosal nodules. The mesentery and mesenteric lymph nodes are often involved. The tumor cells are polymorphic, medium- to large-sized with or without pleomorphism and admixed with inflammatory cells. The cells often extend from the mucosa through the wall into the mesentery (transmural infiltrate). By immunophenotype, EATL is positive for CD3, CD7, and cytotoxic markers and more often expresses αβ T-cell receptor. CD30 is often positive when the lymphoma cells are of larger size. EATL is the most common subtype of the intestinal T-cell lymphomas.

Monomorphic epitheliotropic intestinal T-cell lymphoma (MEITL): This is the entity formerly known as EATL type II or the monomorphic variant of EATL. It has monomorphic medium-sized tumor cells with no inflammatory component. MEITL is not associated with celiac disease. By immunophenotype, MEITL is positive for CD3, CD8, CD56, and cytotoxic markers and more often expresses γδ T-cell receptor. Most cases are also positive for MATK (megakaryocyte-associated tyrosine kinase).

Intraepithelial lesions in the small intestine are typically seen in both EATL and MEITL. T-cell lymphoma involving the intestines, but which cannot be categorized as EATL or MEITL, can be designated intestinal T-cell, NOS.

Lastly, indolent T-cell lymphoproliferative disorder of the gastrointestinal tract is a clonal expansion of T-cells that can affect the small intestine or colon but follows an indolent clinical course. There is involvement of the lamina propria but typically no destruction of glandular epithelium. The immunophenotype is usually CD3, CD8, and TIA1 positive. As might be expected by the indolent course, the proliferation index by Ki-67 is very low. All reported cases so far have been on alpha beta T-cell origin. Despite the indolent course, patients usually do not respond well to chemotherapy and experience multiple relapses with a subset of patients progressing to higher grade lymphoma.

Hepatosplenic T-cell lymphoma

Hepatosplenic T-cell lymphoma (HSTSL) is a T-cell neoplasm derived from gamma delta cytotoxic T-cells. Typically, the tumor cells are medium sized with minimal cytologic atypia and exhibit sinusoidal infiltration of the liver, spleen, and bone marrow. Splenomegaly is usually marked, and the lymphoma involves the splenic red pulp. Lymphadenopathy is absent. HSTCL has several unique features. It often affects adolescent and young adult males (median age about 35 years). Individuals who are chronically immunosuppressed (e.g., solid organ transplant patients, patients on azathioprine and infliximab for Crohn's disease) are at risk of developing this type of lymphoma. Immunophenotypically, HSTCL is usually CD2+, CD3+, CD4−, CD5−, CD7+/−, CD8−/+, CD56+, and TCRγδ+. They are positive for TIA1 and granzyme M but usually negative for perforin and granzyme B. Isochromosome 7q is detected in most cases. Less commonly, additional copies of isochromosome 7q (or 7q) or a ring chromosome 7 may be seen. The prognosis is generally poor.

Subcutaneous panniculitis-like T-cell lymphoma

Subcutaneous panniculitis-like T-cell lymphoma (SPTCL) is a T-cell lymphoma that infiltrates the subcutaneous tissue. The dermis and epidermis are not involved. The lymphoma cells are CD3+, CD8+, and αβ+ cytotoxic T-cells. Cytotoxic markers such as TIA-1, perforin, and granzyme are positive. Cases expressing γδ T-cell receptor are excluded and are instead classified as γδ T-cell lymphoma [10]. This distinction is important because the two diagnoses carry very different prognoses and are managed differently. The tumor cells exhibit a range of size, but in any particular case, tumor cell size tends to be constant. They also tend to have a pale rim of cytoplasm. Tumor cells characteristically surround or rim individual fat cells (adipocyte rimming), but this is not a specific finding. Admixed with the tumor cells are histiocytes. However, other inflammatory cells such as plasma cells are typically relatively few to absent. Lupus-associated panniculitis (LEP), also known as lupus profundus, shares many overlapping clinical and morphologic features with SPTCL, and the distinction can be quite challenging; LEP tends to have more associated plasma cells than SPTCL, but there is still substantial overlap. SPTCL has an indolent clinical course.

EBV positive T-cell and NK-cell lymphoproliferative diseases of childhood

The following two entities are typically seen in the pediatric age group and are more common in people from Asia and Native Americans from Central and South America and Mexico.

Systemic EBV-positive T-cell lymphoma of childhood is an aggressive and life-threatening disease of clonal EBV infected cytotoxic T-cells in the setting of acute EBV infection or chronic active EBV infection (CAEBV). It is rapidly progressive over days to weeks, most often involves the liver and spleen (can also involve the lymph nodes, bone marrow, skin, lungs), and can result in multiorgan failure, hepatosplenomegaly, pancytopenia, sepsis, and death. It is almost always associated with hemophagocytic syndrome. Microscopically, the T-cells are typically small without significant atypia. There is sinusoidal infiltration in the liver and spleen with depletion of the splenic white pulp. Hemophagocytosis is often prominent. Immunophenotypically, the T-cells are CD2+, CD3+, CD56-, and TIA1+. Cases in the setting of acute EBV infection are typically CD8+ whereas cases in the setting of CAEBV infection are CD4+. EBER staining is positive. The T-cells have monoclonal rearrangement of the T-cell receptor. EBV type A has been found in all cases.

Chronic active EBV infection of T- and NK-cell type, systemic form is a systemic EBV positive lymphoproliferative disorder, usually involving the liver, spleen, lymph nodes, bone marrow and/or skin. The diagnostic criteria include: infectious mononucleosis-like symptoms lasting >3 months, increased EBV DNA in the peripheral blood, histological evidence of organ disease, and positivity for EBV in an affected tissue in a non-immunocompromised host. Microscopic examination does not typically show features of a neoplastic process and the cells involved are usually phenotypically normal T-cells (typically CD4 positive) and NK-cells. As expected, EBV encoded RNA (EBER) stain is positive. However, cases which are monoclonal and monomorphic are considered to be overt lymphoma. The clinical course is variable, ranging from indolent to rapidly progressive.

Cutaneous T-cell lymphoma

Cutaneous T-cell lymphoma is characterized by localization of neoplastic T-lymphocytes in the skin. The most common types of cutaneous T-cell lymphomas are mycosis fungoides (MF) and its leukemic variant Sézary syndrome. The incidence of cutaneous T-cell lymphoma in the United States is 7.7 cases per million people per year, whereas combined incidence of MF and Sézary syndrome is 6.4 cases per million persons per year.

Mycosis fungoides

MF is the most common primary cutaneous lymphoma, accounting for approximately 50% of cases. MF is typically confined to skin. Skin lesions include a progression of patches, plaques, and tumors. The diagnosis is highly dependent on the characteristic clinical features and is very much a clinicopathologic diagnosis. Most patients present with limited plaque stage disease. The disease is characterized by epidermal and later on dermal infiltrate of small- to medium-sized T-cells with cerebriform nuclei (resembling the gyri of the brain). In the patch stage, there is a superficial band-like infiltrate in the basal layer epidermis (epidermotropism). The plaque stage has more prominent epidermotropism and Pautrier microabscesses (intraepidermal collection of atypical lymphocytes). The tumor stage (late lesions) exhibits dense dermal infiltrate, and epidermotropism may no longer be apparent. Immunophenotyping shows a mature T-helper cell immunophenotype, positive CD2, CD3, CD4, CD5, and TCR beta. CD7 is frequently negative and CD8 is also negative. Decreased or absence of CD7 and CD26 are useful aberrancies identified by flow cytometry. In the late lesions, large cell transformation may occur (defined as >25% large cells), and these large cells may be CD30 positive. Prognosis is typically excellent for limited disease confined to the skin but worsens significantly if there is extensive skin involvement or disseminated disease.

Sézary syndrome

This is a triad of erythroderma, generalized lymphadenopathy and Sézary cells (neoplastic T-cells with cerebriform nuclei) in the peripheral blood, skin, and lymph nodes. Smaller abnormal cells are referred to as Lutzner cells. At least one of the following criteria for Sézary syndrome is also required:

- 1000 Sézary cells/mm^3 (per microliter)
- CD4:CD8 >10
- Loss of one or more T-cell antigens

Erythroderma refers to diffusely red skin that is often pruritic with exfoliation that may also be accompanied by palmoplantar keratosis. Pruritus is a common manifestation in both MF and SS. It is almost uniformly present in SS, can be severe or incapacitating, and may not respond well to therapy [11]. SS tends to behave aggressively and carries a poor prognosis.

Primary cutaneous CD30 positive T-cell lymphoproliferative disease

This category comprises the second most common group of cutaneous T-cell lymphomas (~30%) and includes:

- Primary cutaneous ALCL (PCALCL)
- Lymphomatoid papulosis (LyP); papules; and spontaneous regression can progress to lymphoma

PCALCL typically presents with a solitary skin lesion (unlike LyP) that can grow rapidly. However, the prognosis is still excellent. There should be no evidence or history of MF.

PCALCL is composed of sheets of large anaplastic T-cells with CD30 expression in >75% of the tumor cells. The morphology of the cells is similar to those seen in systemic ALCL.

The T-cell phenotype is similar to that of systemic ALK-negative ALCL: CD30+ and CD4+ with cytotoxic marker expression and variable loss of CD2, CD3, or CD5. Cutaneous lymphocyte antigen (CLA) is positive (unlike systemic ALCL that is negative for CLA). ALK and EMA are negative (unlike systemic ALK-positive ALCL). PCALCL lacks the characteristic t(2;5)(p23;q35) of systemic ALK-positive ALCL. Rearrangement of the *DUSP22-IRF4* locus on chromosome 6p25.3 is the most common recurrent genetic abnormality in PCALCL, seen in approximately 30% of cases. *DUSP22* rearrangements have also rarely been reported in LyP (see below). A recurrent *NPM1-TYK2* gene fusion has also been reported in the CD30 positive lymphoproliferative disorders (both PCALCL and LyP) and leads to constitutive downstream STAT signaling [12].

LyP is a chronic, spontaneously regressing, and recurring disease with a highly variable number of skin lesions (but usually many) at different stages of development predominantly affecting the trunk and extremities. It has an excellent prognosis. LyP is composed of large cells and may appear anaplastic, immunoblastic, or Hodgkin-like. There is a marked accompaniment of inflammatory cells in the background. There are currently at least six subtypes recognized:

- Type A: Most common; wedge-shaped dermal infiltrate; scattered large multinucleated or Reed—Sternberg-like cells (CD3+, CD4+, CD8−, and CD30+); numerous inflammatory cells (e.g., histiocytes, small lymphocytes, neutrophils, eosinophils).
- Type B: Cerebriform cells in epidermis (resembling MF); few inflammatory cells. The tumor cells are CD3+, CD4+, and CD8− and can be either CD30 positive or negative.

- Type C: Nodular infiltrate composed of sheets of large CD30+, CD3+, CD4+, and T-cells with few inflammatory cells (resembling PCALCL).
- Type D: Epidermotropic but CD8+ with cytotoxic marker expression.
- Type E: Eschar-like ulcers; angiocentric and angiodestructive with necrosis; also CD8+ with cytotoxic marker expression.
- LyP with 6p25.3 (*DUSP22-IRF4*) rearrangement: Typically has a biphasic growth pattern with a dense dermal nodule or infiltrate and extensive atypical lymphoid infiltrates in the epidermis [13].

Primary cutaneous peripheral T-cell lymphomas, rare subtypes

Primary cutaneous CD4+ small/medium T-cell lymphoproliferative disorder: Typically presents as a solitary nodule in the head and neck region and follows an indolent clinical course. The lesion consists of a dense or nodular dermal infiltrate composed of small- to medium-sized pleomorphic lymphocytes admixed with occasional larger cells but with minimal to no epidermotropism. The immunophenotype is typically CD3+, CD4+, CD8−, CD30−, EBV−, and negative for cytotoxic markers. The cells also express the T-cell follicular helper cell markers PD1, BCL6 (variable), and CXCL13.

Primary cutaneous CD8+ aggressive epidermotropic cytotoxic T-cell lymphoma: Very rare (<1%) provisional entity with an aggressive clinical course. Often presents with ulcerated nodules, tumors, plaques, or papules as well as widespread dissemination at diagnosis. There is marked epidermotropism by small- to medium-sized atypical lymphocytes and occasional larger pleomorphic cells as well as epidermal necrosis. There is often angiocentricity and angioinvasion. The immunophenotype is usually CD3+, CD4−, and CD8+ with variable loss of CD2, CD5, and CD7 and negative for CD30, CD56, and EBV. They express TCRαβ and cytotoxic markers.

Primary cutaneous acral CD8+ T-cell lymphoma: This is a new addition to the 2017 WHO classification that preferentially involves acral sites, especially the ear. It has a male predominance and an indolent clinical course. The infiltrate is usually dermal and composed of medium-sized atypical lymphocytes. The immunophenotype is CD3+, CD4−, CD8+, TIA1+ (Golgi dot−like staining), and TCR αβ.

Primary cutaneous gamma delta T-cell lymphoma: As the name implies, it consists of γδ T-cells with a cytotoxic phenotype. The category also now includes cases of SPTCL with a γδ phenotype. Cases that otherwise clinically resemble MF but with a γδ phenotype should still be classified as MF. There can be three patterns of involvement, epidermotropic, dermal, or subcutaneous, but more than one can be present at the same time. The immunophenotype is γδ TCR+, αβ TCR−, CD2+, CD3+, CD5−, CD7+/−, CD56+, and strong cytotoxic marker expression. Most cases are negative for both CD4 and CD8, but some may express CD8. As expected, there is clonal rearrangement of the TRG and TRD genes, and some cases have *STAT5B* mutations. The prognosis is typically poor.

EBV positive T-cell and NK-cell lymphoproliferative diseases of childhood

The following two entities are under the same category as systemic EBV-positive T-cell lymphoma of childhood and chronic active EBV infection of T- and NK-cell type, systemic form described earlier but with primarily cutaneous manifestations.

Hydroa vacciniforme-like lymphoproliferative disorder is a chronic EBV positive lymphoproliferative disorder of cytotoxic (CD8+) T-cells and NK cells in childhood involving the skin with an increased risk of developing systemic lymphoma in the future. The EBV positive T-cells often also express CD30. Patients typically have recurring skin lesions over many years with a subset of patients progressing to systemic involvement, after which the clinical course is aggressive.

Severe mosquito bite allergy is a very uncommon NK cell lymphoproliferative disorder manifesting as fever and severe local skin symptoms. Patients may develop hemophagocytic syndrome or progress to NK cell lymphoma or leukemia. There is usually an angioinvasive and angiodestructive dense dermal infiltrate of small lymphocytes, larger atypical cells, eosinophils and histiocytes. The cells have an NK cell phenotype, and the EBV positive cells are also often CD30 positive. After recovery, patients are typically asymptomatic until the next episode.

Leukemia/disseminated

Mature T-cell leukemias are clonal proliferation of postthymic T-cells that often exhibit systematic manifestation or involvement of extramedullary sites in conjunction with hematological abnormalities. Some types of mature T-cell lymphomas, for example, adult T-cell leukemia/lymphoma, demonstrate particular epidemiological features such as such as association with human T-cell leukemia/lymphoma virus type 1 (HTLV-1).

T-cell prolymphocytic leukemia

T-cell prolymphocytic leukemia (T-PLL) is an aggressive T-cell leukemia with presence of leukemic cells in peripheral blood, bone marrow, liver, spleen, and sometimes skin. There is often marked leukocytosis, often $>100 \times 10^9$/L. The bone marrow is typically effaced by sheets of atypical lymphocytes. The cells are typically small to medium, with basophilic cytoplasm and a prominent single nucleolus. They typically also have cytoplasmic protrusions or blebs. Immunophenotyping indicates positive CD2, cytoplasmic CD3, surface CD3 ($\pm$), CD5, CD7 (frequently bright), CD26 (therapeutic target), CD52 (therapeutic target with drugs such as with alemtuzumab), and TCL1 but negative for CD1a, CD56, CD57, and TdT (terminal deoxynucleotide transferase). Most cases are CD4+ and CD8−, but some are double positive for CD4+ and CD8+. The majority of cases are positive for TCR α/β.

The small cell variant of T-PLL has leukemic cells, which are small, lack prominent nucleoli, and morphologically resemble CLL/SLL, but has the characteristic phenotype and genetic abnormalities of conventional T-PLL. It does not appear to predict clinical behavior [14,15].

Chromosomal abnormalities in the form of inv(14)(q11q32) occurs in up to 80% of cases and involves juxtaposition of *TCL1A* and *TCL1B*, which results in constitutive activation and overexpression of the TCL1 protein. Abnormalities in chromosome 8, including trisomy 8q, t(8;8)(p11-12; q12) and idic(8)(p11), are also very common and occur in 70%−80% of cases. Missense mutations in ataxia—telangiectasia mutated (*ATM*) are other recurring genetic abnormalities. FISH can also often detect deletions in the *ATM* locus (11q23).

T-cell large granular lymphocyte leukemia

T-cell large granular lymphocyte leukemia (T-cell LGL) is an indolent leukemia characterized by >2000 LGL cells/mm^3 (per microliter) of blood, for 6 months or more. T-LGL is usually seen in adults, typically 50−60 years old, and follows an indolent clinical course. It may be accompanied by cytopenia (neutropenia or anemia, most often), bone marrow involvement, and mild splenomegaly. T-LGL is frequently associated with autoimmune diseases (especially rheumatoid arthritis or Felty's syndrome), autoantibodies, circulating immune complexes, and hypergammaglobulinemia. Felty's syndrome is the combined presence of rheumatoid arthritis, splenomegaly, and neutropenia; patients with T-LGL and rheumatoid arthritis or Felty's syndrome often have HLA-DR4.

The LGLs show characteristic morphology: abundant basophilic cytoplasm containing large azurophilic (red-pink) granules, round to slightly irregular nuclei, and condensed chromatin without nucleoli. Electron microscopy shows that the azurophilic granules are composed of numerous microtubules, known as parallel tubular arrays.

Immunophenotyping shows a cytotoxic T-cell phenotype: positive for CD2, CD3, CD8, CD16, CD57, CD94, and TCR alpha-beta along with positive for cytotoxic effector proteins (TIA1, granzyme B and M, cytotoxic granule–associated RNA-binding protein). Aberrant loss of CD5 and/or CD7 is common. T-LGL is typically negative for CD56 in contrast to CLD-NK (see below).

STAT3 (usually Y640 or D661) mutations have been documented in approximately one-third of cases. The *STAT5B* N642H mutation is much rarer but appears to be associated with more aggressive disease.

It is also important to be aware that oligoclonal LGL proliferations can occur in the setting of poststem cell transplant and may represent lymphocyte reconstitution. However, T-LGL leukemia has also rarely been reported as a posttransplant lymphoproliferative disorder. Immunophenotypic aberrancies and the presence of mutations can assist in this difficult distinction.

Chronic lymphoproliferative disorders of natural killer cells

This is an indolent leukemia characterized by >2000 NK cells/mm^3 of blood, for 6 months or more. Unlike aggressive NK-cell leukemia, this is not EBV driven or may exhibit racial or genetic predisposition. Patients may be asymptomatic and may exhibit features of cytopenia or organomegaly (lymphadenopathy, hepatomegaly, and splenomegaly).

Immunophenotyping indicates CD2+, surface CD3−, cCD3ε+, CD8+, CD16+, CD56+, and CD57−, and is also positive for cytotoxic effector proteins (TIA1, granzyme B and M). There is abnormal expression of the killer cell immunoglobulin-like receptor (KIR) family, either lack of KIR or restriction to one isoform. Uniform, bright CD94 or decreased CD161 are other immunophenotypic abnormalities that can be seen.

Karyotype is normal in most cases. Mutations in *STAT3* SH2 domain are seen in about one-third of cases and result in constitutive activation. Because this is a proliferation of NK cells, no clonal rearrangements are detected in the TCR genes (TCR genes are in the germline configuration).

Aggressive natural killer−cell leukemia

This is a highly aggressive leukemia seen more often in young Asians and associated with EBV infection (>90% of cases). Hepatosplenomegaly, fever, and cytopenias are frequently present. Aggressive NK-cell leukemia has a similar immunophenotype as extranodal NK/T-cell lymphoma, nasal type except that CD16 is frequently positive in aggressive NK-cell leukemia. Immunotyping shows CD2+, surface CD3−, cCD3ε+, CD16+, CD56+, and CD57− as well as presence of cytotoxic molecules. TCR genes are in the germline configuration.

Adult T-cell leukemia/lymphoma

Adult T-cell leukemia/lymphoma (ATCL) is a T-cell neoplasm caused by HTLV-1 infection and characterized by the presence of highly pleomorphic cells. HTLV-1 is a retrovirus and, similar to HIV, infects CD4+ T-cells and may result in T-cell immunodeficiency and opportunistic infections (such as *Pneumocystis jiroveci*). HTLV-1 also has similar modes of transmission to HIV (such as intravenous drug use, sexual transmission, breastfeeding, and blood transfusions). Serologic detection of antibodies against HTLV-1 supports the diagnosis. This disease is endemic in southwestern Japan, Caribbean basin, and parts of central Africa. It is seen only in adults with an average age of presentation of 58 years and typically presents with disseminated disease including widespread lymphadenopathy and peripheral blood involvement. However, the bone marrow is frequently negative or only minimally involved.

The leukemic cells are medium to large cells with irregular nuclei and basophilic cytoplasm. There may be many nuclear convolutions and lobules, referred to as flower cells. Immunophenotyping shows that the neoplastic T-cells are typically positive for CD2, CD3, CD4, and CD5 but often lack CD7. Occasional cases are double positive for CD4 and CD8. CD25 is strongly positive in almost all cases and is the most specific immunohistochemical finding to differentiate from other T-cell lymphomas. Because the neoplastic cells are believed to be derived from regulatory T-cells, CCR4 and FOXP3 are also frequently expressed.

The skin is the most frequently involved site of extranodal involvement (approximately 50% of patients). Hypercalcemia with lytic bone lesions is a common and characteristic finding (see variants below). Serum lactate dehydrogenase (LDH) is also often elevated. ATCL carries a poor prognosis.

There are several clinical variants of this disease:

- Acute: most common; leukocytosis, "flower cells," lymphadenopathy, skin lesions, hypercalcemia, lytic bone lesions, widespread disease dissemination, extranodal disease, elevated LDH.
- Lymphomatous: lymphadenopathy without leukemic cells in blood.
- Chronic: exfoliative skin rash, leukemic cells in PB (usually not many) and without hypercalcemia.
- Smoldering: normal white blood cells with >5% circulating leukemic cells with small and normal appearance. Frequently have skin lesions. No hypercalcemia is present.

Key points

- AITL is strongly associated with EBV; however, the tumor cells are EBV negative. There is a polymorphous infiltrate of small- to medium-sized lymphocytes. The tumor cells may be admixed with other cells such as small lymphocytes, plasma cells, eosinophils, and histiocytes. There is increase in HEV, and there is also proliferation of follicular dendritic cells (CD21+). By immunohistochemistry, AITL is positive for CD2, CD3, CD4, and CD5. AITL is thought to arise from TFH cells within the germinal centers; therefore, CD10, Bcl-6, CXCL13, ICOS, and PD1 may be positive.
- FTCL: Neoplasm of TFH cells with a predominantly follicular growth pattern but lacking the characteristic features of AITL (such as HEVs and proliferations of follicular dendritic cells). Approximately 20% of cases harbor a t(5;9)(q33;q22) *ITK-SYK*, which is relatively specific for FTCL and has not been reported in other peripheral T-cell lymphomas.

- Nodal peripheral T-cell lymphoma with a TFH cell phenotype: CD4 positive peripheral T-cell lymphoma with a TFH phenotype. At least two, but preferably three, TFH makers (CD10, BCL6, CXCL13, ICOS, PD1) are required for the diagnosis. It has a diffuse growth pattern but lacks the polymorphous background seen in AITL.
- Peripheral T-cell lymphoma, not otherwise specified is a heterogeneous category of T-cell lymphomas, which do not meet criteria for any other specific category. They exhibit a broad cytological spectrum with nodal and extranodal distribution (bone marrow, peripheral blood, liver, spleen, skin). By immunohistochemistry, T-cell markers are positive with frequent downregulation of CD5 and CD7. Most often CD4 is positive but CD8 is negative.
- ALK-positive ALCL is a T-cell neoplasm characterized by large anaplastic cells with expression of CD30 and ALK protein. Hallmark cells are large cells with horseshoe nucleus/reniform. The patterns of ALCL are common pattern, lymphohistiocytic pattern, Hodgkin-like pattern, and composite pattern. The t(2;5)(p23;q35) results in a fusion gene involving *NPM* (5q35) and *ALK* (2p23), and immunohistochemical staining for ALK protein is both cytoplasmic and nuclear. With variant translocations, staining for ALK is cytoplasmic or membranous. The majority of cases are EMA+, CD4+, and CD8−. ALK-positive ALCL is more common in children and young adults.
- ALK-negative ALCL morphologically resembles ALK-positive ALCL and is also CD30 positive, but, by definition, lacks ALK expression. This tumor is seen more in adults. ALK-negative ALCL with rearrangements of *DUSP22* (6p25.3) has a good prognosis similar to that of ALK-positive ALCL. However, ALK-negative ALCL with rearrangements of *TP63* has a very poor prognosis.
- Breast implant—associated ALCL is uniquely associated with breast implants and usually presents with a unilateral peri-implant effusion approximately 9−11 years after implant placement. It is morphologically and immunophenotypically similar to ALK-negative ALCL. In general, it is slow growing, follows an indolent clinical course, and has an extremely good prognosis following complete capsulectomy (surgical excision of the implant and entire surrounding fibrous capsule).
- Extranodal NK-/T-cell lymphoma, nasal type is an extranodal lymphoma where vascular damage occurs along with consequent necrosis, and this disorder has a very strong association with EBV. Nasal cavities and adjacent areas (nasopharynx, paranasal sinuses) are the most frequent sites of involvement. In this disorder, a diffuse infiltrate with angiocentric and angiodestructive pattern is usually observed. Immunohistochemistry analysis shows CD2+, cytoplasmic CD3ε, CD56+, and EBV+ with cytotoxic marker (e.g., granzyme, perforin, TIA1) expression but negative surface.
- EATL is associated with celiac disease and is composed of medium to large cells with or without pleomorphism and admixed inflammatory cells. The lymphoma cells are positive for CD3 and cytotoxic markers and more often expresses alpha beta T-cell receptor.
- MEITL has monomorphic lymphoma cells with no inflammatory component. MEITL is not associated with celiac disease. MEITL is positive for CD3, CD8, CD56, and cytotoxic markers and more often express gamma delta T-cell receptor. Most cases are also positive for MATK.
- HTSL is derived from cytotoxic T-cells of the gamma delta T-cell receptor type, which exhibit sinusoidal infiltration of liver, spleen, and bone marrow. HSTL is more common in young adults and in individuals who are chronically immunosuppressed (e.g., solid organ transplant patients, patients on azathioprine, and infliximab for Crohn's disease). There is an also an association with isochromosome 7q.

- SPTCL infiltrates the subcutaneous tissue. The dermis and epidermis are not involved. Tumor cells rim individual fat cells. The lymphoma cells are cytotoxic CD8+ and TCRαβ+. Cases expressing γδ T-cell receptor are excluded and instead classified as γδ T-cell lymphoma.
- MF accounts for 50% of all primary cutaneous lymphomas. MF has a characteristic set of three progressive stages: patch, plaque, and tumor. The lymphoma cells themselves often have cerebriform nuclei. The patch stage is the earliest and is predominantly confined to the superficial epidermis. The plaque stage is next and has more pronounced epidermotropism and a dermal infiltrate and Pautrier microabscesses. The tumor stage is last and has a dense dermal infiltrate, but epidermotropism may no longer be apparent. Large cell transformation (>25% large cells which may be CD30+) can occur late in MF. Immunohistochemistry analysis shows CD2+, CD3+, CD5+, CD7− (frequently), CD4+, and CD8−.
- Sézary Syndrome is a triad of erythroderma, generalized lymphadenopathy, and Sézary cells (neoplastic T-cells with cerebriform nuclei) in peripheral blood, skin, and lymph nodes. Smaller abnormal cells are referred to as Lutzner cells.
- Primary cutaneous CD30+ T-cell lymphoproliferative disorders are the second most common group of cutaneous T-cell lymphomas and include PCALCL and LyP.
- PCALCL is usually a solitary lesion composed of sheets of large anaplastic CD30+ (>75%) T-cells with variable loss of CD2, CD5, or CD3. CLA is positive, unlike systemic ALCL, which is negative for CLA. ALK and EMA are negative, unlike systemic ALK-positive ALCL.
- LyP is a chronic, self-resolving, and recurring disease usually with many skin lesions. The cells are large and may appear anaplastic, immunoblastic, or Hodgkin like. There is a marked inflammatory background. There are multiple different subtypes of LyP, with Type A being the most common.
- T-PLL: Aggressive T-cell leukemia with presence of leukemic cells in peripheral blood, bone marrow, liver, spleen, and sometimes skin. Immunophenotyping indicates positive CD2, CD3, CD4, CD7, CD52 (potential target for therapy), and TCL1, but TdT (terminal deoxynucleotide transferase) and CD1a are negative. Inv(14)(q11q32), causing juxtaposition of *TCL1A* and *TCL1B* and constitutive activation, occurs in up to 80% of cases. Abnormalities in chromosome 8 and mutations in *ATM* are also frequent.
- T-cell LGL: indolent, >2000 LGL cells/mm^3 of blood, >6 months duration. This is associated with autoimmune diseases (especially rheumatoid arthritis) and hypergammaglobulinemia. Immunophenotype: CD3, CD8, CD16, CD57, and TCR alpha-beta and cytotoxic markers (TIA1, granzyme B and M). T-LGL is typically negative for CD56 in contrast to CLD-NK. *STAT3* mutations can be seen in approximately one-third of cases.
- Chronic lymphoproliferative disorders of NK cells: indolent, >2000 NK cells/mm^3 of blood, >6 months duration. Not EBV driven. Patients may be asymptomatic, exhibit features of cytopenia or organomegaly (lymphadenopathy, hepatomegaly, and splenomegaly). Immunophenotype: surface CD3−, cCD3ε+, CD16+, CD56+, and also positive for cytotoxic effector proteins (TIA1, granzyme B and M). Karyotype is normal in most cases. Mutations in *STAT3* SH2 domain are seen in about one-third of cases. TCR genes are in the germline configuration.
- Aggressive NK-cell leukemia: highly aggressive leukemia seen more often in young Asians and strongly associated with EBV infection. Immunotyping: CD2+, surface CD3−, cCD3ε+, CD16+, CD56+, CD57− and presence of cytotoxic molecules. CD16 is frequently positive in

aggressive NK-cell leukemia in contrast to extranodal NK-/T-cell lymphoma, nasal type, which otherwise has a nearly identical immunophenotype. TCR genes are in the germline configuration.

- ATCL: caused by HTLV-1 and characterized by the presence of highly pleomorphic cells, including polylobated "flower cells." Endemic in southwestern Japan, Caribbean basin, and parts of central Africa. The leukemic cells are positive for CD25. Hypercalcemia and lytic bone lesions can be seen.

References

[1] Tang T, Tav K, Quek R, Tao M, et al. Peripheral T-cell lymphoma: review and updates of current management strategies. Adv Hematol 2010;2010:624040.

[2] De Leval L, Gaulard P. Pathology and biology of peripheral T-cell lymphomas. Histopathology 2011;58: 49–68.

[3] Warnke RA, Jones D, His ED. Morphological and immunophenotypic variants on nodal T-cell lymphomas and T-cell lymphoma mimics. Am J Clin Pathol 2007;127:511–27.

[4] Willenbrock K, Bräuninger A, Hansmann ML. Frequent occurrence of B-cell lymphomas in angioimmunoblastic T-cell lymphoma and proliferation of Epstein-Barr virus-infected cells in early cases. Br J Haematol 2007;138(6):733–9.

[5] Drexler HG, Gignac SM, von Wasielewski R, Werner M, et al. Pathobiology of NPM-ALK and variant fusion genes in anaplastic large cell lymphoma and other lymphoma. Leukemia 2000;14:1533–59.

[6] Parrilla Castellar ER, Jaffe ES, Said JW, Swerdlow SH, et al. ALK-negative anaplastic large cell lymphoma is a genetically heterogeneous disease with widely disparate clinical outcomes. Blood August 28, 2014; 124(9):1473–80.

[7] Quesada AE, Medeiros LJ, Clemens MW, Ferrufino-Schmidt MC, Pina-Oviedo S, Miranda RN. Breast implant-associated anaplastic large cell lymphoma: a review. Mod Pathol 2019;32(2):166–88.

[8] Miranda RN, Aladily TN, Prince HM, Kanagal-Shamanna R, et al. Breast implant-associated anaplastic large-cell lymphoma: long-term follow-up of 60 patients. J Clin Oncol 2014;32(2):114–20.

[9] Ferrufino-Schmidt MC, Medeiros LJ, Liu H, et al. Clinicopathologic features and prognostic impact of lymph node involvement in patients with breast implant-associated anaplastic large cell lymphoma. Am J Surg Pathol 2018;42(3):293–305.

[10] Willemze R, Jansen PM, Cerroni L, Berti E, et al. EORTC Cutaneous Lymphoma Group. Subcutaneous panniculitis-like T-cell lymphoma: definition, classification, and prognostic factors: an EORTC Cutaneous Lymphoma Group Study of 83 cases. Blood 2008;111(2):838–45.

[11] Vij A, Duvic M. Prevalence and severity of pruritus in cutaneous T-cell lymphoma. Int J Dermatol 2012;51: 930–4.

[12] Velusamy T, Kiel MJ, Sahasrabuddhe AA, Rolland D, et al. A novel recurrent NPM1-TYK2 gene fusion in cutaneous CD30-positive lymphoproliferative disorders. Blood 2014;124(25):3768–71.

[13] Karai LJ, Kadin ME, Hsi ED, Sluzevich JC, et al. Chromosomal rearrangements of 6p25.3 define a new subtype of lymphomatoid papulosis. Am J Surg Pathol 2013;37(8):1173–81.

[14] Rashidi A, Fisher SI. T-cell chronic lymphocytic leukemia or small-cell variant of T-cell prolymphocytic leukemia: a historical perspective and search for consensus. Eur J Haematol 2015;95(3):199–210.

[15] Collignon A, Wanquet A, Maitre E, Cornet E, et al. Prolymphocytic leukemia: new insights in diagnosis and in treatment. Curr Oncol Rep 2017;19(4):29.

Hodgkin lymphoma

14

Introduction

Hodgkin's type lymphoma was first recognized in 1832 by Thomas Hodgkin when he described postmortem findings of in seven patients with enlarged lymph nodes and spleen. In 1865, Samuel Wilks confirmed Thomas Hodgkin's findings in 15 additional patients and named this lymphoma as Hodgkin disease [1]. Later this lymphoma was classified as Hodgkin lymphoma, a hematolymphoid neoplasm primarily of B-cell lineage with unique histological, immunophenotypic, and clinical features. Symptoms include painless enlargement of lymph nodes, spleen, or other immune tissues. Other nonspecific symptoms such as fever, night sweats, weight loss, low appetite, itchy skin, and fatigue may also be present. American Cancer Society estimates that 9190 new cases of Hodgkin lymphoma are expected to be diagnosed in 2014, and an estimated 1180 deaths may occur. Again this is a rare disease with expected number of new cases per year being 2.7 per 100,000 people. The death rate is 0.4 per 100,000 people. The age distribution of this disease is bimodal with first peek occurring between ages of 15−30 years and the second peak in the sixth decade of life. Treatment is based on disease stage, and prognostic factors and therapy is usually multiagent chemotherapy most often using doxorubicin, bleomycin, vinblastine and dacarbazine, or radiotherapy or combination of chemotherapy and radiotherapy.

Overview of Hodgkin lymphoma

Hodgkin lymphoma represents approximately 30% of all lymphomas. Over the years, the absolute incidence remains unchanged. In general, Hodgkin lymphoma arises in lymph nodes, most often the cervical region, spreading to contiguous lymph nodes. Young adults are most often affected. There is a childhood form of Hodgkin lymphoma (0−14 years), which is seen more often in developing countries. There is a male preponderance of Hodgkin lymphoma (male to female ratio of 1.5:1), but this male preponderance is however not seen in nodular sclerosis subtype.

Neoplastic tissues usually contain a small number of tumor cells. Hodgkin lymphoma is characterized by a small number of scattered tumor cells residing in an abundant heterogeneous admixture of nonneoplastic inflammatory and accessory cells. The tumor cells produce cytokines which are responsible for the presence of the background cells, lymphocytes, histiocytes, plasma cells, eosinophils, and neutrophils.

Hematology and Coagulation. https://doi.org/10.1016/B978-0-12-814964-5.00014-0

Classification of Hodgkin lymphoma

Rye classification (1966) was the first classification of Hodgkin lymphoma. This classification includes nodular sclerosis, lymphocyte predominant, mixed cellularity, and lymphocyte depleted form. However, present classification is based on WHO (World Health Organization) guideline where Hodgkin lymphoma is classified into two major types; nodular lymphocyte predominant Hodgkin lymphoma (NLPHL) (about 5% of all cases) and classic Hodgkin lymphoma (CHL) (about 95% cases which can be further subclassified into nodular sclerosis, mixed cellularity, lymphocyte rich, and lymphocyte-depleted type). Both classifications are summarized in Table 14.1.

Various neoplastic cells are observed in Hodgkin lymphoma. The neoplastic cells seen in CHL include

- Reed—Sternberg cell (RS cells): This is a large cell (20—50 μm) with abundant cytoplasm and two mirror image nuclei, each with an eosinophilic nucleolus. The nuclear membrane is thick with chromatin being distributed close to the nuclear membrane. Carl Sternberg (from Germany) provided the first detailed description of these cells in 1898, and in 1902, Dorothy Reed also described these cells independently.
- Mononuclear Hodgkin cell (Hodgkin cell): This cell has same features as RS cell but has only one nucleus.
- Lacunar cell: This cell has cytoplasm that is retracted around the nucleus, creating an empty space. The nucleus is single and hyperlobulated.
- Mummy cell: This cell contains basophilic cytoplasm and a compact nucleus but without the presence of a nucleolus.

In NLPHL, "Popcorn" cells (LP cells formerly known as lymphohistiocytic cells or L&H cells) are observed [2]. These are large cells with hyperlobulated nucleus.

In Hodgkin lymphoma, microscopically, nodules may be seen in the following situations:

- NLPHL
- Nodular sclerosis classical Hodgkin lymphoma
- Lymphocyte-rich classical Hodgkin lymphoma

Table 14.1 Rye and World Health Organization (WHO) classification of Hodgkin lymphoma.

Rye classification (1966 classification)	WHO classification
Nodular sclerosis Lymphocyte predominant Mixed cellularity Lymphocyte depleted	Nodular lymphocyte predominant Hodgkin lymphoma (5% of Hodgkin lymphoma) Classical Hodgkin lymphoma (95% of Hodgkin lymphoma) which can be subclassified into four groups as follow: • Nodular sclerosis (70% of cases) • Mixed cellularity (20% of cases) • Lymphocyte rich (5% of cases) • Lymphocyte depleted (5% of cases)

Nodular lymphocyte predominant Hodgkin lymphoma

NLPHL consists of 5% of all reported cases of Hodgkin lymphoma. This disease is a predominately male disease with a 3:1 ratio of male to female in Caucasian and a 1.2:1 ratio in African individuals. In adults, median age of onset is 30–35 years [2]. Most patients present with localized peripheral lymphadenopathy, which develops slowly and is responsive to therapy. NLPHL tends to spare the mediastinum, spleen, or bone marrow. This lymphoma was considered to be analogous to "low-grade" B-cell lymphomas, but disseminated disease is not usually observed. In addition, NLPHL is typically negative for Epstein–Barr virus (EBV). Progressively transformed germinal centers are seen in association with NLPHL, but it is uncertain whether these lesions are preneoplastic or not. However, most patients with reactive hyperplasia and progressive transformation of germinal centers (PTGC) do not develop Hodgkin lymphoma. NLPHL may progress to large B-cell lymphoma in 2%–3% of cases.

Sites of involvement

Cervical, axillary, and inguinal lymph nodes are common sites of involvement.
Involvement of mediastinum is rare
Sometimes the Waldeyer's ring may be involved (not seen in CHL)

Histology
The architecture of NLPHL may be nodular or nodular and diffuse.

The nodules are composed of small lymphocytes, histiocytes (giving rise to a moth-eaten appearance), and tumor cells (LP cells).

Immunophenotype
The neoplastic cells, also known as popcorn cells, and the background cells are CD20 positive. CD20 also highlights the nodularity. The neoplastic cells are positive for CD45 and negative for CD15 and CD30.

In addition, the LP cells are also positive for CD79a, PAX5, Bcl6, J chain, BOB1, OCT2, MEF2B, and sometimes EMA.

The LP cells are surrounded by T cells (forming rosette) and are positive for CD3, CD4, CD57, and PD1.

The background cells are typically B cells and are predominantly composed of IgD-positive mantle zone B cells.

If the background lymphocytes are rich in T cells, clinical outcome is poorer.

With diffuse pattern, there may be confusion with T cell/histiocyte-rich large B-cell lymphoma (THRLBL). In THRLBL, the large tumorous B cells do not form aggregates. Rosette formation is not seen.

Classic Hodgkin lymphoma

This is 95% of Hodgkin lymphomas with a bimodal age distribution. EBV has been postulated to play a role in CHL. The prevalence of EBV in RS cells varies according to the histological subtype, highest in mixed cellularity (75%), and lowest in nodular sclerosis (10%–40%). Cervical lymph nodes are the most common area of involvement, and 60% patients have mediastinal involvement. However, bone marrow involvement is rare and observed in only 5% patients. The disease is characterized by the presence of RS cells in the appropriate cellular background. The neoplastic cells are B cells. However, they are usually not CD20 positive. They are weakly positive for PXA5, proving that they are B cells. The staining is characteristically weak. The background lymphocytes are T cells (CD20 negative).

Immunophenotype of CHL

The tumor cells are weakly positive for PAX5. Other B-cell markers such as CD20 and CD79a are typically negative.
The tumor cells are positive for CD15, CD30, and negative for CD45.
The tumor cells in CHL are also positive for MUM1 and fascin.
EBER positivity favors CHL.
In CHL, CD79a, PAX5, Bcl6, J chain, BOB1, OCT2, and MEF2B are negative in the tumor cells.
In CHL, the background lymphocytes are mostly T cells.

Nodular sclerosis classic Hodgkin lymphoma

The most common subtype of classical Hodgkin lymphoma is nodular sclerosis classical Hodgkin lymphoma, which in contrast to other forms of Hodgkin lymphoma occurs more often in young adults than elderly. This subtype is more frequently seen in developed countries (resource-rich countries) in patients who belong to high-socioeconomic group. This subtype is also less frequently associated with EBV [3].

Nodular sclerosis classical Hodgkin lymphoma accounts for approximately 70% of all classical Hodgkin lymphoma, and mediastinal involvement occurs in about 80% of cases.

Nodular sclerosis classical Hodgkin lymphoma is characterized by the presence of nodules and broad bands of sclerosis, and the lymph node capsule may also be thickened. The collagen bands surround at least one nodule. This disease shows a cellular phase and fibrotic phase. Lacunar cells are seen more often in this subtype which may form aggregates. These aggregates may be associated with necrosis and histiocytes. This may resemble necrotizing granulomas. There is also a syncytial variant that is an extreme form of the cellular phase. Nodular sclerosis classical Hodgkin lymphoma can be further subdivided into two types: NS1 (nodular sclerosis type I) and NS2.

NS2 is characterized by

>25% of the nodules show pleomorphic or reticular lymphocyte depletion or
>80% of the nodules show features of the fibrohistiocytic variant or
>25% of the nodules show numerous bizarre anaplastic appearing Hodgkin cells without lymphocyte depletion

Mixed cellularity classic Hodgkin lymphoma

Mixed cellularity subtype of classical Hodgkin lymphoma is more frequently seen in patients with HIV infection and also in developing countries. A bimodal age distribution is not seen in this type of lymphoma. Mixed cellularity subtype demonstrates an interfollicular growth pattern with presence of typical RS cells within an inflammatory background.

Lymphocyte rich classic Hodgkin lymphoma

Lymphocyte rich subtypes of classical Hodgkin lymphoma is characterized by nodular (common) and diffuse architecture. The nodules (which represent expanded mantle zones) are composed of small lymphocytes which may harbor germinal centers. The RS cells are found within the nodules but not in the germinal centers. Some of the HRS cells (Hodgkin and RS cells) may resemble LP cells and with the nodularity may resemble NLPHL. Intact germinal centers are infrequent in NLPHL. Thus presence of nodules with germinal centers and the difference in immunophenotype will help to resolve between the diagnosis of lymphocyte rich classical Hodgkin lymphoma and NLPHL.

Lymphocyte depleted classic Hodgkin lymphoma

Lymphocyte depleted classical Hodgkin lymphoma is the rarest subtype that is also observed more frequently in patients with HIV infection and also in developing countries. As the name implies, few background lymphocytes are seen and two subtypes are described, diffuse fibrosis (with a few HRS cells) and reticular (increased number of HRS cells) cells.

Major characteristics of these subtypes of classical Hodgkin lymphoma are listed in Table 14.2.

Immunostains for diagnosis of Hodgkin lymphoma

Immunological and molecular studies have shown that most HRS cells of classical Hodgkin lymphoma are derived from germinal center B cells with rearranged immunoglobin genes bearing

Table 14.2 Major characteristics of subtypes of classical Hodgkin lymphoma.

Nodular sclerosis	Mixed cellularity	Lymphocyte rich	Lymphocyte depleted
Most common subtype and observed in developed countries Capsular fibrosis and broad collagen bands; often lacunar and mummified cells are present Lacunar cells may form aggregates that may be associated with necrosis and histiocytes. Two subtypes are NS1 and NS2 Sites are often cervical, axillary, and mediastinal, and this type has intermediate prognosis.	More frequent in patients with HIV infection and in developing countries Demonstrates an interfollicular growth pattern with presence of typical Reed—Sternberg cells within an inflammatory background Sites are peripheral lymph node, spleen and prognosis is intermediate.	Nodular (common) and diffuse architecture The Reed—Sternberg cells are found within the nodules but not in the germinal centers. Some of the Hodgkin and Reed—Sternberg (HRS) cells may resemble LP cells, and immunophenotyping will help to resolve the diagnosis. Sites are often peripheral lymph nodes, and this type has a good prognosis.	Rarest subtype seen; more frequent in patients with HIV infection and in developing countries Diffuse fibrosis (with a few HRS cells) and reticular subtypes (increased number of HRS cells) may be observed Sites are often retroperitoneal and abdominal (in advanced stage), and this lymphoma has an aggressive course.

crippling mutations. Immunohistological studies have detected B-cell markers in HRS cells including CD20 and CD79a (although less often expressed). Rassidakis et al. reported that CD20 was expressed by HRS cells in 22% of patients with classical Hodgkin disease [4]. Eberle et al. reviewed histopathology of Hodgkin lymphoma and immunostaining available for diagnosis in clinical settings [5]. Important points regarding immunostains in Hodgkin lymphoma diagnosis include the following:

- NLPHL is a B-cell neoplasm, and thus the LP cells are CD20 positive. The predominant cell population of the nodules is also B cells. In CHL (in most cases also derived from B cells), CD20 may be detectable in 30%—40% of cases but is usually of varied intensity and usually is present in a minority of cases. In NLPHL, the LP cells are also positive for CD79a (a B-cell marker) in most cases. In classical Hodgkin lymphoma, CD79a is less often expressed. Another B-cell marker PAX 5 (paired box family of transcription factor 5) is weakly expressed by HRS cells. This is a very useful marker in CHL as the large tumor cells are easily picked by the weak staining. In contrast, the background smaller B lymphocytes exhibit bright staining.
- LP cells (Popcorn cells) are CD15 negative (in nearly all cases) and CD30 negative (in nearly all cases). These cells are positive for CD45 (in nearly all cases) and positive for EMA (epithelial membrane antigen) in 50% of cases. HRS cells are positive for CD15 (in approximately 80% of cases) and positive for CD30 (in nearly all cases). These cells are usually negative for CD45 and negative for EMA. CD15 and CD30 stain the membrane of the tumor cells in CHL with accentuation of the Golgi area.
- In NLPHL, the tumor cells are ringed by T cells in a rosette-like manner. Most of LP cells are ringed by T cells and less often by CD57-positive T cells. T cell rosettes may be seen in lymphocyte-rich CHL.
- Immunostain by BOB 1 (B-lymphocyte specific coactivator of octamer-binding transcription factors OCT 1 and OCT 2) and OCT 2 typically fails to stain HRS cells but typically and consistently stains LP cells. The plasma cell transcription factor IRF4/MUM1 (multiple myeloma 1/interferon regulatory factor 4) is consistently seen in classical Hodgkin lymphoma.
- EBV has been postulated to play a role in the pathogenesis of CHL. Presence of EBER (EBV encoded RNA) is indicative of CHL.
- Bcl 6 (B-cell lymphoma 6 protein) is positive in nearly all cases of NLPHL, and CD68 is typically negative in classical Hodgkin lymphoma.
- The LP cells express IgD in 9%—27% of cases.
- Most tumor cells of NLPHL and classical Hodgkin lymphoma express KI 67, a protein associated with cellular proliferation.
- The nodules of NLPHL contain expanded meshwork of follicular dendritic cells which are stained by CD21 and CD23.

More recently, flow cytometry has been applied for immunophenotyping of classical Hodgkin lymphoma. Fromm and Wood recently demonstrated that six-color flow cytometry has acceptable sensitivity and specificity for clinical application allowing immunophenotyping by this method [6]. Therefore, in the near future clinical laboratories may use flow cytometry for diagnosis of Hodgkin lymphoma [7].

Differential diagnosis

Differential diagnosis of Hodgkin lymphoma includes the following:

- Non-Hodgkin lymphoma
- NS2, reticular variant of lymphocyte-depleted Hodgkin lymphoma, anaplastic large cell lymphoma), and T cell—rich diffuse large B-cell lymphoma may look histologically similar to Hodgkin lymphoma.

Staging of Hodgkin lymphoma

Hodgkin lymphoma is a potentially curable malignancy with a 5-year survival of 81%. However, this disease may relapse in up to 30% of patients. The commonest sites of disease are cervical, supra-clavicular, and mediastinal lymph node, while subdiaphragmatic presentation with bone marrow and hepatic involvement is less common. Splenic involvement is usually associated with liver disease. The staging of disease is important and is usually achieved by workup including physical examination, chest X-ray, chest and abdominal CT scan, and bone marrow biopsy. More recently it has been shown that 18 FDG (18-fluordeoxyglucose) positron emission tomography is useful for staging of Hodgkin lymphoma [8]. Staging of Hodgkin lymphoma includes the following:

- Stage I: Involvement of a single lymph node region or lymphoid structure (e.g., Waldeyer's ring, thymus, spleen)
- Stage II: Involvement of two or more lymph node regions on the same side of the diaphragm
- Stage III: Involvement of lymph node regions or structures, both side of the diaphragm
- Stage IV: Diffuse or disseminated involvement of one or more extralymphatic organs, including any involvement of the liver or bone marrow.

Letters are often associated with staging where A indicates absence and B represents presence of symptoms such as fever, night sweats, and weight loss (10% or more) In addition, E indicates if the disease is extranodal or the disease has spread from lymph nodes to adjacent tissue, X denotes bulky disease if the largest deposit is >10 cm or if the mediastinum is wider than one-third of the chest on X-ray The presence of letter S indicates spleen involvement. The treatment and prognosis of CHL typically depends on the stage of the disease rather than the histologic classification.

Key points

- Hodgkin lymphoma is 30% of all lymphomas. Young adults are most often affected, and there is a bimodal age distribution (15—40 years and over 55 years most commonly in sixth decade of life).
- Hodgkin lymphoma is characterized by a small number of scattered tumor cells residing in an abundant heterogeneous admixture of nonneoplastic inflammatory and accessory cells. The tumor cells produce cytokines that are responsible for the presence of the background cells, lymphocytes, histiocytes, plasma cells, eosinophils, and neutrophils.

- WHO classification of Hodgkin lymphoma include NLPHL (5% of Hodgkin lymphoma) and classical Hodgkin lymphoma (95% of Hodgkin lymphoma), which can be further subdivided into nodular sclerosis (70%), mixed cellularity (20%), lymphocyte-rich (5%), and lymphocyte-depleted (5%) type.
- The neoplastic cells are seen in Hodgkin lymphoma. In CHL, RS cell is seen, which is a large cell (20–50 μm) with abundant cytoplasm and two mirror image nuclei, each with an eosinophilic nucleolus. The nuclear membrane is thick with chromatin being distributed close to the nuclear membrane; mononuclear Hodgkin cell has same features as RS cells but with one nucleus. Lacunar cell has cytoplasm that is retracted around the nucleus, creating an empty space. The nucleus is single and hyperlobulated. Mummy cell is a cell with basophilic cytoplasm and a compact nucleus with no nucleus.
- In NLPHL, Popcorn cells (LP cells formerly known as lymphohistiocytic cells or L&H cells) are observed. These are large cells with hyperlobulated nucleus.
- CHL represents 95% of Hodgkin lymphomas with a bimodal age distribution. EBV has been postulated to play a role in this disease. The prevalence of EBV in RS cells varies according to the histological subtype, highest in mixed cellularity (75%), and lowest in nodular sclerosis (10%–40%). Cervical lymph nodes are the most common area of involvement, and 60% have mediastinal involvement. Bone marrow involvement is rare (5%).
- NLPHL is typically negative for EBV. Progressively transformed germinal centers are seen in association with NLPHL. It is uncertain whether these lesions are preneoplastic or not. However, most patients with reactive hyperplasia and PTGC do not develop Hodgkin lymphoma. NLPHL may progress to large B-cell lymphoma in 2%–3% of cases.
- NLPHL is a B-cell neoplasm, and thus the LP cells are CD20 positive. The predominant cell population of the nodules is also B cells. In classical Hodgkin lymphoma (in most cases also derived from B cells), CD20 may be detectable in 30%–40% of cases but is usually of varied intensity and usually is present in a minority of cases. In NLPHL, the LP cells are also positive for CD79a (a B-cell marker) in most cases. In classical Hodgkin lymphoma, CD79a is less often expressed. Another B-cell marker PAX 5 is weakly expressed by HRS cells.
- LP cells are CD15 negative (in nearly all cases) and CD30 negative (in nearly all cases). They are positive for CD45 (in nearly all cases) and positive for EMA in 50% of cases. HRS cells are positive for CD15 (in approximately 80% of cases) and positive for CD30 (nearly all cases). They are usually negative for CD45 and negative for EMA.
- In NLPHL, the tumor cells are ringed by T cells in a rosette-like manner. Most of LP cells are ringed by T cells and less often by CD57-positive T cells. T-cell rosettes may be seen in lymphocyte-rich CHL.
- The treatment and prognosis of CHL typically depends on the stage of the disease rather than the histologic classification.

References

[1] Tamaru J. Pathological diagnosis of Hodgkin lymphoma. Nihon Rinsho 2014;72:450–5.
[2] Goel A, Fan W, Patel AA, Devabhaktuni M. Nodular lymphocyte predominant Hodgkin lymphoma: biology, diagnosis and treatment. Clin Lymphoma Myeloma Leuk 2014;14:261–70.

[3] Mani H, Jaffe ES. Hodgkin lymphoma: an update on its biology with new insights into classification. Clin Lymphoma Myeloma 2009;9:206−16.

[4] Rassidakis G, Medeiros LJ, Viviani S, Bonfante V, et al. CD 20 expression in Hodgkin and Reed-Sternberg cells of classical Hodgkin disease: association with presenting features and clinical outcome. J Clin Oncol 2002;20:1278−87.

[5] Eberle FC, Mani H, Jaffe ES. Histopathology of Hodgkin's lymphoma. Cancer J 2009;15:129−37.

[6] Fromm JR, Wood BL. A six color flow cytometry assay for immunophenotyping classical Hodgkin lymphoma in lymph node. Am J Clin Pathol 2014;141:388−96.

[7] Fromm JR, Thomas A, Wood BL. Characterization and purification of neoplastic cells of nodular lymphocyte predominant Hodgkin Lymphoma from lymph nodes by flow cytometry and flow cytometric cell sorting. Am J Pathol 2017;187:304−17.

[8] Gobbi PG, Ferreri AJ, Ponzoni M, Levis A. Hodgkin lymphoma. Crit Rev Oncol Hematol 2013;85:216−37.

Lymphoproliferative disorders associated with immune deficiencies, histiocytic and dendritic cell neoplasms, and blastic plasmacytoid dendritic cell neoplasm

Introduction

The immune-deficient state predisposes a patient not only to infectious diseases but also to cancer, particularly cancer of the immune system [1]. Immune deficiencies are associated with a range of lymphoproliferative disorders that range from benign reactive hyperplasia to atypical hyperplasias to frank lymphomas. Of the lymphomas, non-Hodgkin lymphomas are the most common type seen in such situations. These lymphomas are known to be aggressive and resistant to therapy. Patients infected with human immunodeficiency virus (HIV) are also at a higher risk of developing lymphoproliferative disorder. Histiocytic and dendritic cell tumors are rare diseases, and their pathogenesis is still not completely understood [2].

Lymphoproliferative disorders associated with immune deficiency

This chapter will be dealt in three sections, lymphoproliferative disorders associated with primary immune deficiency, lymphoproliferative disorders associated with HIV infection, and posttransplant lymphoproliferative disorders (PTLD).

Lymphoproliferative disorders associated with primary immune deficiency

This category includes lymphoid proliferations that arise in the setting of a primary immune deficiency. The risk of developing lymphoma in individuals with primary immune deficiency is 10- to 200-folds higher than in immunocompetent people, and the risk and frequency of lymphomas depend on the specific type of immune deficiency [3]. There usually is a latency period before the lymphoma develops. The primary immune deficiencies most often associated with lymphomas are ataxia telangiectasia, Wiskott–Aldrich syndrome, common variable immunodeficiency, severe combined immunodeficiency, X-linked lymphoproliferative disorder, Nijmegen breakage syndrome, hyper-IgM syndrome, and autoimmune lymphoproliferative syndrome (ALPS). Epstein–Barr virus (EBV) is involved in the majority of cases. The lymphomas often present in extranodal sites. The gastrointestinal

(GI) tract, lungs, and the central nervous system (CNS) are most often involved. Overall, diffuse large B-cell lymphomas (DLBCL) are most common. Burkitt lymphoma, Hodgkin lymphomas, and T-cell leukemias and lymphomas are also seen. In general, the lymphomas that occur in this setting are similar in morphology and immunophenotype to those described in immunocompetent people. In ataxia telangiectasia, T-cell leukemias and T-cell lymphomas are more common than B-cell neoplasms [4]. ALPS is associated with mutations in *FAS* or *FASLG* genes, which may directly contribute to lymphoid proliferations due to a failure of these cells to undergo apoptosis. ALPS is also characterized by a proliferation of CD3+, CD4−, CD8−, and alpha beta T-cells in the peripheral blood, bone marrow, or involved tissues.

Lymphoproliferative disorders associated with human immunodeficiency virus infection

HIV-positive patients are at increased risk (60−200 times) of all types of non-Hodgkin lymphoma. HIV-associated lymphomas often involve extranodal sites such as the GI tract, CNS, liver, and bone marrow. The most common HIV-associated lymphomas are Burkitt lymphoma, DLBCL (often of the CNS), primary effusion lymphoma (PEL), and plasmablastic lymphoma (PBL). Of note, PEL, PBL, and HHV8-positive DLBCL occur more specifically in patients with HIV. With the use of combination antiretroviral therapy (cART), the incidence of non-Hodgkin lymphoma has decreased; however, the risk of classic Hodgkin lymphoma has been relatively increased [5]. EBV is detected in about 40% of HIV-related lymphomas but varies depending on the site and subtype of lymphoma.

Posttransplant lymphoproliferative disorders

PTLDs are lymphoid or plasmacytic proliferations seen in individuals who have undergone solid organ or stem cell transplantations that develop as a consequence of immunosuppression. The disorder may range from EBV-driven polyclonal proliferations (infectious mononucleosis type) to frank lymphomas. The lymphomas may or may not (20−40%) be positive for EBV. However, most PTLDs are associated with EBV infection, and the most important risk factor for the EBV-driven PTLDs is EBV seronegativity before transplant. The vast majority (>90%) of PTLDs in solid organ transplants are of host origin. However, the majority of PTLD cases in stem cell transplants are of donor origin [6].

The broad categories of PTLDS are as follow:

- Nondestructive PTLDs (previously known as early lesions): lymphoid proliferations usually forming a mass lesion but with architectural preservation of the involved tissue and without overt features of lymphoma; includes plasmacytic hyperplasia, florid follicular hyperplasia, and infectious mononucleosis; EBV is typically positive in all three types.
- Polymorphic PTLD (P-PTLD): consists of a heterogeneous proliferation of immunoblasts, plasma cells, and small lymphocytes that efface the normal architecture of the involved lymph node or tissue but do not meet the criteria for any subtype of lymphoma. Unlike most lymphomas, the full range of B-cell maturation is observed. B-cells may or may not exhibit light chain restriction. Prominent CD30 expression is also common but in CD20+ B-cells (unlike Hodgkin lymphoma). Most cases of P-PTLD are EBER positive, which is a very useful tool in the distinction between PTLD and allograft rejection.

- Monomorphic PTLD (M-PTLD): meets criteria for a B-cell or T-cell/NK-cell neoplasms that are seen in immunocompetent people; they make up the majority of PTLDs (60–80%). M-PTLDs should be designated as PTLD but also further subcategorized into the lymphoma that they meet WHO diagnostic criteria for. In general, M-PTLDs are clinically, morphologically, and immunophenotypically similar to the lymphomas that they resemble. The most common B-cell M-PTLDs are DLBCLs (usually EBV-positive and nongerminal center phenotype), but Burkitt lymphoma and plasma cell neoplasms can also occur; clonal IG gene rearrangements are present in almost all cases. T-cell M-PTLD encompasses almost the entire spectrum of T-cell and NK-cell neoplasms, but the most common is PTCL, NOS followed by HSTCL. An exception is that the cases that meet the criteria for EBV-positive mucocutaneous ulcer (see later in this chapter) should be classified as such and not as M-PTLD.
- Classical Hodgkin lymphoma (CHL) PTLD: least common of the four categories of PTLD. It fulfills diagnostic criteria for classic Hodgkin lymphoma but is almost always EBV-positive. They are most commonly the mixed cellularity type of CHL.

Other iatrogenic immunodeficiency–associated lymphoproliferative disorders

Lymphoid proliferations or lymphomas arise following immunosuppression or autoimmune disease (such as rheumatoid arthritis, systemic lupus erythematosus, psoriasis, and inflammatory bowel disease) but specifically not in the posttransplant setting. Otherwise, they can resemble P-PTLD or a specific lymphoma, such as DLBCL, other B-cell lymphomas, T-cell or NK-cell lymphomas, or classical Hodgkin lymphoma. In general, they are also clinically, morphologically, and immunophenotypically similar to the lymphomas that they resemble.

Methotrexate immunosuppressive therapy for rheumatoid arthritis was the first reported agent associated with lymphoproliferative disorders in this setting. Methotrexate-associated cases are usually DLBCL (usually EBV-positive, nongerminal center phenotype, and often expressing CD30) or CHL (usually mixed cellularity) and typically occur at extranodal sites. Spontaneous remission of the lymphoproliferative disorder after removal of the immunosuppressive drug also can sometimes occur.

EBV-positive mucocutaneous ulcer is a specific type of immunosuppression (either iatrogenic or age-related)-associated lymphoproliferative disorder that often has Hodgkin-like features but is typically self-limited and follows an indolent clinical course.

Histocytic and dendritic cell neoplasms

Neoplasms of this category are very rare, representing less than 1% of tumors presenting in lymph nodes or soft tissues. Histocytes are derived from bone marrow–derived monocytes. Dendritic cells may be myeloid-derived or derived from mesenchymal stem cells.

Histiocytic sarcoma

As defined in the World Health Organization classification, this is a category of malignant tumors with morphologic and immunophenotypic features of mature tissue histiocytes. Histiocytic sarcomas are aggressive tumors. Most cases occur at extranodal sites, typically the GI tract, skin, and soft tissue. Presentation may include a mass with systemic symptoms (e.g., fever and weight loss).

The tumor typically has a diffuse proliferation of large cells with abundant eosinophilic cytoplasm. The chromatin is vesicular, and atypia may be mild to marked. A spindle cell pattern may be seen. There is sometimes a background of reactive inflammatory cells, consisting of neutrophils, small lymphocytes, plasma cells, eosinophils, and benign histiocytes. Multinucleated giant cells may be seen. Erythrophagocytosis and emperipolesis may be seen [7,8]. The tumor cells are positive for one or more histiocytic markers (CD68/KP1/PGM1, CD163, and lysozyme) as well as CD4, CD11c, CD45, and HLA-DR. The tumor cells are negative for follicular dendritic cell (FDC) markers (CD21, CD23, and CD35), Langerhans cells markers (CD1a, langerin), and myeloid markers (CD13, CD33, and myeloperoxidase [MPO]). Some cases may be associated with a mediastinal germ cell tumor (usually malignant teratoma) and may also show an isochromosome 12p, identical to the germ cell tumor. A subset of histiocytic sarcomas has a concomitant and clonally related lymphoid neoplasm; in most of these cases, the lymphoid neoplasm precedes (or occurs simultaneously with) the histiocytic neoplasm. This is a concept is known as transdifferentiation and suggests that the neoplastic lymphoid cells can either dedifferentiate or redifferentiate to a histiocyte [9].

Dendritic cell neoplasms

Dendritic cell neoplasms are neoplasms derived from antigen-presenting cells and include tumors derived from Langerhans cells, interdigitating dendritic cells (IDCs), and FDCs.

Tumors derived from Langerhans cells

There are two major subtypes within this category, which are both neoplastic proliferation of Langerhans cells: Langerhans cell histiocytosis (LCH) and Langerhans cell sarcoma. The distinction is primarily made based on the degree of cytologic atypia and clinical aggressiveness. Langerhans cells are also myeloid derived.

Langerhans cell histiocytosis

LCH is a clonal neoplastic proliferation of Langerhans cells with typically minimal cytologic atypia and a relatively good prognosis (excellent prognosis for unifocal disease). Most cases of LCH occur in children, and there is a male predominance. LCH can be either localized or multifocal, and it often involves the bone (skull, vertebrae, ribs, pelvis, or femur). LCH can also involve lymph nodes, skin, and lungs. Cases that involve the lungs are almost always associated with smoking.

LCH cells are oval cells with grooved, folded, indented or lobulated nuclei, fine chromatin, indistinct nucleoli, a thin nuclear membrane, and moderate cytoplasm. The LCH cells are often distributed amidst a background of eosinophils, neutrophils, histiocytes, and small lymphocytes.

Electron microscopy demonstrates the hallmark Birbeck granules, which resemble tennis rackets. Birbeck granules can also be confirmed by langerin expression. LCH cells are positive for CD1a, langerin (also known as CD207), and S100. LCH cells may also be positive for CD68, HLA-DR, PDL1, and vimentin. LCH cells are negative for FDC markers (CD21, CD23, and CD35). Immunophenotype characteristics of histiocytic and dendritic cell neoplasms are summarized in Table 15.1.

Approximately 50% of LCH cases have a *BRAF* V600E mutation. Approximately 25% of cases have a *MAP2K1* mutation, but these mutations are almost always seen in BRAF germline cases.

Table 15.1 Immunophenotypic characteristics of histiocytic and dendritic cell neoplasms.

Histiocytic sarcoma	The tumor cells are positive for CD45, CD4, CD11c, CD68, and lysozyme. They are negative for myeloid markers (CD13, CD33 and MPO), follicular dendritic cell markers (CD21, CD23 and CD35) and CD1a.
Follicular dendritic cell sarcoma	The tumor cells are positive for CD21, CD23, and CD35. There is a variable positivity for S100, CD68, and epithelial membrane antigen (EMA). They are negative for CD1a, CD34, lysozyme, and MPO.
Interdigitating dendritic cell sarcoma	The tumor cells are positive for S100 and vimentin. They are negative for FDC markers, CD21, CD23, and CD35.
Langerhans cell histiocytosis and Langerhans cell sarcoma	Tumor cells are positive for CD1a, langerin, and S100. Cells may also be positive for vimentin and CD68.

LCH is also sometimes referred to as histiocytosis X. The term eosinophilic granuloma is no longer used but was historically the name given due to its appearance and if the disease was solitary and localized to the bone. Hand—Schuller—Christian disease is also no longer used but was historically applied when there were multiple sites of involvement. Letterer—Siwe disease is another term that is no longer used but describes disseminated disease or visceral involvement.

Langerhans cell sarcoma

Langerhans cell sarcoma is a rare high-grade neoplasm in which the tumor cells display overt malignant cytologic features (pleomorphism, prominent nucleoli, high mitotic rate) and an immunophenotype identical to LCH. It occurs in adults, usually involves multifocal extranodal sites (skin and bone), and has a poor prognosis.

Indeterminate dendritic cell tumor

This is a neoplastic proliferation of spindle to ovoid cells with an immunophenotype similar to normal indeterminate cells, which are thought to be the precursor cells of Langerhans cells. These neoplasms often present as skin lesions that are usually dermal in distribution. The neoplastic cells resemble Langerhans cells with nuclear grooves and clefts. The neoplastic cells are positive for CD1a and S100; however, in contrast to LCH, and by definition, they lack Birbeck granules and thus will be negative for langerin. They are also negative for CD21, CD23, CD35, and CD163.

Interdigitating dendritic cell sarcomas

IDC sarcoma is an extremely rare and aggressive neoplasm of spindle to ovoid cells with an immunophenotype similar to IDCs. IDCs are myeloid derived. Patients usually present with an asymptomatic mass. Solitary lymph node involvement is common and, when affected, demonstrates a tumor in the paracortical area with residual follicles present. The tumor cells are spindle to ovoid and form whorls, fascicles, or storiform pattern. The morphologic features are thus very similar to FDC sarcoma, and immunohistochemistry is necessary to make that distinction. The tumor cells are positive for S100, vimentin, fascin (usually) and occasionally with strong nuclear p53 expression. In contrast to FDC sarcoma, markers for FDCs (CD21, CD23, and CD35) are negative.

Follicular dendritic cell sarcoma

FDC sarcoma is a neoplastic proliferation of cells, typically spindle to ovoid, with immunophenotypic features similar to nonneoplastic FDCs. FDCs are mesenchymal in origin and are located within the lymphoid follicle, being most numerous in the light zone of the germinal center. Most tumors present as a mass, with lymphadenopathy being a frequent finding. The tumor cells are spindle to ovoid with eosinophilic cytoplasm and form fascicles, whorls, or a storiform pattern. Diffuse sheets may also be seen. Binucleated and multinucleated cells may be seen. The mitotic rate is generally relatively low. The tumor cells are positive for one or more FDC markers (CD21, CD23, and CD35), as well as CXCL13, D2-40 (podoplanin), and clusterin. There a is variable positivity for CD68, S100, and epithelial membrane antigen (EMA). It is negative for CD1a, CD34, lysozyme, and MPO.

Inflammatory pseudotumor-like follicular/fibroblastic dendritic cell sarcoma

It usually occurs in young to middle-aged adults with a marked female predominance and typically involves the spleen and/or liver [10]. Microscopically, the cells are spindled with vesicular chromatin and are small, but they have distinct nucleoli in a lymphoplasmacytic infiltrate. The neoplastic cells are usually positive for FDC markers (CD21, CD23, and CD35), but the staining can range from diffuse to only focal. However, the neoplastic cells are associated with EBV in the large majority of cases, making EBER staining a very useful marker [11]. This tumor is indolent, but local recurrences are common.

Fibroblastic reticular cell tumor: This is very rare and histologically resembles FDC sarcoma or IDC sarcoma, but lacks the immunophenotype of either. The cells are variably reactive CD68, smooth muscle actin (SMA), desmin, and cytokeratin (dendritic pattern).

Disseminated juvenile xanthogranuloma: This is a benign histiocytic proliferation of small, oval, occasionally slightly spindled, bland cells without grooves and often with accompanying foamy macrophages and Touton-type giant cells. It usually presents as a solitary dermal lesion and is associated with neurofibromatosis type 1. The neoplastic cells express markers that are also seen on macrophages, such as CD14, CD68 (PGM1), CD163, and vimentin. The cells are negative for CD1a and langerin. There are no documented *BRAF* or *MAP2K* mutations in these tumors.

Erdheim—Chester disease

Erdheim—Chester disease (ECD) is a rare systemic clonal proliferation of histiocytes often with foamy cytoplasm and with interspersed Touton giant cells. The diagnosis is based on histopathologic findings in the appropriate clinical and radiological context. There is also a background of small lymphocytes, plasma cells, and neutrophils. The ECD histiocytes are positive for macrophage markers (CD14, CD68, CD163) as well as factor XIIIa and fascin; they are negative for Langerhans cells markers (CD1a, S100, langerin). Cases with mutated *BRAF* V600E are positive by BRAF immunostain.

Mutations in *BRAF* V600E are seen in more than 50% of cases as well as other MAPK pathway genes, such as *NRAS* mutations.

ECD occurs in adults (typical age 55–60 years) and has a male predominance (M:F = 3:1). Almost any organ can be involved, but the vast majority (>95%) of cases have skeletal involvement. Cardiovascular involvement occurs in approximately half of the cases, and about one-third have retroperitoneal involvement. CNS involvement causing diabetes insipidus or exophthalmos can also occur. Cutaneous involvement usually manifests as xanthelasma of the eyelids or periorbital areas. The diagnosis is greatly facilitated by characteristic radiology findings seen in bones. ECD is a chronic disease, and the prognosis depends on the sites and extent of involvement [12].

Blastic plasmacytoid dendritic cell neoplasm

Blastic plasmacytoid dendritic cell neoplasm (BPDCN) is derived from precursors to plasmacytoid dendritic cells. BPDCN was previously classified as a subtype of acute myeloid leukemia in the 2008 WHO classification system but has now been made a separate and distinct entity. It can be present at any age but typically affects older adults (~60s) and involves the skin and bone marrow in the vast majority of cases.

Histologic examination by H&E shows a diffuse proliferation of monomorphous blasts, which may resemble myeloid or lymphoid blasts. By Giemsa stain, the BPDCN cells may also resemble blasts with irregular nuclei, fine chromatin, small nucleoli, and scant cytoplasm; however, the cytoplasm may often appear blue gray and be eccentrically displaced, imparting a "hand mirror" or "tadpole" appearance.

The BPDCN cells are negative for myeloperoxidase and butyrate esterase stains by cytochemistry. By immunohistochemistry, the BPDCN cells express CD4, CD56, CD123, CD303, and TCL1A. Although nonspecific, BPDCN also often expresses CD7, CD33, and CD68. The neoplastic BPDCN cells may also aberrantly express BCL2, BCL6, and IRF4 (MUM1), which should be negative in normal (nonneoplastic) plasmacytoid dendritic cells.

Patients with BPDCN often have an abnormal karyotype and commonly a complex karyotype. However, the abnormalities are relatively nonspecific but include abnormalities of 5q, 6q, 12p, 13q, and 15q. Mutations in *TET2* are the most common in BPDCN. Mutations in *ASXL1*, *NRAS*, and *KRAS* have also been described.

BPDCN can be associated with concomitant myeloid neoplasms including myelodysplastic syndrome, chronic myelomonocytic leukemia, and acute myeloid leukemia. The clinical course of BPDCN is aggressive, and relapses with chemoresistance are common.

Key points

- The risk of developing lymphoma in individuals with primary immune deficiency is 10- to 200-fold, and the risk and frequency of lymphomas depend on the specific type of immune deficiency.
- Overall, in primary immune deficiency patients, DLBCLs are most common. Burkitt lymphoma, Hodgkin lymphomas, and T-cell leukemias and lymphomas are also seen. In ataxia telangiectasia, T-cell leukemias and T-cell lymphomas are more common than B-cell neoplasms.

- ALPS is associated with mutations in *FAS* or *FASLG* genes and characterized by a proliferation of CD3+, CD4−, CD8−, alpha beta T cells.
- The most common HIV-associated lymphomas are Burkitt lymphoma, DLBCL (often of the CNS), PEL, and PBL. With the use of combination antiretroviral therapy (cART), the incidence of non-Hodgkin lymphoma has decreased.
- PTLDs are seen in individuals who have undergone solid organ or stem cell transplantations and develop as a consequence of immunosuppression.
- Majority of PTLDs in solid organ transplants are of host origin, whereas the majority of stem cell transplant cases are of donor origin.
- There are four broad categories of PTLD:
  (1) Nondestructive PTLDs (plasmacytic hyperplasia, florid follicular hyperplasia, and infectious mononucleosis) without significant architectural distortion.
  (2) P-PTLD that consists of a heterogeneous proliferation of immunoblasts, plasma cells, and small lymphocytes that efface the normal architecture of the tissue but do not meet criteria for any subtype of lymphoma.
  (3) M-PTLD that meets criteria for a specific B-cell or T-cell/NK-cell neoplasms that are seen in immunocompetent people. They are clinically, morphologically, and immunophenotypically similar to the lymphomas that they resemble. The most common M-PTLD is DLBCL (usually EBV positive and nongerminal center phenotype).
  (4) Classical Hodgkin lymphoma PTLD fulfills diagnostic criteria for classic Hodgkin lymphoma (typically mixed cellularity type) but is almost always EBV-positive.
- Iatrogenic immunodeficiency−associated lymphoproliferative disorders arise following immunosuppression in the nontransplant setting or autoimmune disease. They are also clinically, morphologically, and immunophenotypically similar to the lymphomas that they resemble. A classic scenario is methotrexate immunosuppressive therapy for rheumatoid arthritis. Spontaneous remission of the lymphoproliferative disorder after removal of the immunosuppressive drug also can sometimes occur.
- EBV-positive mucocutaneous ulcer is a specific type of immunosuppression (either iatrogenic or age-related)-associated lymphoproliferative disorder that often has Hodgkin-like features but is typically self-limited and follows an indolent clinical course.
- Histiocytic sarcoma typically has a diffuse proliferation of large cells with a background of reactive cells. The tumor cells are positive for histiocytic markers (CD68/KP1/PGM1, CD163, lysozyme) as well as CD4, CD11c, CD45, and HLA-DR; they are negative for CD21, CD23, CD35, CD1a, langerin, CD13, CD33, and MPO.
- A subset of histiocytic sarcomas has a concomitant and clonally related lymphoid neoplasm, which usually precedes the histiocytic neoplasm (transdifferentiation).
- Langerhans cell histiocytosis (LCH) is a clonal neoplastic proliferation of Langerhans cells with grooved, folded, indented, or lobulated nuclei and typically minimal cytologic atypia.
- Most cases of LCH occur in children, and there is a male predominance. It often involves the bone (skull, vertebrae, ribs, pelvis, or femur), and cases that involve the lung are almost always associated with smoking.
- Electron microscopy demonstrates the hallmark Birbeck granules of LCH, which resemble tennis rackets. Birbeck granules can also be confirmed by expression of langerin expression.
- LCH cells are positive for CD1a, langerin (aka CD207), and S100.

- Approximately 50% of LCH cases have a *BRAF* V600E mutation. Approximately 25% of cases have a *MAP2K1* mutations, but these mutations are almost always seen in BRAF germline cases.
- Langerhans cell sarcoma is a rare high-grade neoplasm in which the tumor cells display overt malignant cytologic features (pleomorphism, prominent nucleoli, high mitotic rate) and an immunophenotype identical to LCH.
- IDC sarcoma is composed of spindle to ovoid cells and forms whorls, fascicles, or storiform pattern, very similar morphologically to FDC sarcoma. The tumor cells are positive for S100, vimentin, fascin (usually) but are negative for FDCs (CD21, CD23, and CD35).
- FDC sarcoma consists of spindle to ovoid cells that form fascicles, whorls, or a storiform pattern. The tumor cells are positive for CD21, CD23, and CD35 as well as CXCL13, D2-40 (podoplanin), and clusterin; they are negative for CD1a, CD34, lysozyme, and MPO.
- Inflammatory pseudotumor-like follicular/fibroblastic dendritic cell sarcoma usually occurs in young to middle-aged adults with a marked female predominance and typically involves the spleen and/or liver. The cells are positive for FDC markers (CD21, CD23, and CD35) and EBER (strongly associated with EBV). This tumor is indolent, but local recurrences are common.
- ECD is a rare systemic clonal proliferation of histiocytes often with foamy cytoplasm and with interspersed Touton giant cells.
- The vast majority (>95%) of ECD cases have skeletal involvement; cardiovascular, cutaneous, CNS, and retroperitoneal involvement are also common. The diagnosis is greatly facilitated by characteristic radiology findings seen in bones.
- Mutations in *BRAF* V600E are seen in more than 50% of ECD cases as well as other *MAPK* pathway genes, such as *NRAS* mutations. Cases with mutated *BRAF* V600E are positive by BRAF immunostain.
- BPDCN is an aggressive neoplasm derived from precursors to plasmacytoid dendritic cells. The cells resemble blasts, but the cytoplasm may be eccentrically displaced, imparting a "hand mirror" or "tadpole" appearance on Giemsa stain.
- BPDCN cells express CD4, CD56, CD123, CD303, and TCL1A. The neoplastic BPDCN cells may also aberrantly express BCL2, BCL6, and IRF4 (MUM1). Mutations in *TET2* are the most common in BPDCN.

References

[1] Van Krieken J. Lymphoproliferative disease associated with immune deficiency in children. Am J Clin Pathol 2004;122(Suppl. 1):S122—7.
[2] Said J. Follicular lymphoma and histocytic/dendritic neoplasm related? Blood 2008;111:5418—9.
[3] Filipovich AH, Mathur A, Kamat D, et al. Primary immunodeficiencies genetic risk factors for lymphoma. Cancer Res 1992;52:5465s—7s.
[4] Taylor AM, Metcalfe JA, Thick J, Mak YF. Leukemia and lymphoma in ataxia telangiectasia. Blood 1996; 87:423—38.
[5] Clifford GM, Polesel J, Rickenbach M, et al. Cancer risk in the Swiss HIV Cohort study: associations with immunodeficiencies, smoking, and highly active retroviral therapy. J Natl Cancer Inst 2005;97:425—32.
[6] Zutter MM, Martin PJ, Sale GE, et al. Epstein-Barr virus lymphoproliferation after bone marrow transplantation. Blood 1988;72:520—9.

[7] Pileri SA, Grogan TM, Harris NL, et al. Tumors of histiocytes and accessory dendritic cells: an immuno-histochemical approach to classification from the International Lymphoma study Group based on 61 ceases. Histopathology 2002;41:1−29.

[8] Vos JA, Abbondanzo SL, Barekman CL, et al. Histiocytic sarcoma: a study of five cases including the histiocytic marker CD163. Mod Pathol 2005;18:693−704.

[9] Feldman AL, Arber DA, Pittaluga S, et al. Clonally related follicular lymphomas and histiocytic/dendritic cell sarcomas: evidence for transdifferentiation of the follicular lymphoma clone. Blood 2008;111(12): 5433−9.

[10] Chen Y, Shi H, Li H, et al. Clinicopathologic features of inflammatory pseudotumor-like follicular dendritic cell tumor of the abdomen. Histopathology 2016;68:858−65.

[11] Cheuk W, Chan JK, Shek TW, et al. Inflammatory pseudotumor-like follicular dendritic cell tumor: a distinctive low-grade malignant intra-abdominal neoplasm with consistent Epstein-Barr virus association. Am J Surg Pathol 2001;25(6):721−31.

[12] Haroun F, Millado K, Tabbara I. Erdheim-Chester Disease: comprehensive review of molecular profiling and therapeutic advances. Anticancer Res 2017;37:2777−83.

Essentials of coagulation

16

Introduction

Blood clotting (coagulation) is initiated within seconds after vascular injury and is considered one of fastest tissue repair system in human body. The main purpose of coagulation is to seal an injured vessel, which is accomplished by aggregation of platelets at the site of injury. First, a loose platelet plug is formed, which is then stabilized by the formation of fibrin network. Both events known as primary and secondary hemostasis, respectively, not only prevent blood loss but also trigger wound healing and tissue regeneration. Because a bleeding site is also a potential entry point of invading microorganism, coagulation is also one of the first humoral regulatory systems that encounters invading microorganism. Antimicrobial peptides are released from platelet when coagulation is activated. In addition, an intact platelet–fibrinogen plug can provide an active surface that allows the recruitment, attachment, and activation of phagocytosing cells. Moreover, coagulation factors are able to induce pro- and antiinflammatory reactions by activating protease-activated receptors on immune cells [1].

Normal hemostasis

Hemostasis consists of three steps:

- Vasoconstriction: this is mediated by reflex neurogenic mechanisms. The vasoconstriction is augmented by endothelin, which is released from damaged endothelial cells. Vasoconstriction reduces flow of blood, thus reducing extent of blood loss.
- Platelet plug (primary hemostasis): platelets adhere to the subendothelial collagen along with shape change and release of platelet granule contents. Additional platelets are recruited and a platelet plug is formed. The primary platelet plug that is formed is reversible, but with the help of fibrin, secondary and irreversible platelet plug is formed.
- Activation of the coagulation cascade (secondary hemostasis): activation of the clotting cascade results in formation of fibrin and cross-linking of fibrin with resultant arrest of bleeding.

Hematology and Coagulation. https://doi.org/10.1016/B978-0-12-814964-5.00016-4

Platelets and platelet events

Platelets are anucleated discoid-shaped blood cells derived from megakaryocytes and have glycoproteins attached to the outer surface, which serves as receptors. Morphologically, a platelet has three zones:

- A peripheral zone responsible for adhesion and aggregation and also contain platelet membrane
- Sol-gel zone responsible for contraction and support of microtubule system and
- Organelle zone where platelet granules, mitochondrion, and glycogen are located.

Platelets have glycoproteins and important ones are

- Platelet glycoprotein Ib (GpIb), which is a disulfide-linked alpha-beta heterodimer (molecular weight: 160,000) that forms a complex with GpIX (molecular weight: 22,000). Absence of GpIb results in Bernard—Soulier syndrome.
- Platelet glycoprotein V (GpV; molecular weight 82,000) is a major membrane protein, which is a substrate for thrombin and also forms a noncovalent complex in the platelet membrane with other platelet glycoproteins (GpIb/IX).
- GpIb/IX/V plays a central role in attachment of platelets with von Willebrand factor (vWF) in the subendothelium of damaged vessel wall [2]. GpIb/IX/V has a central GpV and to each side of this is one GpIb alpha, GpIb beta, and GpIX. It is the GpIb alpha that binds to vWF, specifically the A1 domain. Binding of vWF to GpIb alpha initiates platelet adhesion.
- GpIIb/IIIa is responsible for attachment with fibrinogen. Platelet aggregation is mediated by GpIIb/IIIa. Absence of GpIIb/IIIa results in Glanzmann's thrombasthenia.
- GpIa/IIa and GpVI are also attached to collagen where GpIa/IIa acts as a receptor for platelet adhesion to collagen. Binding of collagen to GpVI initiates platelet aggregation and platelet degranulation.

Adenosine diphosphate (ADP) is an important physiological agonist that plays an important role in normal hemostasis and thrombosis. ADP causes platelets to undergo shape change, release granule content, and aggregate. Platelets also at that point hydrolyze arachidonic acid (AA) from phospholipids and convert it into thromboxane A2 via cyclooxygenase and thromboxane A2 synthase. It is well established that ADP activates platelet through three purinergic receptors namely P2Y1, P2Y12, and P2X1 [3]. Major characteristics of various membrane receptors present in platelets include the following:

- ADP receptors (P2Y1, P2Y12, and P2X1): P2Y1 receptors once activated are involved in platelet shape change. P2Y1 receptors are widely distributed in many tissues. P2Y12 receptors are only found in platelets and once activated is involved in platelet aggregation, secretion, and thrombus stabilization.
- Thrombin receptors (PAR-1 and PAR-4): thrombin can activate platelets. This is done via receptors, PAR-1 and PAR-4.
- Thromboxane receptors: once activated stimulates platelet activation
- Adrenergic receptors

The canalicular system in platelets represents a reservoir of membranes connected to the outer surface, which allows release of platelet granule contents to the exterior. Chemicals released from platelets are capable of activating other platelets. Peripheral microtubules are responsible for shape change and release of chemicals.

Platelets are capable of secreting granules, which are critical to normal platelet function. Among the three types of platelet secretory granules are

- Alpha granule
- Dense granule and
- Lysosomes.

Alpha granules are most abundant (50–80 per platelets ranging in size from 200 to 500 nm). Alpha granules are stained by the Wright-Giemsa stain and contain fibrinogen, fibronectin, factor V, vWF, platelet factor 4 (PF-4), platelet-derived growth factors (PDGF), transforming growth factor beta (TGF-β), and thrombospondin. The contents of the alpha granules are released into the canalicular system. Deficiency of alpha granules results in platelets, which appear pale and gray on peripheral smear, and this is referred to as gray platelet syndrome. It is transmitted in autosomal recessive manner.

Dense granules (also known as dense bodies or delta granules) are electron dense due to presence of calcium and appear as dark bodies under electron microscope. Dense granules are present in low numbers, are less than 10 per platelet, and contain adenosine triphosphate (ATP), ADP, ionized calcium, histamine, serotonin (5-hydroxytryptamine, 5-HT), and epinephrine. The dense granule contents are released directly through fusion with the plasma membrane. Lysosomes contain acid hydrolases.

Platelets circulate for about 10 days in the circulation. Approximately one-third of platelets are normally sequestered in the spleen.

The three main platelet events that take place in primary hemostasis:

- Adhesion and shape change: with exposure of subendothelial collagen to vWF, there occurs conformational change in high molecular weight multimers followed by interaction between vWF and GP Ib/IX/V receptors. This allows adhesion of platelets to subendothelial collagen via interaction with vWF. The vWF, GpIb/IX/V, collagen, and GpIa/IIa as well as GpVI are all involved in this step. In von Willebrand's disease and Bernard–Soulier syndrome (where GpIb/IX/V is lacking), abnormal platelet function (thrombocytopathia) is observed.
- Platelet activation: during this process, there is an increase in cytoplasmic calcium concentration with change in shape of platelets, extension of pseudopodia, and release of chemicals (release reaction). Phosphatidylserine is translocated to the external surface. Release of ADP, activation of thrombin receptors, and production of thromboxane A2 (from AA with the help of cyclooxygenase enzyme) are all involved in platelet activation. Activation of ligand binding site on GpIIb/IIIa also takes place at this time. GpIIb/IIIa then interacts with GpIIb/IIIa on other platelets through fibrinogen, causing platelet aggregation. If there is deficiency of enzymes responsible for synthesis of chemicals normally stored and released from platelets or abnormalities of receptors through which they act, these may result in defective platelet function, also known as thrombocytopathia.
- Aggregation: fibrinogen-mediated binding of activated GpIIb/IIIa receptors on adjacent platelets, which is augmented by thrombospondin, a component of α-granules. Thus, with lack of GpIIb/IIIa (known as Glanzmann's thrombasthenia, transmitted as autosomal recessive) and hypofibrinogenemia, abnormal platelet function is observed.

Thrombocytopenia and thrombocytopathia

Bleeding due to platelet disorders is mainly due to thrombocytopenia (low platelet count) or thrombocytopathia (dysfunctional platelets) or both. Manifestation of bleeding is most often in the form of purpuras, mucosal bleeding (e.g., epistaxis, gum bleeding, GI bleed), prolonged bleeding from superficial cuts and abrasions, and menorrhagia.

Major causes of thrombocytopenia include the following:

- Decreased production: generalized bone marrow failure or selective megakaryocyte depression results in decreased platelet formation. Congenital diseases associated with reduced platelet production include hereditary thrombocytopenias, macrothrombocytopenia with neutrophilic inclusions (MYH9 disorders), Wiskott—Aldrich syndrome (X-linked recessive), and Bernard—Soulier disease
- Increased breakdown: idiopathic thrombocytopenic purpura (ITP), heparin-induced thrombocytopenia (HIT), drug-induced thrombocytopenia, neonatal and posttransfusion purpura
- Increased utilization: disseminated intravascular coagulation (DIC), thrombotic thrombocytopenic purpura, and hemolytic uremic syndrome
- Increased sequestration: Kasabach—Merritt syndrome (platelets sequestered in a hemangioma)

Hereditary thrombocytopenias

Hereditary thrombocytopenia includes the following:

- Thrombocytopenia with absent radii (TAR syndrome);
- Congenital amegakaryocytic thrombocytopenia (in this disorder, thrombopoietin receptor is deficient or defective; thrombopoietin is required for maturation of megakaryoblasts to megakaryocytes);
- Familial thrombocytopenia-leukemia (thrombocytopenia combined with thrombocytopathy with increased incidence of acute myeloid leukemia); and
- Fanconi's anemia (autosomal recessive transmission).
- Macrothrombocytopenia with neutrophilic inclusions (MYH9 disorders) is a group of disorders characterized by mutations in the MYH9 gene. This gene encodes the nonmuscle myosin heavy chain class IIA protein. These disorders are characterized by thrombocytopenia, large/giant platelet, and Dohle-like bodies in neutrophils. This group of disorders includes May—Hegglin anomaly (autosomal dominant), Sebastian syndrome, Fechtner syndrome (nephritis, ocular defects, and sensorineural hearing loss), and Epstein syndrome.
- Wiskott—Aldrich syndrome is due to defect in the WASP gene, and this syndrome is characterized by immune deficiency, eczema, and thrombocytopenia with small platelets.
- Bernard—Soulier syndrome is due to deficiency of GpIb/IX/V receptor. There is also thrombocytopenia with large/giant platelets. Some cases of Bernard—Soulier syndrome are due to defects of the GpIb-β gene located on chromosome 22. This gene may be affected in velocardiofacial syndrome or DiGeorge syndrome associated with deletion of 22q11.2.

Pseudothrombocytopenia

It is important to be aware of causes of pseudothrombocytopenia, which as the name implies is not true thrombocytopenia. Pseudothrombocytopenia may be due to

- Platelet clumps: this phenomenon occurs when blood is collected in EDTA (ethylenediamine tetraacetic acid) containing blood collection tube. Hematology analyzers can flag a sample when it detects clumps. Clumps may be seen on peripheral smear examination. Recollection of blood should be done with heparin or citrate tube for accurate platelet count [4].
- Platelet satellitism: this phenomenon also occurs when blood is collected in EDTA. Here, platelets surround neutrophils resulting in satellitism. The platelets are not counted resulting in thrombocytopenia.
- Large platelets: large platelets may be counted as red blood cells by the hematology analyzer. If these large platelets are numerous, this may falsely lower the actual platelet count.
- Traumatic venipuncture may result in activation of clotting process, resulting in thrombocytopenia.

Thrombocytopathia

Thrombocytopathia can be congenital or acquired. Examples of congenital thrombocytopathia include

- von Willebrand disease
- Bernard–Soulier syndrome
- Chediak–Higashi syndrome
- Hermansky–Pudlak syndrome
- Glanzmann's syndrome
- Gray platelet syndrome

Acquired thrombocytopathia may be due to

- Nonsteroidal antiinflammatory drugs such as aspirin
- Uremia or
- Acquired von Willebrand disease: acquired von Willebrand disease is an acquired bleeding disorder that may suddenly manifest in an individual without any family history of bleeding disorder. Usually this disease is frequently associated with cardiopulmonary bypass, monoclonal gammopathy, lymphoproliferative, myeloproliferative, and autoimmune disorder, and pathogenic mechanism involves autoantibodies against vWF, resulting in inactivation of plasma vWF or rapid clearance of this factor [5]. Various causes of thrombocytopathia are listed in Table 16.1.

Table 16.1 Congenital and acquired causes of thrombocytopathia.

Congenital thrombocytopathia	Acquired thrombocytopathia
• Disorders of platelet adhesion: von Willebrand's disease, Bernard–Soulier syndrome • Disorders of platelet activation: storage pool disorders, Chediak–Higashi syndrome, and Hermansky–Pudlak syndrome • Disorders of platelet aggregation: Glanzmann's syndrome	• Drugs: aspirin and various other nonsteroidal antiinflammatory drugs • Uremia • Acquired von Willebrand disease • Myeloproliferative diseases • Antiplatelet antibodies

Tests for platelet function

Platelets are main regulator of hemostasis and also interact with a large variety of cell types including monocytes, neutrophils, endothelial cells, and smooth muscle cells. Platelets are sensitive to manipulation and are prone to artifactual in vitro activation. Therefore, care must be exercised during performing platelet function tests. Platelet functions tests are performed in a variety of patients including patients with bleeding disorders, and in most cases, a platelet-mediated hemostatic disorder cannot be established by just a single function defect but rather by a combination of platelet function abnormalities [6]. Applications of platelet function tests are listed in Table 16.2.

Bleeding time

Bleeding time, the time taken for bleeding to stop after a defined incision, is made into the skin was introduced by Duke in 1910. Ivy made the method more reliable by introducing a blood pressure cuff on the upper arm, which was inflated to 40 mmHg and placing the incision into the anterior surface of the forearm. This protocol is still followed, and drops of blood are absorbed with filter paper disks every 30s. The time taken for bleeding to stop is noted. This test has been used most often to detect qualitative defects of platelets, vascular defects, or von Willebrand disease, but this test has poor clinical correlation.

Capillary fragility test

Capillary fragility test (also known as a Rumpel–Leede capillary fragility test or tourniquet test) determines capillary fragility and is a clinical diagnostic method to determine hemorrhagic tendency of a patient. This test assesses fragility of capillary walls and is used to identify thrombocytopenia or thrombocytopathia. The test is defined by the World Health Organization as one of the necessary requisites for diagnosis of dengue fever. A blood pressure cuff is applied and inflated to a point between the systolic and diastolic blood pressures for 5 min. The test is positive if there are 10 or more petechiae per square inch. In dengue hemorrhagic fever, the test usually provides a definite positive result with 20 petechiae or more. This test does not have high specificity.

PFA-100

PFA-100 system is a platelet function analyzer designed to measure platelet-related primary hemostasis. The instrument uses two disposable cartridges that are coated with platelet agonist.

Table 16.2 Application of platelet function tests.

- Patients with bleeding disorders
- Monitoring response of a patient receiving antiplatelet therapy
- Platelet function tests to screen donors
- Assessment of platelet function in platelet concentrates
- Assessment of platelet function following platelet transfusion in a patient
- Perioperative assessment of platelet function in a patient

For analysis, whole blood is collected from the patients in a citrate tube, and testing should be performed within 4 h of collection. Blood is transferred into a sample cup. Blood is aspirated from the sample cup by the analyzer and passes through an aperture in a membrane that is already coated with platelet agonists. When platelet aggregation takes place, the aperture closes and the blood flow stops. This is the closure time. One membrane is coated with collagen/epinephrine (CEPI), and the other membrane is coated with collagen/adenosine diphosphate (CADP).

Interpretation of PFA-100

- If the closure time of both CEPI and CADP is normal, this means no evidence of platelet dysfunction. However, PFA-100 is insensitive to von Willebrand disease type 2N, clopidogrel, ticlopidine, and storage pool disease.
- If the CEPI closure time is prolonged but CADP closure time is normal, this is most likely due to aspirin.
- If both CEPI and CADP closure time are prolonged or CEPI closure time is normal and CADP closure time is abnormal, this denotes platelet dysfunction or von Willebrand disease.

PFA-100 is very sensitive to von Willebrand disease type I and GpIIb/IIIa antagonists. PFA-100 results are affected by thrombocytopenia, low hematocrits. It is not affected by heparin or deficiencies of clotting factors other than fibrinogen.

VerifyNow

Patients with inadequate response to antiplatelet medications remain at risk for myocardial infarction, stent thrombosis, and death. Up to 40% of patients on antiplatelet medications may not exhibit adequate platelet-inhibiting effect. Causes include

- Genetic differences
- Noncompliance
- Drug interactions (e.g., with proton pump blockers)
- Preexisting medical conditions (e.g., diabetes)

VerifyNow (VFN) is a rapid, turbidimetric whole blood assay capable of evaluating platelet aggregation. This assay is based on the ability of activated platelets to bind with fibrinogen. There are three different assays:

- VFN IIb/IIIa assay
- VFN P2Y12 assay
- VFN aspirin assay

The VFN IIb/IIIa assay uses fibrinogen-coated microparticle and thrombin receptor activating peptide is used as an agonist to maximally stimulate platelets to determine platelet function. If GpIIb/IIIa antagonists are present in the patients' blood, then platelet aggregation should be reduced. In the VFN aspirin assay, AA is used as the agonist to measure the antiplatelet effect of aspirin. Results are expressed in aspirin reactive units (ARU). The VFN P2Y12 assay is similarly used to assess the antiplatelet effect of clopidogrel, prasugrel, and ticagrelor, and result is expressed in P2Y12 reactive units (PRU).

Interpretation of results:

- PRU values less than 180: clinically significant effect of P2Y12 inhibitor
- PRU values greater than 220: clinically insignificant effect of P2Y12 inhibitor
- PRU values between 180 and 220: gray zone
- ARU values greater than 550: clinically insignificant effect of aspirin
- ARU values less than 550: clinically significant effect of aspirin

PRU results may be affected by prior exposure to GpIIb/IIIa antagonists. Results may be affected by exposure to eptifibatide or tirofiban exposure within 48 h or abciximab exposure with 14 days.

VFN IIb/IIIa assay:

This test assesses response to abciximab and eptifibatide. Before administration of either of the drugs, a baseline value is obtained. The baseline value for abciximab is 125–330 PAU (Platelet Aggregation Units) and the baseline value for eptifibatide is 136–288 PAU. Ten minutes after IV infusion of these drugs, the test is repeated. The two numbers are used to calculate % inhibition.

VFN assays may be used to assess efficacy of the abovementioned drugs, checking patient compliance, and also to assess residual effect of these drugs if patient is to undergo surgery or invasive procedures.

Plateletworks

Plateletworks is a test for the assessment of platelet function test using whole blood and kits for the test, and impedance cell counter (ICHOR II analyzer) is commercially available from the Helena Laboratory so that this test can be used as a point of care test. This test assesses platelet function by comparing the platelet count before and after exposure with a specific platelet agonist. For this test, blood is collected in EDTA tubes and in other platelet agonist containing tubes, for example, ADP or AA or collagen. In the agonist tube, functional platelets should aggregate and the nonfunctional platelets should not aggregate. Then a hematology analyzer such as ICHOR II is used to count the number of platelets in the EDTA tube and also the number of unaggregated platelets in the agonist tube. The unaggregated platelets are dysfunctional. To calculate the number of functional platelets, the platelet count in the presence of a platelet agonist should be subtracted from platelet count obtained from using blood collected in the EDTA tube. Campbell et al. commented that point of care test platform Plateletworks is useful in monitoring platelet response in patients receiving antiplatelet agents including aspirin and clopidogrel [7].

Platelet aggregation

Platelet aggregation test using platelet aggregometry is a widely used laboratory test to screen patients with inherited or acquired defect of platelet function. Platelet aggregometry measures the increase in light transmission through platelet-rich plasma that occurs when platelets are aggregated due to addition of an agonist. For this test,

- Blood should be collected in citrate tubes
- The test should be performed within 4 h of blood collection

- Before analysis, specimen should be stored at room temperature
- The sample cannot be tubed to the laboratory
- Platelet count should ideally be 100,000 or greater

Platelet-rich plasma is obtained from the sample by centrifugation. Ideally, the platelet count of the platelet-rich plasma should be approximately 200,000 to 250,000, and if the platelet count is higher, it can be adjusted by saline. Before actual testing, the platelet-rich plasma should be left at room temperature for approximately 30 min because the test is performed at 37°C. If the original platelet count of the patient is less than 100,000, then the test might be invalid. If the test needs to be performed, then the platelet count of the control should also be lowered. Platelet aggregation can also be performed using whole blood instead of platelet-rich plasma.

Various agonists used for this test include AA, collagen, ristocetin, ADP, and epinephrine. The agonists induce platelet aggregation. This will result in increased light transmission as aggregated platelets settle to the bottom. Strictly speaking, ristocetin does not cause aggregation, rather agglutination.

In general, agonists such as ADP and epinephrine are considered as weak agonists. These two weak agonists, in low concentration in a normal person, demonstrate two waves of aggregation. The primary wave of aggregation is due to activation of the GpIIb/IIIa receptor. The secondary wave is due to platelet granule release. Lack of secondary wave implies a storage pool disorder due to reduced number of granules or defective release of granule contents. These agonists (ADP and epinephrine) if used in higher concentration results in the two waves merging into one wave of aggregation. Collagen characteristically demonstrates an initial shape change before the wave of aggregation. This is seen as a transient increase in turbidity. Effective aggregation is typically considered as 70%—80% aggregation. However, values should be compared with the control. Values of 60% or more are generally considered to be adequate.

Various patterns of platelet aggregation include

- Normal: adequate aggregation (>60%) with ADP, collagen, epinephrine, AA, and ristocetin at high dose but not at low dose (0.6 mg/mL or less). Low-dose ristocetin will demonstrate aggregation values close to zero, less than 10%.
- von Willebrand/Bernard—Soulier pattern: adequate aggregation with ADP, collagen, epinephrine, and AA but not at higher dose of ristocetin.
- von Willebrand type IIB pattern/pseudo-von Willebrand pattern: increased aggregation with low dose of ristocetin. Normal individuals have aggregation values 0%—10%. Here, the values will be much higher.
- Glanzmann's thrombasthenia/hypofibrinogenemia pattern: adequate aggregation with ristocetin, but impaired aggregation with all other agonists is observed. Uremia and simultaneous use of multiple antiplatelet medications can also produce similar results.
- Disorder of activation (storage pool disorder): loss of secondary wave of aggregation with ADP and epinephrine at lower doses is the characteristic of this disorder. Storage pool disease is the most common inherited platelet function defect. It is subdivided into alpha granule and dense granule storage pool diseases.
- Aspirin effect: Significant impairment with aggregation with AA is observed in patients taking aspirin. There is usually impairment of aggregation (not as much as AA) with ADP, collagen, and epinephrine.

- Plavix (clopidogrel) effect/ADP receptor defect: significant impairment of aggregation with ADP at high concentration. At our institute, we use ADP at a concentration of 50uM/mL to elicit effect of Plavix. If there is impairment of aggregation at this concentration, we conclude presence of inhibitory effect of Plavix. There may be variable degree of impairment of aggregation with AA, collagen, and epinephrine.
- In chronic myeloproliferative disorder (CMPD), there may be impairment of aggregation with epinephrine.

Thromboelastography

The whole blood thromboelastography (TEG) is a method of assessing global hemostasis and fibrinolytic function that includes interaction of primary and secondary hemostasis, and subsequently, defect in one component of hemostasis can affect the other to certain extent. This technique has existed for more than 60 years, but improvement of technology has led to increased utilization of this test in clinical practice for monitoring hemostatic and fibrinolytic rearrangements [8]. TEG is a visualization of viscoelastic changes that occur during in vitro coagulation and provides a graphical representation of the fibrin polymerization process, but during interpretation of TEG data/tracing, it is most important to focus on the most significant defect. In classical thrombelastography, a small sample of blood (typically 0.36 mL) is placed into a cuvette (cup), which is rotated gently through $4°45'$ (cycle time 6/min) to imitate sluggish venous flow and activate coagulation. When a sensor shaft is inserted into the sample, a clot forms between the cup and the sensor. Once a clot has formed, the sensor now moves, and this movement results in a tracing. The speed and strength of clot formation depends on various factors including the activity of the plasmatic coagulation system, platelet function, fibrinolysis, and other factors that can be affected by illness, environment, and medications.

TEG analysis provides four basic parameters including R (reaction time), K value, angle alpha, and maximum amplitude (MA). Reaction time is measured in seconds and represents initial latency from start of the test until the initial fibrin formation (usually amplitude of 2 mm). K (denotes clot kinetics) value is also measured in seconds and indicates time taken to achieve a certain level of clot strength (usually amplitude of 20 mm). Alpha angle (degree) measures the speed of fibrin build up and cross-linking taking place and thus assesses the rate of clot formation. MA (measured in millimeter; mm) represents the ultimate strength of the fibrin clot. Possibly, the most important information provided by the TEG is clot strength, which may help to resolve if the bleeding is related to coagulopathy or a mechanical bleeding. Clot strength is measured by MA value and a low MA value indicates thrombocytopenia or platelet dysfunction. G is a computer-generated value reflecting the strength of the clot from initial fibrin blast to fibrinolysis.

$$G = (5000 \times \text{amplitude})/100 - \text{amplitude (normal } 5.2 - 12.4)$$

Therefore, MA and G value represent a direct function of the maximum dynamic properties of fibrin and platelet bonding via GPIIb/IIIa and represents the ultimate strength of the fibrin clot. G is the best measurement of clot strength [9]. CI (clot index) represents hemostasis profile and is calculated based on R, K, alpha angle, and MA. Ly30 (clot lysis at 30 min) indicates percentage decrease in amplitude at 30 min after MA, the stability, or degree of fibrinolysis. Various parameters obtained from TEG analysis are listed in Table 16.3. These parameters are also illustrated in Fig 16.1.

Table 16.3 Various parameters of thromboelastography (TEG).

Parameter	Comment
R: Reaction (measured in seconds)	The value indicates the time until the first evidence of a clot is detected.
K: Clot kinetics (measured in seconds)	K value is the time from the end of R value until the clot reaches 20 mm, and this value represents the speed of clot formation.
Angle Alpha (measured in degree)	Measures the rapidity of fibrin build-up and cross-linking (clot strengthening) and is the tangent of the curve made as the K is reached.
MA (Maximum amplitude; measured in millimeter) and G: measures clot strength	Direct measure of highest point of the TEG curve and represents clot strength.
CI: clotting index	CI is a mathematic equation calculated from R, K, alpha angle, and MA values.
Ly30 (clot lysis at 30 min)	Percentage of clot lysis 30 minutes after MA.

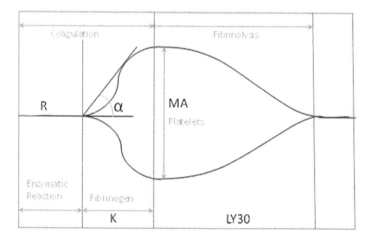

FIGURE 16.1

Different components of thromboelastography (TEG).

Use of TEG in management of bleeding patient

TEG may be used to dictate use of blood products and certain medication in a bleeding patient. Blood products that may be used in bleeding patients include packed red blood cells (PRBC), fresh frozen plasma (FFP), cryoprecipitate, and platelets. Medications that may be used include protamine and antifibrinolytic agents (e.g., epsilon amino caproic acid, tranexamic acid). However, use of PRBC is dictated by hemoglobin and hematocrit, and analysis TEG is not used for this purpose.

- Use of FFP based on TEG: FFP is indicated if the R time is prolonged. If the prolonged R is not due to anticoagulants (such as heparin), then the prolongation is due to clotting factor deficiency. In such patients, prothrombin time (PT) and partial thromboplastin time (PTT) should also be prolonged. If the volume of FFP that needs to be given is too high (for example, in patients with right ventricular strain or failure), then prothrombin complex concentrate may be used instead of FFP.

- Use of cryoprecipitate based on TEG: cryoprecipitate is used in bleeding patients with hypofibrinogenemia and uremic thrombocytopathia. Cryoprecipitate is also useful in von Willebrand disease patients who are bleeding. In hypofibrinogenemia, the angle alpha is low with a normal MA. These patients also have prolonged TT (thrombin time) and low levels of fibrinogen.
- Use of platelets based on TEG: platelet transfusion is valuable in thrombocytopenia and thrombocytopathia, and in both conditions MA should be low. Complete blood count (CBC) in addition will also document thrombocytopenia.
- Use of protamine based on TEG: protamine is used to neutralize heparin, and TEG must be performed with and without heparinase (enzyme which destroys heparin). If the R value in the TEG analysis without heparinase is >50% longer than the R with heparinase, protamine is indicated. In these patients, TT and PTT are also prolonged.
- Use of antifibrinolytics based on TEG: fibrinolysis may be primary or secondary. Causes of primary include physiological (due to thrombolytic therapy) or pathological. Secondary fibrinolysis is seen in DIC. In primary and secondary hyperfibrinolysis, Ly30 is prolonged. However, in primary fibrinolysis, MA value from the TEG analysis is low and CI value is also low. In secondary fibrinolysis, MA is normal or high with high CI. Antifibrinolytic agents are only used in patients with primary pathologic fibrinolysis.

Platelet mapping

Platelet mapping is a special TEG assay to measure effects of antiplatelet drug therapy on platelet function. Antiplatelet drugs, whose efficacy can be tested, include

- ADP receptor inhibitors such as clopidogrel and ticlopidine
- AA pathway inhibitors such as aspirin
- GPIIb/IIIa inhibitors such as abciximab, tirofiban, eptifibatide

The platelet mapping assay specifically determines the MA reduction present with antiplatelet therapy and reports the percent inhibition and aggregation.

The platelet mapping assay measures platelet function in the presence of antiplatelet drugs in a patient's blood sample. The results obtained by the TEG 5000 will be reported as % inhibition and % aggregation. The platelet mapping assay measures the presence of platelet-inhibiting drugs using whole blood and four different steps:

- No additive sample to measure total platelet function and the contribution of fibrin to MA and yields $MA_{THROMBIN}$. Thrombin overrides inhibition at other platelet activation pathways and indicates complete activation
- Activator F is added to measure the contribution of fibrin only to MA and yields MA_{FIBRIN}, which indicates no activation
- Activator F is added to the sample along with ADP to measure MA due to ADP receptor uninhibited platelets to yield the value of MA_{ADP}, which indicates activation of noninhibited platelets

- Activator F is added to the sample along with AA to measure MA due to TxA2 pathway to yield the value of MA_{AA}, which also indicates activation of noninhibited platelets.

The presence of platelet-inhibiting drugs is reflected in a reduction in the MA values. The % inhibition is a derived by the following equation:

$$\text{Percent MA Reduction} = 100 - [\{(MA_P - MA_F)/(MA_T - MA_F)\} \times 100]$$

where MA_P represents MA_{ADP} or MA_{AA}, MA_F represents MA_{FIBRIN}, and MA_T represents $MA_{THROMBIN}$. The optimum time, from the time the blood enters the syringe to the time it is placed in the TEG instrument, is 4 min.

Idiopathic thrombocytopenic purpura

ITP is immune-mediated destruction of platelets causing a low platelet count. Bone marrow shows increased megakaryocytes, and these megakaryocytes release relatively immature platelets, which are larger than normal. It is thought that in ITP, antibodies are formed against pathogens, which cross-reacts with platelet GpIb/IX, IIb/IIIa. Acute ITP is self-limiting. Chronic ITP lasts for more than a year, and 10% of individuals with chronic ITP have splenomegaly. ITP may be seen alone or sometimes as part of Evan's syndrome (with autoimmune hemolytic anemia).ITP may also occur in patients with systematic lupus erythematosus, chronic lymphocytic anemia, patients with HIV infection, and following stem cell transplantation.

Heparin-induced thrombocytopenia

HIT is a potential life-threatening complication arising from exposure to heparin. It may occur with exposure to unfractionated heparin or low molecular weight heparin. It occurs in up to 5% of individuals exposed to heparin, regardless of dose, schedule, or route of administration.
There are two major types of HIT:

- HIT I: here fall in platelet within first 2 days of heparin administration is observed. This disorder is nonimmune and due to direct effect of heparin on platelet activation. There is no need to discontinue heparin therapy.
- HIT II: affecting 0.2%−5% of patients who are receiving heparin for more than 4 days. Clinical features include heparin administration for more than 4 days with reduction of platelet count by 50% or more. Patients who are at risk for HIT II are females, surgical patients (especially cardiac and orthopedic surgery), patients with previous exposure to heparin, and patients receiving unfractionated heparin. In HIT II, antibodies are formed, which binds to PF-4 and results in platelet aggregation causing thrombosis. It is thought that the antibodies are initially formed against antigens mimicking PF4/heparin complex (e.g., PF4 complex with polysaccharide on surface of bacteria).

HITT stands for heparin-induced thrombocytopenia with thrombosis.

Pathophysiology

The PF4 protein is found in the alpha granules of platelets. PF4 when released on platelet activation binds and neutralizes heparin. Heparin−PF4 complexes are found on platelet surfaces. Binding of heparin with PF4 creates a neoantigen, and antibodies are formed against this neoantigen.

HIT antibodies bind to the heparin—PF4 complexes on the platelet surfaces. The Fc portion of the HIT antibody binds with Fc receptors on the same or adjacent platelets, which leads to further platelet activation. Further platelet activation causes further release of PF4.

HIT antibodies are of IgG, IgM, and IgA subclasses. Only IgG HIT antibodies are thought to be pathological.

Patient with HIT develop thrombocytopenia as platelets are consumed with thrombus formation. In addition, IgG-coated platelets are removed by macrophages of the reticuloendothelial system.

Risk factors for heparin-induced thrombocytopenia

- HIT occurs more commonly with unfractionated heparin exposure than LMW heparin exposure
- A therapeutic dose of heparin is thought to result in greater likelihood of HIT than prophylactic doses.
- HIT is seen more in females than males
- Surgical patients have increased risk of HIT. This is especially true for patients undergoing cardiac surgery.

Assessment of a patient with heparin-induced thrombocytopenia

The 4T's score is an easy-to-use score system that quantifies the clinical findings associated with HIT. The score assesses

- Degree of thrombocytopenia
- The timing of thrombocytopenia relative to heparin exposure
- The presence of thrombosis
- Other causes of thrombocytopenia

Calculating the 4T score

Thrombocytopenia:

- Platelet count fall >50% and nadir >/ = 20,000: 2 points
- Platelet count fall 30%—50% or nadir 10—19,000: 1 point
- Platelet count fall <30% or nadir <10,000: 0 point

Timing:

- Clear onset of thrombocytopenia between 5 and 10 days or platelet count fall at = /<1 day if there is prior exposure to heparin within the last 30 days: two points
- Consistent with fall at 5—10 days but unclear (e.g., missing platelet count), onset after 10 days or fall = /<1 day with prior heparin exposure within 30—100 days: one point
- Platelet count fall at <4 days without recent exposure: 0 point

Thrombosis:

- Confirmed new thrombosis, skin necrosis, or acute systemic reaction after IV unfractionated heparin bolus: two points
- Progressive or recurrent thrombosis, nonnecrotizing (erythematous) skin lesions, or suspected thrombosis: one point

- None: 0 point

Other causes of thrombocytopenia:

- None apparent: two points
- Possible: one point
- Definite: 0 point

Probability of HIT, based on the 4T score:

- 0–3 points: low
- 4-5 points: intermediate
- 6-8 points: high

The tests for HIT include

- ELISA-based test: heparin PF4 complexes are coated on a plate to detect antibody in the patient's plasma/serum. The test is very sensitive but with low specificity. Results are published as positive or negative based on the OD (optical density) value of the patient compared with the control. It is important to appreciate that individuals with positive ELISA do not necessarily have platelet-activating antibody. However, individuals with OD > 1.0 are more likely to be associated with platelet-activating antibody.
- Heparin-induced platelet aggregation: heparin induced-platelet aggregation is performed using donor platelets and patient's serum. The control test has no heparin added to it. Another test will be performed with low concentration of heparin being added to it (0.1–0.3 U/mL). Another test will be performed with high concentration of heparin added to it (10–100 U/mL). A positive test is one where there is platelet aggregation (more than 25%) with low concentration of heparin, but not with high or in the control.
- Serotonin release assay: platelets from a normal donor are incubated with radiolabeled serotonin (^{14}C), and then labeled platelets are incubated with patient's serum in the absence and presence of heparin in therapeutic (0.1 U/mL) or high concentration of heparin (100 U/mL). For a test to be positive, there must be at least 20% of radioactivity release in the presence of 0.1 U/mL of heparin, but release of radioactivity must be substantially reduced in the presence of high heparin concentration. Serotonin release assay is considered the gold standard. However, this is often a send out test, and thus results are available after considerable delay.

Secondary hemostasis

Secondary hemostasis takes place with formation of fibrin clot. The clotting pathways consist of extrinsic pathway, intrinsic pathway, and the common pathway. The concept of the coagulation as a series of stepwise enzymatic conversion was first proposed in 1964 and was described under the headings of intrinsic pathway (dependent on contact activation by a negatively charged surface and involving coagulation factors V, VIII, IX, XI, and XII) and extrinsic pathway (dependent on tissue factor being exposed to circulation and involving tissue factor and factor VIII), then converging to a common pathway to activate factor X.

Extrinsic pathway

Tissue factor (factor III) and factor VII together activates factor X.

Intrinsic pathway

The intrinsic pathway starts with activation of factor XII, which in turn activates factor XI. Activated factor IX and activated factor VIII activate factor X. Activation of factor XII requires kallikrein and high molecular weight kininogen (HMWK) for activation. Kallikrein is derived from prekallikrein.

Common pathway

Activated factor X with the help of factor V converts prothrombin (factor II) to thrombin (factor IIA). Thrombin converts fibrinogen (factor I) to fibrin. Fibrin is stabilized by factor XII.

Important points to note regarding the clotting pathway include

- PT is used to assess integrity of the extrinsic and common pathway
- PTT is used to assess integrity of the intrinsic and common pathway
- All clotting factors are produced by the liver with the exception of factor VIII. Factor VIII is produced by endothelial cells
- Calcium (factor IV) is required in multiple steps
- Factors V and VIII are cofactors
- Factor VII has the shortest half-life (4–6 h). With acute liver dysfunction, levels of factor VII are reduced early. Factor VII is required for the extrinsic pathway, and therefore, PT is a very good test to assess liver function in the acute setting
- Individuals with factor XII deficiency or HMWK or kallikrein deficiency do not bleed, but PTT may be prolonged.
- Individuals with factor XIII deficiency have normal PT and PTT but bleed

All deficiencies of clotting factors are transmitted as autosomal recessive with the exception of factor VIII deficiency (hemophilia A) and factor IX deficiency (hemophilia B), which is transmitted as X-linked recessive. Dysfibrinogenemia meaning fibrogen is dysfunctional and is transmitted as autosomal dominant.

Tests for secondary hemostasis

Various tests for evaluating secondary hemostasis are addressed in this section.

Prothrombin time

PT is a widely used test to evaluate secondary hemostasis. In this test, platelet poor plasma from a patient (collected in a blood collection tube containing sodium citrate) is mixed with thromboplastin and calcium, and then clotting time is determined at 37°C using a variety of methods including photooptical and electromechanical. Automated coagulation analyzers are commercially available for measuring PT along with other coagulation parameters. PT is a functional measure of the extrinsic pathway, and common pathway and reference range is 8.8–11.6 s. Therefore, PT is a useful test to detect inherited or acquired defects in coagulation related to extrinsic pathway. However, often PT is reported as INR (International Normalized Ratio). The thromboplastin used may vary from laboratory to laboratory and from country to country. However, reporting results as INR ensures results are comparable between different laboratories.

$$INR = [Patient\ PT/Mean\ normal\ PT]^{ISI}$$

In this equation, ISI stands for international standardized index, which is available from the reagent package insert. Normal value of INR is 0.8−1.2. Causes of isolated prolonged PT (i.e., PTT is normal) include

- Coumarin (warfarin)
- Vitamin K deficiency (dietary deficiency, defective absorption such as seen in cholestatic jaundice)
- Liver disease

Vitamin K is responsible for carboxylation of glutamic acid residues of factors II, VII, IX, and X. Dietary vitamin K is vitamin K1 (phytonadione) and vitamin K that is absorbed from the gut by bacterial activity is vitamin K2 (menaquinone). Vitamin K2 has 60% activity of vitamin K1. Vitamin K is converted to the epoxide form by the enzyme epoxidase. The epoxide form is inactive. The epoxide form is converted back to active vitamin K by the reductase enzyme. The full name of the enzyme complex is vitamin K epoxide reductase complex I (VKROCI). Warfarin blocks the reductase enzyme. There is genetic variation in the gene for VKROCI amongst individuals. Warfarin is metabolized by the P450 CYP system, including CYP2C9. Polymorphism of the CYP2C9 affects CYP2C9 activity. Variation in activity could result variation in the half-life of warfarin amongst different patients. Drugs may also affect activity of CYP2C9. Thus, genetic variation of VKROC I gene, polymorphism of CYP2C9, and effect of other drugs all result in variation in warfarin activity amongst individuals. The other vitamin K dependant proteins are protein C and protein S.

Partial thromboplastin time

PTT (also known as activated partial prothrombin time) is another useful test for evaluation of secondary homeostasis. In this test, patient's platelet poor plasma (citrated plasma but oxalate can also be used), surface activating agent (silica, kaolin, Celite, or ellagic acid), calcium, and platelet substitute (crude phospholipid) are mixed, and clotting time is determined most likely using an automated coagulation analyzer. It is a functional measure of the intrinsic pathway as well as common pathway and can detect hereditary or acquired defects of the coagulation factors XII, XI, X, IX, VIII, and V, prothrombin, and fibrinogen. PTT or APTT is called partial thromboplastin time because of the absence of tissue factor (thromboplastin) in the tests and the normal value varies from laboratory to laboratory but usually between 25 and 39 s. Causes of isolated prolonged PTT (i.e., PT is normal) include

- Heparin, direct thrombin inhibitors (DTIs)
- Factor deficiency of the intrinsic pathway (including HMWK and prekallikrein deficiency)

Important examples are hemophilia A due to factor VIII deficiency and hemophilia B due to factor IX deficiency.

Table 16.4 Interpretation of prothrombin time (PT) and partial thromboplastin time (PTT) results in various clinical scenarios.

PT result	PTT result	Clinical scenario
Normal	Prolonged	Factor deficiency of the intrinsic pathway, above the common pathway (e.g., factor VIII, factor IX deficiency) Heparin therapy, direct thrombin inhibitor therapy Von Willebrand disease Inhibitors (e.g., lupus anticoagulant, factor VIII or IX inhibitor)
Prolonged	Normal	Liver disease, vitamin K deficiency, warfarin therapy
Prolonged	Prolonged	Factor deficiency of the common pathway High dose of heparin, direct thrombin inhibitor, warfarin Strong lupus anticoagulant

- Inhibitors: VIII and IX inhibitors, lupus anticoagulant, or lupus antibody (LA)
- von Willebrand disease

Please note deficiency of factor XII, HMWK, and prekallikrein may result in prolongation of PTT, but there is no bleeding in these individuals.

Various scenarios where PT, PTT, or both may be prolonged are listed in Table 16.4.

Heparin and DTIs typically cause prolonged PTT. They will also cause prolonged TT. They typically do not cause significant prolongation of PT, but with higher dose, both may cause some prolongation of PT. Currently, four parenteral DTIs including lepirudin, desirudin, bivalirudin, and argatroban are FDA-approved [10].

There are three types of hemophilias, A, B, and less well-known C. Hemophilia A is due to inherited deficiency of factor VIII. Hemophilia B is due to inherited deficiency of factor IX. Hemophilia C is due to inherited deficiency of factor XI. Hemophilia A and B are transmitted as X-linked recessive, and hemophilia C is transmitted as autosomal recessive.

Hemophilia A occurs in approximately 1 in 10,000 individuals, and up to 30% of cases may be due to spontaneous mutation. Over 100 mutations in the factor VIII gene have been described. Hemophilia A is clinically divided into mild (>10% factor VIII activity), moderate (2%−10% factor VIII activity), and severe (<1% factor VIII activity). Treatment includes use of desmopressin acetate for mild cases (causes 2- to 10-fold increase in factor VIII level) and factor VIII replacement (1 unit/kg raises factor VIII level by 2%). Factor VIII has a half-life of 8 h. Typical target values of factor VIII activity in various situations include 100% for surgery, central nervous system bleeding, gastrointestinal and genitourinary bleeds, and 40%−80% for bleeding into joints and muscle. Factor VIII levels in female carriers are about 50% of normal, but levels of vWF are normal. Thus, most hemophilia carriers have factor VIII to von Willebrand antigen ratio less than 0.5.

Hemophilia B is less frequent than hemophilia A and occurs in about 1 in 25,000 males. A variation of hemophilia B is known as Leyden phenotype where features of hemophilia in childhood improve after puberty. In these patients, a mutation in the promoter region of the gene, which disrupts hepatocyte nuclear factor 4, takes place, but the site for androgen response element is intact.

Hemophilia C occurs in about one in 1 million individuals. However, Hemophilia C is the second most common inherited bleeding disorder in women. It also has a high prevalence in Ashkenazi Jews. Unlike factor VIII and factor IX, factor XI deficiency has to be treated with FFP. Interestingly, hemophiliacs are protected against coronary artery disease, and coinheritance of factor V Leiden improves symptoms.

Inhibitors to factor VIII and factor IX may develop spontaneously (autoantibodies) or posttreatment (alloantibodies) with factor replacement. Incidence of alloantibody formation in hemophiliacs is approximately 1%. Autoantibodies may be associated with pregnancy, autoimmune diseases, malignancy, or allergy. Autoantibodies may respond to immune suppression medication. Tests to detect the presence of these antibodies are factor VIII and factor IX inhibitor screen and when positive followed by factor VIII or IX inhibitor assay. The amount of inhibitor quantified by the assay is expressed as Bethesda unit (BU) for both, where one BU of inhibitor is that amount of inhibitor that destroys half the factor VIII or IX:C activity in an equal mixture of patient and normal plasma. If the result is 5 or less, it is considered that the antibody is in low titers. In such cases, increasing doses of factor VIII or IX may be useful. If the result is greater than 5, then the individual has high titers. In such cases, factor VIII inhibitor bypass activity will be required. This may be done by use of prothrombin complex concentrate or activated factor VII.

Spurious causes of prolonged PTT include high hematocrit, underfilling of citrate tube and EDTA contamination, which may happen if purple top tube is collected before blue top tube. Delay in transport and processing (testing should be done within 4 h of collection; sample should be stored at room temperature) may also cause spurious result.

Thrombin time

TT is a test where patient's plasma is mixed with thrombin and clotting time is determined. It is a measure of functional fibrinogen. Heparin produces prolonged TT. Patients on heparin, however, have a normal reptilase time. Causes of prolonged TT include

- Heparin
- Hypofibrinogenemia
- Dysfibrinogenemia
- Thrombolytic therapy

PT/PTT mixing study

Individuals with prolonged PT or prolonged PTT or both may undergo PT/PTT mixing study. The objective of this test is to determine if the cause of prolonged PT or PTT is due to a factor deficiency or due to the presence of inhibitor. In this test, patient's plasma is mixed with an equal volume of normal plasma, and PTT is measured at 0 and 1–2 h. Failure of correction of prolonged PTT means presence of inhibitors. If results at 0 and 1–2 h are similarly prolonged, it implies lupus anticoagulant. If results show time-dependent prolongation, it implies coagulation factor antibody (e.g., factor VIII or IX inhibitor). If PT and PTT are both prolonged and mixing study shows correction, then most likely there is deficiency of factor in the common pathway. Hemophiliacs typically have prolonged PTT with normal PT. Mixing study should show correction.

Causes of prolonged PT and PTT, not showing correction of either include

- Heparin, DTI
- Lupus anticoagulant
- Factor inhibitor (factor II, factor V, and factor X inhibitor; factor V inhibitor is the most common factor inhibitor of the common pathway).

Factor assays

PTT-based factor assay performed for factors VIII, IX, XI, and XII. PT-based factor assay is performed for factors II, V, VII, and X. For these tests, commercially available normal plasma is used as reference plasma, while factor deficient plasmas are also available, which are completely deficient of factor of interest (for example, factor VIII deficient plasma for factor VIII assay). The major steps in the procedure include

- Serial dilution of the reference (normal plasma) is made, and each dilution is mixed with an equal volume of factor deficient plasma and PTT is measured.
- The clotting times (y axis) are plotted against dilution (x axis) on Log-Lin graph. A best fit line is drawn. This is the reference line/curve (first line)
- The patient plasma is treated the same way as the reference plasma (i.e., serial dilution and mixed with equal volume of factor deficient plasma), and PTT is measured.
- The clotting times (y axis) are plotted against dilution (x axis) on Log-Lin graph. A best fit line is drawn. This is the reference line/curve (second line). The second line should be parallel to the first line or be superimposed to the first line.

If the second line is not parallel, then maybe an inhibitor is present. Often with inhibitor, the clotting time shortens with increasing dilution as the inhibitor is diluted out. If all the clotting time is grossly prolonged and does not change with dilution, then maybe the clotting factor is less than 1%. The two lines are used to calculate factor activity in the patient plasma, for example, if a 1:20 dilution of normal plasma shortens the clotting time of deficient plasma to the same extent as a 1:5 dilution of the patient's plasma, the latter sample has 25% of normal activity. Assay results are expressed as %. For fibrinogen, level is expressed as mg/dL.

Important points in interpretation of coagulation tests:

- Abnormal PT or PTT results may be explained by medication effect or clotting factor deficiency or inhibitors to clotting factors
- Clotting factor deficiency may be inherited or acquired
- Most common medication prolonging only PTT is heparin
- Most common medication prolonging only PT is warfarin
- Most common clotting factor deficiency resulting in isolated prolonged PT is factor VII deficiency, due to vitamin K deficiency or liver disease
- Inherited causes of prolonged PTT are hemophilia A and B and von Willebrand disease
- Causes of prolonged PTT, which does not show correction, are factor VIII inhibitor, factor IX inhibitor, and LA
- Individuals with factor VIII or IX inhibitor may have history of bleeding
- Individuals with LA may have history of thrombosis

- If a patient has prolonged PTT, mixing study does not show correction, and factor VIII assay is low, then the patient must have a factor VIII inhibitor. If the patient had factor VIII deficiency, mixing study should have corrected.
- If a patient has prolonged PTT, mixing study does not show correction and factor IX assay is low, then the patient must have a factor IX inhibitor

Tests for bleeding patients

Initial tests for a bleeding patient that are typically done are

- CBC
- PT
- PTT
- +/− TT (TT would pick up dysfibrinogenemia that causes prolonged TT with normal PT and PTT)

In most cases, these yield a result. However, in certain situations, additional testing is required. In these situations, we need to test for

- Platelet dysfunction: the gold standard for platelet dysfunction is platelet aggregation test. PFA-100 may be considered as a screen test.
- Factor XIII deficiency or factor XIII inhibitors: factor XIII screen or factor XIII levels may be ordered. Factor XIII screen is performed by observing clot solubility in 5M urea solution.
- Primary fibrinolysis: euglobulin clot lysis time is a test to document primary fibrinolysis. This test is not easily available. DIC screen would exhibit low fibrinogen with high D dimmers. TEG would demonstrate high Ly30 values.

Von Willebrand disease

Von Willebrand disease is the most commonly inherited bleeding disorder in the general population. An estimated 1%−2% of the population is affected by this disease (estimated between 60−120 million people worldwide). This disease is 150 times more prevalent than hemophilia and affects men and women equally including all racial, ethnic, and socioeconomic groups and is found worldwide.

Von Willebrand disease is a group of genetically heterogenous disorders, resulting in abnormal function of the vWF, and more than 100 mutations have been described. It is predominantly transmitted as an autosomal dominant condition. However, only a small fraction of individuals inheriting the gene suffer from a clinically significant diathesis.

vWF is an unusual extremely large multimeric glycoprotein composed of repeating units that are polymerized from dimer subunits by disulfide bonds. Propeptides are synthesized in the endothelial cells and megakaryocytes and processed. Then vWFs are released as series of multimers with larger multimeric compounds being more functional. vWF acts as a carrier protein for factor VIII and prevents its inactivation by protein C. It also serves as the ligand that binds GpIb receptor on platelets to initiate platelet adhesion to damaged blood vessels. Factor VIII is brought into storage granules (Weibel−Palade bodies in endothelial cells and platelet alpha granules) by vWF. In the absence of this factor, factor VIII survival is drastically reduced resulting in secondary factor VIII deficiency [11]. There are three main types of von Willebrand disease:

- Type 1 is the most common type where 75% of all people with von Willebrand disease are accounted for. These patients have partial quantitative deficiency with 20%−50% reduction of vWF, but the factor is structurally normal. This disorder is transmitted as autosomal dominant with reduced penetrance and variable expressivity. Clinical symptoms are usually mild.
- Type 2: these patients there have qualitative defect of vWF, and these patients account for 20%−25% of all cases. The type 2 disorder can be subclassified into four subtypes: 2A, 2B, 2M (M for multimer), and 2N (N for Normandy). Type 2N is transmitted as autosomal recessive. All others are transmitted as autosomal dominant
- Type 3: this is the rarest type where patients have total deficiency of vWF resulting in sever form of disease. This disorder is transmitted as autosomal recessive
- Pseudo (platelet-type) von Willebrand disease
- Acquired von Willebrand disease

  Testing for von Willebrand disease include following tests:

- CBC (in certain subtypes, type 2B and platelet thrombocytopenia may be present).
- Bleeding time should be prolonged as vWF is required for platelet adhesion.
- PTT (PTT should be prolonged due to low levels of factor VIII). Mixing study should show correction.
- von Willebrand panel: this panel of tests consists of factor VIII levels, von Willebrand antigen assay (measures quantity of vWF), and ristocetin factor activity (vWF:RcoF), which is a functional test (ability to aggregate normal platelets in the presence of ristocetin). Typically all three should be low.
- Ristocetin-induced platelet aggregation: ristocetin induces von Willebrand and Gp1b interaction causing platelet aggregation. Normal healthy individuals should show adequate aggregation with higher dose of ristocetin. These normal individuals should not show aggregation with low dose ristocetin (0.6 and below).
- Multimer analysis (by electrophoresis; allows analysis of multimer according to size).
- Factor VIII to vWF binding assay (assesses ability of binding of vWF and factor VIII).

Diagnosis of various types of von Willebrand disease

It is important to establish the diagnosis of von Willebrand disease by classifying the subtype. Following approach is usually helpful:

- Type I: CBC does not show thrombocytopenia, but bleeding time may be prolonged. PTT should be prolonged. Abnormal tests on von Willebrand panel (all decreased) are also other characteristics. No aggregation with low-dose ristocetin (just like normal individuals) is observed, but ristocetin at higher dose also fails to induce platelet aggregation. Multimer analysis will demonstrate presence of all various multimers; however, quantity will be reduced. Type I responds to desmopressin. Desmopressin induces release of stored vWF and improves clinical condition.

- Type III: CBC does not show thrombocytopenia, but bleeding time may be prolonged. PTT should also be prolonged. Abnormal von Willebrand panel (all severely decreased) and ristocetin at higher dose fail to induce platelet aggregation. No aggregation with low-dose ristocetin (just like normal individuals) is also observed. Multimer analysis should demonstrate absence of all mutimers.
- Type IIA: In this disorder, reduced levels of large and intermediate sized multimers are observed. CBC does not show thrombocytopenia, but bleeding time may be prolonged. PTT should be prolonged along with abnormal von Willebrand panel (all decreased). Ristocetin at higher dose fails to induce platelet aggregation. No aggregation with low-dose ristocetin (just like normal individuals). There are two subtypes of type 2A: 2A-1 (defect in intracellular transport and 2A-2: vWF more susceptible to proteolysis by the vWF cleaving protease: ADAMTS13 which stands for a disintegrin and metalloprotease with a thrombospondin type 1, motif 13).
- Type IIB: In this disorder, there is increased affinity of the large vWF multimers for platelet binding (to GpIb). vWF binds to platelets in the bloodstream, and these large bundles of platelets are removed from circulation with resultant thrombocytopenia. Thus, CBC does show thrombocytopenia, but bleeding time may be prolonged. PTT may be normal or prolonged. Von Willebrand panel may be normal or decreased. Ristocetin induces aggregation with low dose, which is a reflection of increased affinity of vWF to platelet GpIb. Multimer analysis should demonstrate reduction of large multimers because large multimers along with the platelets are consumed.

An identical clinical and laboratory phenotype is seen in the platelet GPIb receptor abnormally, where there is increased affinity of GpIb to bind to normal vWF. This is called "pseudo" or platelet-type von Willebrand disease. To distinguish these two disorders, repeating the test using normal (donor) fixed washed platelets and incubating them with the patient's plasma and with low or higher dose of ristocetin will not result in aggregation. This is because the donor (normal) platelets are normal with no increased affinity for vWF.

- Type 2M (M for multimer): In this disorder, there is mutation in the A1 region resulting in decreased platelet-dependent function. The multimers are present but dysfunctional. Therefore, von Willebrand disease panel will show normal vWF antigen assay but low functional assay. Multimer analysis should demonstrate normal multimer pattern.
- Type 2N (N for Normandy): This disorder is due to mutation in the factor VIII binding region of vWF resulting in rapid turnover of unbound factor VIII, causing a reduced plasma level of factor VIII. It is transmitted as autosomal recessive pattern. Factor VIII levels are quite low ($<10\%$) and clinically it mimics hemophilia A (autosomal hemophilia). In cases of apparent hemophilia A with poor factor VIII recovery and survival in vivo and normal mixing studies, von Willebrand disease type 2N should be considered in the differential diagnosis. vWF to factor VIII binding assay is necessary to confirm the diagnosis.
- Acquired von Willebrand disease: This condition typically presents as a sudden onset of mucocutaneous bleeding in a previously asymptomatic patient. Mechanisms include antibody formation against vWF with resultant increased clearance of the factor from circulation or inhibition of function of vWF. Another mechanism may be due to adsorption of vWF by tumor cells. Tumor cells may have aberrant GP Ibα receptor expression.

Management of von Willebrand disease includes use of

- Desmopressin: Twofold to tenfold increase in plasma vWF level, but this therapy is contraindicated in type 2B and platelet-type von Willebrand disease because release of abnormal vWF may induce thrombocytopenia
- Cryoprecipitate
- Factor VIII concentrates

Antiplatelets and anticoagulants

Antiplatelets used in clinical practice include the following:

- Aspirin: This drug inhibits cyclooxygenase enzyme. Efficacy of aspirin can be assessed by VFN test where effective inhibition is indicated by values less than 550 ARU. No antidotes available. Bleeding due to aspirin requires platelet transfusion.
- Clopidogrel (Plavix): This drug inhibits P2Y12 ADP receptors. Other members of the same group are prasugrel and ticagrelor. Efficacy can be assessed by VFN test where effective inhibition is indicated by values less than 210 PRU. No antidote is available and bleeding due to clopidogrel requires platelet transfusion. This drug should be stopped typically 5 days before surgery.
- Ticagrelor (Brilinta): This drug also inhibits P2Y12 ADP receptors. Literature suggests that efficacy can be assessed by VFN test. No antidote is available and bleeding due to ticagrelor requires platelet transfusion. This drug is typically stopped 3—5 days before surgery. Other P2Y12 inhibitors that are FDA approved include prasugrel and ticlopidine.
- Eptifibatide (Integrilin): This drug inhibits GpIIb/IIIa receptors. No antidotes available. Bleeding due to eptifibatide requires platelet transfusion, and this drug is typically stopped 3—6 h before surgery. Other GpIIb/IIa inhibitors include abciximab and tirofiban.
- Phosphodiesterase inhibitors: These agents inhibit the enzyme phosphodiesterase and cause accumulation of cAMP resulting in impaired platelet function. In this group, dipyridamole and cilostazol are commercially available.

Various anticoagulants are also used in therapy. These drugs include the following:

- Warfarin: This drug is a vitamin K antagonist and efficacy is assessed by measuring PT (INR). If patient is bleeding while on warfarin, effective management includes FFP infusion. If there is no bleeding, but PT is prolonged and requires reversal, vitamin K administration in dosage of 1—10 mg oral or intravenous is recommended. The goal is to bring the INR to less than 1.5.
- Unfractionated heparin: This agent inhibits factors IX, X, XI, and XII. Antithrombin III is required for effective function and a low level of ATIII is a cause of heparin resistance. Efficacy is measured by PTT (ideally after 6 h of dose adjustment). Antidote of heparin is protamine. Patients on heparin should have prolonged TT. During surgery using heparin (e.g., as in cardiopulmonary bypass), heparin effect is monitored by activated clotting time). Heparin efficacy can also be monitored by anti-Xa assay. This is applicable when PTT measurement can be relied on, for example, when a patient has a lupus anticoagulant (prolongs PTT) and is on heparin.

- Low molecular weight heparin: This agent inhibits factor Xa, and efficacy is monitored by antifactor Xa assay (ideally after 4 h after dose adjustment). Antidote is protamine (66% efficient when compared with neutralization of unfractionated heparin).
- DTI: Examples of these agents in use are bivalirudin (Angiomax), dabigatran (Pradaxa), argatroban, and lepirudin. None of these agents have antidotes. Bivalirudin has the shortest half-life (approximately 25 min). Impaired renal function results in impaired excretion in all except argatroban (hepatic clearance). Efficacy can be monitored by PTT. The Ecarin clotting time can also be used to monitor DTIs. Venom extract of the *Echis carinatus* snake coverts prothrombin to meizothrombin. Meizothrombin is an intermediate product in the conversion of prothrombin to thrombin. DTIs inhibit the conversion of prothrombin to meizothrombin. DTIs may be immunogenic and cause formation of antibodies. This will reduce efficacy of the drugs. This is least likely with bivalirudin. DTIs are used for HIT.
- Direct factor Xa inhibitors: Examples include apixaban, rivaroxaban, idraparinux, fondaparinux, dalteparin, and enoxaparin. All of these drugs are mainly excreted by the kidneys except apixaban, which is excreted by the kidneys (25%) and by the liver (75%). Enoxaparin and dalteparin may be countered partially with protamine. Others do not have any antidote.

Key points

- Hemostasis consists of three steps: vasoconstriction, which is mediated by reflex neurogenic mechanisms, platelet plug (primary hemostasis), and activation of the coagulation cascade (secondary hemostasis).
- Platelet glycoproteins: GpIb/IX/V attaches to vWF. GpIIb/IIIa attaches to fibrinogen and GpIa/IIa and GpVI attaches to collagen. GpIa/IIa acts as a receptor for platelet adhesion to collagen. Binding of collagen to GpVI initiates platelet aggregation and platelet degranulation.
- Platelet alpha granules are granules stained by the Wright-Giemsa stain and contain fibrinogen, fibronectin, factor V, vWF, platelet factor (PF-4), PDGF, TGF-β, and thrombospondin. Deficiency of alpha granules could result in platelets, which appear pale and gray on peripheral smear, and this is referred to as gray platelet syndrome. It is transmitted as autosomal recessive.
- Platelet dense bodies/delta granules are electron dense due to presence of calcium, which appears as dark bodies under electron microscope. These dense bodies are present in low numbers, less than 10 per platelet, and contain ATP, ADP, ionized calcium, histamine, serotonin (5-HT), and epinephrine.
- The three main platelet events are (a) adhesion and shape change, which occurs due to interaction between vWF and GP Ib/IX/V receptors. In von Willebrand disease as well as Bernard–Soulier syndrome (where GpIb/IX/V is lacking), abnormal platelet function (thrombocytopathia) is observed; (b) platelet activation where there is rise in cytoplasmic calcium with change in shape of platelets, extension of pseudopodia, and release of chemicals (release reaction); and c) platelet aggregation, which is fibrinogen-mediated binding of activated GP IIb/IIIa receptors on adjacent platelets, which is augmented by thrombospondin, a component of α-granules. Therefore, with lack of GpIIb/IIIa (known as Glanzmann's thrombasthenia, transmitted as autosomal recessive) and hypofibrinogenemia, abnormal platelet function is observed.

- Pseudothrombocytopenia may be due to platelet clumps that occurs when blood is collected in EDTA, platelet satellitism, large platelets, or traumatic venipuncture.
- Macrothrombocytopenia with neutrophilic inclusions (MYH9 disorders): This group of disorders is characterized by mutations in the MYH9 gene, which encodes the nonmuscle myosin heavy chain class IIA protein. These are characterized by thrombocytopenia, large/giant platelet, and Dohle-like bodies in neutrophils. This group of disorders includes May–Hegglin anomaly (autosomal dominant), Sebastian syndrome, Fechtner syndrome (nephritis, ocular defects, and sensorineural hearing loss), and Epstein syndrome.
- Wiskott–Aldrich syndrome: This disorder is due to defect in the WASP gene. It is characterized by immune deficiency, eczema, and thrombocytopenia with small platelets.
- Bernard–Soulier syndrome: This disorder is due deficiency of GpIb/IX/V receptor. There is also thrombocytopenia with large/giant platelets. Some cases of Bernard–Soulier syndrome are due to defects of the GpIb beta gene located on chromosome 22. This gene may be affected in velocardiofacial syndrome or DiGeorge syndrome associated with deletion of 22q11.2.
- HIT II: About 0.2%–5% of patients on heparin for more than 4 days experience this type of thrombocytopenia. Clinical clues for HIT II include heparin administration for more than 4 days with reduction of platelet count by 50% or more. Patients who are at risk for HIT TT are use of unfractionated heparin, previous exposure to heparin, females, and surgical patients, especially cardiac and orthopedic surgery.
- Tests for HIT are ELISA, heparin-induced platelet aggregation, and serotonin release assay. Serotonin release assay is considered as the gold standard.
- PFA-100: If the CEPI membrane closure time is prolonged, but ADP closure time is normal; this is most likely due to aspirin. If both CEPI and ADP closure time are prolonged or CEPI closure time is normal and ADP closure time is abnormal, this denotes platelet dysfunction or von Willebrand disease. However, PFA-100 test is insensitive to von Willebrand disease type 2N, patients receiving clopidogrel and ticlopidine, and storage pool disease. It is very sensitive to von Willebrand disease type I and therapy with GpIIb/IIIa antagonists.
- VFN is a rapid, turbidimetric whole blood assay that assesses platelet aggregation. It is based on the ability of activated platelets to bind with fibrinogen.
- Normal platelet aggregation: Indicates adequate aggregation with ADP, collagen, epinephrine, AA, and Ristocetin at high dose but not at low dose (0.6 mg/mL or less).
- Platelet aggregation in von Willebrand disease/Bernard–Soulier pattern: Adequate aggregation with ADP, collagen, epinephrine, and AA, but not at higher dose of ristocetin.
- Platelet aggregation in von Willebrand disease type IIB pattern/pseudo vWD pattern: Increased aggregation with low dose of ristocetin.
- Platelet aggregation in Glanzmann's thrombasthenia/hypofibrinogenemia pattern: Adequate aggregation with ristocetin, but impaired aggregation with all other agonists. Uremia and antiplatelet medication can also produce similar results.
- Platelet aggregation in disorder of activation (storage pool disorder): Loss of secondary wave of aggregation with ADP and epinephrine at lower doses. Storage pool disease is the most common inherited platelet function defect. It is divided into alpha granule and dense granule storage pool diseases.
- Aspirin effect on platelet aggregation: Significant impairment with aggregation with AA. There may be impairment of aggregation (not as much as AA) with ADP, collagen, and epinephrine.

- Plavix effect/ADP receptor defect: Significant impairment with aggregation with ADP.
- In CMPD, there may be impairment with aggregation with epinephrine.
- Thrombelastography (TEG): TEG is a global test for hemostasis, which includes interaction of primary and secondary hemostasis. Parameters of TEG include R (reaction time measured in seconds representing initial latency from start of the test until the initial clot formation), K (value measured in seconds representing time taken to achieve a certain clot strength usually 20 mm), alpha angle (measures the speed of fibrin build up), and MA (MA measured in millimeter and represents clot strength). TEG also produces certain other calculated parameters such as G (computer-generated value representing clot strength), CI (a value calculated based on R, K, alpha angle, and MA), and Ly30 (clot lysis 30 min indicates percentage of decrease in amplitude 30 min after MA and indicates stability of fibrinolysis).
- Platelet mapping is a special TEG assay to measure effects of antiplatelet drug therapy on platelet function.
- Regarding the clotting pathway: PT is used to assess integrity of the extrinsic and common pathway; PTT is used to assess integrity of the intrinsic and common pathway.
- All clotting factors are produced by the liver with the exception of factor VIII. Factor VIII is produced by endothelial cells.
- Factor VII has the shortest half-life (4−6 h). With acute liver dysfunction, levels of factor VII are reduced early. Factor VII is required for the extrinsic pathway. Thus, PT is a very good test to assess liver function in the acute setting.
- Individuals with factor XII deficiency or HMWK or kallikrein deficiency do not bleed. PTT may be prolonged.
- All deficiencies of clotting factors are transmitted as autosomal recessive with the exception of factor VIII deficiency (hemophilia A) and factor IX deficiency (hemophilia B), which is transmitted as X-linked recessive. Dysfibrinogenemia, which means fibrogen is dysfunctional, is transmitted as autosomal dominant.
- $INR = [Patient\ PT/Mean\ normal\ PT]^{ISI}$.
- Causes of prolonged PT include coumarin (warfarin), vitamin K deficiency (dietary deficiency), failure of absorption of vitamin K (e.g., cholestasis), liver disease, and factor deficiency.
- Vitamin K is converted to the epoxide form by the enzyme epoxidase. The epoxide form is inactive. The epoxide form is converted back to active vitamin K by the reductase enzyme. The full name of the enzyme complex is VKROCI. Warfarin blocks the reductase enzyme. There is genetic variation in the gene for VKROC I amongst individuals. Warfarin is metabolized by the P450 CYP system, including CYP2C9. Polymorphism of the CYP2C9 affects CYP2C9 activity. Variation in activity will result variation in the half-life of warfarin amongst individuals. Drugs may also affect activity of CYP2C9. Thus, genetic variation of VKROC I gene, polymorphism of CYP2C9, and effect of other drugs all result in variation in warfarin activity amongst individuals.
- Causes of prolonged PTT include heparin, DTIs, factor deficiency, inhibitors (VIII and IX inhibitors, LA), and von Willebrand disease.
- Inhibitors to factors VIII and IX may develop spontaneously (autoantibodies) or posttreatment (alloantibodies) with factor replacement.
- One BU of inhibitor is that amount of inhibitor that destroys half the factor VIII or IX:C activity in an equal mixture of patient and normal plasma. If the result is 5 or less, it is considered that the antibody is in low titers.

- Causes of prolonged TT include heparin, hypofibrinogenemia, dysfibrinogenemia, and thrombolytic therapy.
- Mixing study: Individuals with prolonged PT or prolonged PTT or both may undergo PT/PTT mixing study. The objective of this test is to determine if the cause of prolonged PT or PTT is due to a factor deficiency or due to the presence of inhibitor. Here, patient's plasma is mixed with an equal volume of normal plasma. PTT is measured at 0 and 1–2 h. Failure of correction of prolonged PTT means presence of inhibitors. If results at 0 and 1–2 h are similarly prolonged, it implies lupus anticoagulant. If results show time-dependent prolongation, it implies coagulation factor antibody (e.g., factor VIII or IX inhibitor). If PT and PTT are both prolonged and mixing study shows correction, then most likely there is deficiency of factor in the common pathway.
- Causes of bleeding with normal PT/PTT: dysfibrinogenemia, factor XIII deficiency, and primary fibrinolysis.
- Von Willebrand disease is the most commonly inherited bleeding disorder in the general population. Type 1 von Willebrand disease is the most common type; 75% of all people with this disease are type 1. In this disorder, partial quantitative deficiency (20%–50% reduction) of vWF is observed, but the factor is structurally normal. This disease is transmitted as autosomal dominant with reduced penetrance and variable expressivity. Clinical symptoms are usually mild. In type 2 disorder, there is qualitative defect of vWF, and this type represents 20%–25% of all cases. There are four subtypes 2A, 2B, 2M, and 2N. Type 2N is transmitted as autosomal recessive, but others are transmitted as autosomal dominant. Type 3 is the rarest type with total deficiency of vWF and manifests as most severe forms of von Willebrand disease. This disorder is transmitted as autosomal recessive. Pseudo (platelet type)-von Willebrand disease and acquired von Willebrand disease are also observed.

References

[1] Van der Poll T, Herwald H. The coagulation system and its function in early immune defense. Thromb Haemostasis 2014;112:640–8.
[2] Modderman PW, Admiral LG, Sonnenberg A, von dem Borne AE. Glycoprotein V and Ib-IX form a noncovalent complex in platelet. J Biol Chem 1992;267:364–9.
[3] Murugappa S, Kunapuli SP. The role of ADP receptors in platelet function. Front Biosci 2006;11:1977–86.
[4] Tan GC, Stalling M, Dennis G, Nunez M, Kahwash SB. Pseudothrombocytopenia due to platelet clumping: a case report and brief review of the literature. Case Rep Hematol 2016;2016:3036476.
[5] Van Genderen PJ, Michiels JJ. Acquired von Willebrand disease. Baillieres Clin Haematol 1998;11:319–30.
[6] Kehrel BE, Brodde MF. State of the art in platelet function testing. Transfus Med Hemotherapy 2013;40: 73–86.
[7] Campbell J, Ridgway H, Carville D. Plateletworks: a novel point of care platelet function screen. Mol Diagn Ther 2008;12:253–8.
[8] Chen A, Teruya J. Global hemostasis testing thromboelastography: old technology, new application. Clin Lab Med 2009;29:391–407.
[9] Luz D, Nascrimento B, Rizoli S. Thrombelastography (TEG): practical considerations and its clinical use in trauma patients.
[10] Lee CJ, Ansell JE. Direct thrombin inhibitors. Br J Clin Pharmacol 2011;72:581–92.
[11] Swami A, Kaur V. von Willebrand disease: a concise review and update for the practicing physician. Clin Appl Thromb Hemost 2017;23:900–10.

Thrombophilia and their detection

Introduction

Thrombophilia is defined as an abnormality of the coagulation or fibrinolytic system that results in a hypercoagulable state and increases the risk of an individual for thrombotic event where intravascular thrombus formation may be arterial or venous. The predisposition to form such blood clots may arise from genetic or acquired factors. Venous thrombosis has an overall incidence of less than 1 person per 1000 people and in pediatric population approximately one in 100,000. However, the frequency increases with advancing age [1]. Certain populations such as individuals with varicose veins and venous ulcers have higher prevalence of thrombophilia than normal population [2]. Although studies have suggested that coagulation disorders are the major cause of ischemic stroke only in 1%—4% of cases, the prevalence of thrombophilic markers is increased in children with stroke compared with control subjects. Specifically factor V Leiden and antiphospholipid antibodies contribute significantly to stroke occurrence and specifically factor V Leiden and presence of antiphospholipid antibodies contribute significantly to stroke occurrence [3].

Thrombophilia: inherited versus acquired

Thrombophilia may be inherited or acquired, and the hypercoagulability state may arise from an excess or hyperfunction of a procoagulant or a deficiency of an anticoagulant moiety. The risk factors for thrombophilia may be broadly divided into inherited and acquired categories, which are summarized in Table 17.1. Established genetic factors associated with thrombophilia include

- Factor V Leiden
- Prothrombin gene mutation
- Protein C or S deficiency, and
- Antithrombin III (AT III) deficiency

  Less common or rare genetic defects are also associated with thrombophilia and include

- Hyperhomocysteinemia
- Dysfibrinogenemia
- Elevated factor VIII, factor IX, and XI
- Elevated lipoprotein (a) (Lp [a])

Table 17.1 Common inherited and acquired risk of thrombophilia.

Common inherited causes	Common acquired causes
Factor V Leiden	Trauma
Prothrombin gene mutation	Surgery
Protein C deficiency	Immobilization
Protein S deficiency	Pregnancy
Antithrombin III deficiency	Use of oral contraceptive pill
Hyperhomocysteinemia (e.g.,	Hormone replacement therapy
methylenetetrahydrofolate reductase mutation)	Paroxysmal nocturnal hemoglobinuria
Increased factor VIII activity	Heparin-induced thrombocytopenia
Dysfibrinogenemia	Lupus anticoagulant (Lupus antibody)
	Anticardiolipin antibodies

Being a critical factor for clot formation process, platelet and platelet glycoprotein gene polymorphism have received increased attention as possible inherited determinant of prothrombotic tendency. However, their role in genetic susceptibility to thrombotic disease is still controversial. A deficiency in tissue type plasminogen activator or plasminogen could, however, reduce the capacity to remove excessive clot and contribute to thromboembolic disease [4]. Various acquired causes of thrombophilia include trauma, surgery, pregnancy, use of various medications such as oral contraceptives, antiphospholipid syndrome, paroxysmal nocturnal hemoglobinuria (PNH), and heparin induced thrombocytopenia (HIT) [5].

Factor V Leiden

In 1993, Dahlback et al. reported that a poor anticoagulant response to activated protein C (APC) was associated with the risk of thrombosis [6]. Further studies demonstrated that a single base pair substitution at nucleotide location 1691 (guanine to adenine) in factor V gene results in arginine being substituted by glutamine at amino acid location 506 (R506Q) in the factor V protein, the cleavage site of factor V by APC, and a natural anticoagulant. Factor V cleaved at position 506 also functions as a cofactor (along with protein S) for APC mediated inactivation of factor VIIIa. Loss of this anticoagulant activity of factor V may contribute to fibrin generation. The mutant factor V is referred as factor V Leiden as Dutch investigators from the city of Leiden first described this mutation. Factor V Leiden can be seen in 4%−10% of Caucasian population, and the risk for venous thromboembolism increases five- to tenfold for heterozygous states and 50-to 100-fold for homozygous states. The carrier state of factor V Leiden is highest in individuals from Greece, Sweden, and Lebanon. It is significantly seen less in African Americans and Asians (<1%). Other variants of Factor V are Factor V Cambridge, Factor V Liverpool, Factor V Hong Kong, and HR2 haplotype. The thrombotic risks for these mutations are not clear. Routine clinical testing for factor V Leiden do not detects these variants.

If an individual inherits factor V Leiden (heterozygous state) and is heterozygous for deficiency of Factor V, this is referred to as pseudohomozygous state. The activate protein C resistance (APCr) test is the recommended screening test for detection of factor V Leiden, followed by a confirmatory test such as DNA analysis of factor V gene, which encodes the factor V protein. Polymerase chain reaction

(PCR)−based assay is available for identifying factor V Leiden mutation. The first acute thrombosis in patients carrying factor V Leiden is treated according to standard protocol, and duration of anti-coagulation therapy is based on clinical assessment. In the absence of a history of thrombosis, long anticoagulation therapy is not recommend in asymptomatic patients who are factor V Leiden heterozygotes [7].

Activated protein C resistance test

In the original test, activated partial thromboplastin time (APTT) in the plasma of a patient is measured with and again without addition of APC. In normal individuals, the APC capable of cleaving V should inactivate it and then the cleaved factor V (along with protein C and S) should also inactivate factor VIII; as a result, APTT should be prolonged. However, if the patient has factor V Leiden, then it is not be inactivated by APC, and as a result, clotting time should be shorter than a normal patient. The APC ratio is then determined using the formula:

$$APC\ Ratio = [APTT\ in\ the\ presence\ of\ APC]/[APTT]$$

If the ratio is less than 2.2, then the individual is likely to have factor V Leiden (however, in some laboratories, normal ratio may be 2.0 or 2.3; values 1.5−2.0 may be considered borderline). However, there is a considerable overlap between healthy individuals and heterozygotes. A limitation of this test is that it requires a normal APTT in the patient and so cannot be used in cases in which there is a prolongation of the APTT (such as patients on oral anticoagulants or with a lupus anticoagulant).

Second generation APCr screens use a protocol where patient's plasma is diluted with commercially available factor V−deficient plasma. This modified assay reduces the number of exogenous confounding factors that might affect the APTT such as high factor VIII levels and makes the test specific for mutations within factor V. However, if lupus anticoagulant (lupus antibody [LA]) is present in patient's plasma, it could compete for phospholipid and can prolong partial thromboplastin time (PTT) measurements. Therefore, presence of LA is a major source of false-positive results in this assay. This modified assay is specific only for mutations within factor V, whereas the original APTT assay without factor V−deficient plasma predilution protocol measures APC resistance from any cause. Polybrene (hexadimethrine bromide) may be added, which makes it insensitive to unfractionated heparin and low molecular weight heparin. Confirmatory test for factor V Leiden is PCR testing, which is positive in 90% of APC resistance cases.

Prothrombin gene mutation

The prothrombin gene G2010A mutation is due to a single base pair substitution at nucleotide position 20210 (guanine to adenine substitution; chromosome 11p-q12) that results in increased prothrombin level with increased risk for venous thrombosis because prothrombin is the precursor of thrombin, a serine protease, and a key enzyme in the process of hemostasis and thrombosis. This disorder is an autosomal dominant inherited defect. This prothrombin gene mutation abnormality is the second most frequent factor after factor V Leiden mutation found in patients with thromboembolism (prevalence varies between 5% and 19%). The prevalence of this mutation is between 1% and 5% of individuals in Caucasian population, where the risk of venous thrombosis is typically increased three- to fivefold (heterozygotes). However, like factor V Leiden, this mutation is rarely found in individuals from Asia or Africa. Jebeleanu and Procopciuc reported that out of 20 patients with venous leg ulcers, the authors studied two patients showing heterozygous G2010A mutation [8]. Laboratory testing includes factor II assay and PCR testing for G20210A.

Protien C deficiency

Protein C is a vitamin K−dependent coagulation inhibitor that is synthesized by the liver and responsible for inactivating factor Va and factor VIIIa after protein kinase C is activated by thrombin. Protein C deficiency is less common than factor V Leiden or mutation in prothrombin gene (G2010A), and this disorder is transmitted as autosomal dominant and is seen in 0.2%−0.5% of population who also has increased risk of venous thrombosis by 6.5- to 8-fold. Protein C deficiency is found in 6%−10% of hypercoagulation cases. Protein C deficiency may be inherited or acquired (Table 17.2). Inherited protein C deficiency may due to decreased production of protein C, which represents most of the cases of inherited protein C deficiency (Type I; approximately 85% cases). A less common inherited disorder is due to dysfunctional protein (Type II). The gene for protein C is located on chromosome 2, and more than 160 mutations have been described. The Type I deficiency is more common, where most affected patients are heterozygous and usually have a 50% reduction in plasma protein C concentration compared with normals. In patients with Type II disorder, plasma protein C concentration is normal, but the protein has decreased functional activity.

Protein C deficiency can also be acquired and can be induced by certain conditions such as a particular malignancy. Because protein C is a vitamin K−dependent protein (molecular weight 62,000), produced in the liver, liver disease may cause reduced level. In addition, patients receiving warfarin therapy or patients who are deficient in vitamin K may also show protein C deficiency. Antineoplastic drugs may impair either the absorption or metabolism of vitamin K that may indirectly cause protein C deficiency. For example, L-asparaginase, which is used in treating patients with acute lymphoblastic leukemia, is known to cause protein C deficiency. Increased clearance of protein C may occur during acute thrombosis or disseminated intravascular coagulation (DIC). Protein C deficiency may also occur postplasmapheresis because of dilution effect. Acquired protein C deficiency is more common than inherited deficiency.

Clinical manifestation of protein C deficiency is recurrent deep vein thrombosis, pulmonary embolism, neonatal purpura fulminans (in homozygotes), and warfarin (Coumadin)-induced necrosis

Table 17.2 Inherited and acquired of protein C deficiency.

Inherited protein C deficiency	Acquired protein C deficiency (more common)
Type I: Decreased production of functional protein C (approximately 85% of inherited cases). Type II: Dysfunctional protein with reduced activity.	Decreased synthesis: Liver disease, vitamin K deficiency or presence of vitamin K antagonist (e.g., Coumadin), L-asparaginase therapy. Increased clearance: Disseminated intravascular coagulation, acute thrombosis, etc. Dilution: Postplasmapheresis, IV administration of crystalloids, etc.

of the skin. Warfarin-induced skin necrosis, a rare complication of warfarin therapy, is observed mainly in obese perimenopausal middle-aged women who are being treated with warfarin for deep vein thrombosis or pulmonary embolism. Most cases of warfarin-induced necrosis were reported in patients with hereditary protein C deficiency, but this may also be due to secondary acquired protein C deficiency. This complication of warfarin therapy is usually observed during the first few days of treatment (onset within first 1—10 days, but most commonly within first 5 days) and is due to shorter half-life of protein C compared with other vitamin K—dependent factors such as factors IX, X, and prothrombin, which may cause temporary exaggeration of the imbalance between procoagulation and anticoagulation pathway. However, protein S and antithrombin II deficiency may also cause this episode. However, this phenomenon can be prevented by carefully adjusting initial dosage of warfarin (progressively increasing warfarin dosage oven and extended period of time) and use of fresh frozen plasma when starting warfarin therapy [9]. Laboratory testing to diagnose this disorder include function assay for protein C and immunological testings.

Protein S deficiency

Protein S (originally discovered and purified in Seattle, hence the name), a vitamin K—dependent protein, is synthesized in the liver and also in megakaryocytes. Protein S serves as a cofactor of protein C in inactivating factor Va and factor VIIIa. In plasma, 60% protein S is bound to C4b-binding protein (an acute-phase reactant) and is functionally inactive. Protein S deficiency is transmitted as autosomal dominant and is observed in 0.7% of population, and this deficiency increases the risk of venous thrombosis by 1.6- to 11.5-folds compared with normal population. Protein S deficiency is found in 5%—10% of hypercoagulation cases. Protein S deficiency may be inherited or acquired (Table 17.3).

Inherited protein S deficiency could be Type I, II, or III. There are two homologous genes for protein S: PROS1 and PROS2, both located in chromosome 3. The Type I is the most common type inherited disorder of protein S accounting for approximately 50% reduction in protein S level in plasma with marked reduction in free protein S level, which is the active form. Type II disorder is characterized by normal total and free protein S level but diminished protein S functional activity. The Type II (also known as Type IIa) is characterized by normal total S level but selectively reduced free protein S level and protein S functional activity (approximately 40% reduction) [4]. This phenomenon is due to increased binding of protein S to C4b-binding protein. However, acquired cases of protein S deficiency are more common than inherited cases of protein C deficiency. Because protein S, similar to

Table 17.3 Inherited and acquired protein S deficiency.

Inherited protein S deficiency	Acquired protein S deficiency (more common)
Type I: Decreased production of functional protein S Type II: Dysfunctional protein Type III: Increased binding to C4b-binding protein	Decreased synthesis: Liver disease, vitamin K deficiency or presence of vitamin K antagonist (e.g., Coumadin), L-asparaginase therapy Increased clearance: Disseminated intravascular coagulation, acute thrombosis Dilution: Postplasmapheresis, IV administration of crystalloids

protein C, is a vitamin K–dependent protein, acquired causes of protein S deficiency are as expected similar to acquired causes of protein C deficiencies. Common causes of acquired protein S deficiency include liver disease, vitamin K deficiency, therapy with warfarin (vitamin K agonist), or L-asparaginase therapy. Increase clearance of protein kinase S as observed in DIC or acute thrombosis may also cause protein S deficiency. Clinical manifestations of protein S deficiency include recurrent deep vein thrombosis, pulmonary embolism, and neonatal purpura fulminans. Laboratory testings are immunological assays (free and total) and functional assay.

Assays for protein C and protein S

Available assays for protein C include quantifying protein C antigen (estimate quantity of protein C in plasma but not functional protein C) using an immunoassay (such as enzyme-linked immunosorbent assay [ELISA] immunodiffusion or immunoelectrophoresis method), where a monoclonal antibody against protein C is used for antigen binding. Functional assays for protein C (determines protein C activity) could be clot based or chromogenic (spectrophotometric detection). Multiple immunoassays to determine protein C concentration and both clot-based and chromogenic-based functional assays for protein C are commercially available.

Protein C functional assay is useful in determining protein C deficiency and should be the first test to be performed in a patient with suspected protein C deficiency [10]. If value is abnormal, then quantity of protein C in serum may be estimated using an immunoassay. In clot-based functional protein C assay, patient's plasma is incubated with Southern Copperhead snake venom (Agkistrodon contortrix; Protac), which activates protein C. Then an activator (such as APTT reagent or Russell viper venom reagent which induces clot formation) is added. If protein C is present and is activated, then factors Va and VIIIa are inactivated, thus clotting time is longer, which can be measured by APTT (if APTT reagent is used) or Russel venom clot time. For chromogen-based functional assay, APC is capable of cleaving a chromogen that produces a color that can be measured photometrically. Conditions that may shorten clotting time may cause falsely low levels of protein C. These include high levels of factor VIII, low levels of protein S, and factor V Leiden. Also, vitamin K deficiency and warfarin may also result in falsely low levels because in both situations protein C is not adequately carboxylated and thus not functional. Situations where the clotting time is prolonged may result in falsely high protein C levels such as heparin therapy, DTI therapy, and lupus anticoagulant.

The protein S functional assay, which measured the biological activity of free protein S, is based on the principle of adding protein S–depleted plasma and APC to test plasma (patient's plasma). Then an activator (Russell viper venom reagent) is added, and APTT is measured. As with protein C assay, clotting time is longer with adequate levels. Causes of false low or high are similar to protein C functional assay. However, immunoassays are commercially available for determining total concentration of protein S or free concentration of protein S in plasma. For determining total concentration of protein S (antigenic assay), ELISA or latex particle–based assays are commercially available. For measuring free concentration of protein S, C4b-bound protein S must be precipitated using a precipitation agent such as polyethylene glycol and then after centrifuging, protein S concentration in the supernatant, which represents free protein S, could be measured. Alternatively, monoclonal antibody directed to protein S epitopes not accessible in the bound form can be used in an immunological assay for direct determination of free protein S.

Antithrombin III deficiency

AT III inactivates thrombin and other factors (Xa, IXa, XIa, XIIa, kallikrein), and this process is accelerated by heparin. AT III deficiency is transmitted as autosomal dominant and is seen in 0.17% of the population. The risk of thrombosis is increased by 5- to 8.1-folds and accounts for 5%−10% of hypercoagulation cases. AT III deficiency may be inherited or acquired. Inherited AT III deficiency is of two types:

- Type I: Reduced production of functionally normal AT III and values are reduced by approximately 50% in heterozygote. More than 80 mutations have been reported in patients with Type I deficiency.
- Type II; Dysfunctional AT III: There are three subtypes: in reactive site subtype, there is altered AT interaction with Xa and thrombin active sites; in heparin-binding site subtype, there is defective heparin binding to antithrombin; and in pleomorphic subtype (PE), there are multiple defects.

Acquired AT III deficiency may be seen in patients with liver disease (reduced synthesis), nephrotic syndrome (renal loss of AT III and reduced synthesis by the liver), DIC, and therapy with L-asparaginase.

Clinical manifestations of AT III deficiency include recurrent deep vein thrombosis and pulmonary embolism. Laboratory testing involve functional assay (chromogenic) and immunologic (antigenic) assay.

Hyperhomocysteinemia

Homocysteine is a sulfhydryl-containing amino acid produced from the metabolism of essential amino acid methionine. Hyperhomocysteinemia may be inherited or acquired and is a risk factor for cardiovascular diseases and thrombophilia. Inherited causes include deficiencies of 5,10-methylenetetrahydrofolate reductase (MTHFR) or cystathionine beta synthase enzymes due to defective genes that encode such enzymes. MTHFR is an enzyme in folate-dependent homocysteine remethylation, catalyzing the reduction of 5,10-methylenetetrahydrofolate into 5-methyltetrahydrofolate. Single base pair substitution at nucleotide location 677 (thymine replaces cytosine: 677 C > T) in the MTHFR gene results in the substitution of alanine by valine at location 223 (A223V) in the 5,10-MTHFR enzyme with decreased activity causing hyperhomocysteinemia. As a result, these individuals are at higher thrombotic risk (mechanism: blood vessel injury, coagulation activation, fibrinolysis inhibition, platelet activation). Approximately 11% of the Caucasian population is homozygote for this mutation with threefold increased risk of thrombosis. Although this mutation is most common, at least 40 different mutations of MTHFR gene have been described in individuals with homocystinuria.

Acquired causes of hyperhomocysteinemia include vitamin B_6, B_{12}, and folate deficiencies as well as renal failure. Combined folic acid vitamin B therapy can reduce elevated homocysteine level in blood [11]. Laboratory testing includes plasma homocysteine assay using an automated analyzer, but PCR testing to detect genetic mutation is currently not fully accepted as part of the routine battery.

Increased factor VIII activity

Factor VIII is a plasma sialoglycoprotein that plays an important role in hemostasis, and for a long time, it has been recognized that factor VIII deficiency in patients with hemophilia A results in bleeding episodes. Research studies in recent years, especially epidemiological studies, have demonstrated that converse is also true as elevated plasma factor VIII coagulant activity is now accepted as an independent marker of increased thrombotic risk. The Leiden thrombophilia study was the first one that reported association between high plasma levels of factor VIII with venous thromboembolism. Factor VIII activity is higher in non-O blood group individuals. It is also well recognized that levels of factor VIII increase with advancing age and may be related to increased risk of thrombotic event with advanced age. Although thrombosis is rare in children, elevated factor VIII levels (>90th percentile) were observed in 19.5% children with venous thrombosis compared with only 4.4% in the control group. At present, no polymorphism of factor 8 gene is associated with increased level of factor VIII in plasma [12]. Factor VIII activity may also be elevated in pregnancy, malignancies, infection, and inflammation.

Acquired causes of thrombophilia

Acquired causes of thrombophilia are more common than inherited causes. Trauma, surgery, immobilization, pregnancy, hormone replacement therapy, use of oral contraceptive, paroxysmal nocturnal hemoglobinuria, and HIT are all acquired causes of thrombophilia. Lupus anticoagulant (LA) and anticardiolipin antibodies (ACA) are commonly encountered in patients with higher risk of thrombotic events. The use of oral contraceptives increases the risk of venous thromboembolism and arterial thrombosis, but third generation oral contraceptives appears to increase the risk of venous thromboembolism compared with the second generation pill, but such medication usually do not increase the risk of arterial thrombosis. The effect of oral contraceptives in increasing the risk of venous thromboembolism is more pronounced in the first year therapy, and as expected, such risks are higher in patients with an inborn or acquired tendency for thrombophilia [12]. Although hormone replacement therapy can increase the risk of thromboembolism, the risk is lower in users of estrogen-only hormone replacement therapy than users of estrogen-progestin hormone replacement therapy [13,14].

Antiphospholipid syndrome

Antiphospholipid syndrome is a systemic autoimmune disorder characterized by:

- Arterial or venous thrombosis and/or adverse pregnancy outcome and
- Laboratory evidence of antiphospholipid antibodies

The condition may be primary or secondary to autoimmune diseases such as systemic lupus erythematosus (SLE).

Pathogenesis of antiphospholipid syndrome: The antibodies develop in susceptible individuals following incidental exposure to infectious agents, and it is thought that a second hit is required for clinical effects. These antibodies effect coagulation pathways.

Clinical effects of antiphospholipid syndrome:

- Arterial or venous thrombosis especially in young
- Fetal death after 10 weeks of gestation, premature birth due to preeclampsia, placental insufficiency, multiple embryonic losses

- Thrombocytopenia
- deep vein thrombosis (DVT), PE
- Stroke, transient ischemic attack (TIA)
- Livedo reticularis
- Superficial thrombophlebitis

  Lab testing for antiphospholipid syndrome:

- Anticardiolipin antibodies IgG, IgM by ELISA
- Anti-beta2 GP I antibodies by ELISA
- Lupus anticoagulant (LA) testing

Of the laboratory tests, we need at least one of these to be positive. The test needs to be positive again 12 weeks later. Please note, LA testing is PTT-based, and blood should be collected before heparin treatment.

Laboratory testing for lupus anticoagulant

Individuals with lupus anticoagulant should have a prolonged PTT. Mixing study should not demonstrate correction. In such cases, there are a number of tests that can be performed for detection of lupus anticoagulant antibodies:

- Dilute Russell viper venom time (dRVVT),
- Platelet neutralization procedure (PNP), and
- Hexagonal phase phospholipid neutralization test (Staclot LA).

The dilute Russell viper venom test is a commonly used screening test for LA. It is called dRVVT test because the phospholipid reagent was diluted such that its concentration becomes rate limiting and thus it becomes a sensitive assay. The principle of the assay is that Russell's venom directly activates factor X, which then activates prothrombin (factor II) in the presence of factor V and phospholipid, thus initiating coagulation (clot formation in 23−27 s). If lupus coagulation antibodies are present in patient's sample, then these antibodies should bind with phospholipids, thus reducing their availability to participate in coagulation, and as a result, clotting time should be prolonged (positive tests).

If the screening test is positive, then mixing study may be performed, where patient's plasma is mixed with normal plasma in a 50: 50 ratio. A prolonged bleeding time, which is not corrected by adding normal serum, indicates the presence of LA. In the mixing study, no phospholipid is added, and the ratio is calculated as following:

dRVVT mix ratio = dRVVT mix (50: 50 patient plasma to normal)/dRVVT screen

For confirmation assay, phospholipid is added, which should correct the prolonged bleeding time. dRVVT confirmation ratio = dRVVT screen/dRVVT with added phospholipid.

The principle of DRVVT confirmation ratio, PNP, and Staclot LA is the same. If excess phospholipid is present in the test media, all antibodies present must bind to added phospholipids, and bleeding time should be reduced because enough phospholipids should be present to participate in coagulation process. In PNP, a platelet lysate is the source of the phospholipid, and in Staclot LA, the source of phospholipid is hexagonal phase phospholipids reagent.

In Staclot LA assay, test plasma is incubated with and without hexagonal phase phospholipid reagent, and PTT is measured. It is a confirmatory test for detecting the presence of lupus anticoagulant antibody. The PNP assay is based on the ability of platelets to bypass the effect of lupus anticoagulant antibody by correcting the prolonged clotting time in various phospholipid-dependent assay system. In this assay, freeze-thawed platelet suspension (platelet lysate), which is added to reaction mixture, can neutralize lupus antibodies by binding with them.

For testing ACA, ELISA (for IgG, IgM, and IgA types of antibodies) assays are commercially available. As cardiolipin is the antigen used for screening tests for syphilis (e.g., venereal diseases research laboratory [VDRL], rapid plasma reagin [RPR]), false-positive tests for syphilis may be seen in individuals positive for ACA. Individuals with syphilis infection may also result in false-positive test for ACA.

Key points

- Factor V Leiden can be seen in 4%−10% of Caucasian population, and the risk for venous thromboembolism increases 5- to 10-folds for heterozygous states and 50- to 100-folds for homozygous states. Factor V Leiden is poorly inhibited by APC, which can lead to venous thrombosis. However, factor V activity in these patients is normal.
- In the original test, APC test, APTT in the plasma of a patient is measured with and again without addition of APC. The APC ratio is then determined using the formula:

$$\text{APC Ratio} = [\text{APTT in the presence of APC}]/[\text{APTT}]$$

If the ratio is less than 2.2, then the individual is likely to have factor V Leiden. A limitation of this test is that it requires a normal APTT in the patient and so cannot be used in cases in which there is a prolongation of the APTT (such as patients on oral anticoagulants or with a lupus anticoagulant). Second generation APCr screens use a protocol where patient's plasma is diluted with commercially available factor V−deficient plasma. This modified assay reduces the number of exogenous confounding factors that might affect the APTT such as high factor VIII levels and makes the test specific for mutations within factor V.

- Prothrombin gene mutation: Single base pair substitution at nucleotide position 20210 (guanine to adenine substitution: G20210A). This results in increased prothrombin level with increased risk for venous thrombosis. Laboratory testing include actor II assay and PCR testing for G20210A.
- Inherited Protein C and S deficiencies are transmitted as autosomal dominant. These proteins inhibit factor Va and VIIIa. Protein C and S deficiencies may be inherited or acquired. Laboratory testing includes immunological assays and functional assays.
- Principle of protein C functional assay: In this test, patient's plasma is incubated with Southern Copperhead snake venom (Protac), which activates protein C. Then an activator such as APTT is added, and clotting time is measured. Alternatively, Russell viper venom reagent is added to activate clotting, and clotting time is measured. If protein C is present and is activated, then factor Va and VIIIa are inactivated, thus clotting time is longer.
- Principle of protein S functional assay: In this test, protein S−depleted plasma and APC are added to test plasma. Then, an activator (e.g., APTT reagent or Russell viper venom reagent) is added, and clotting time is measured.

- AT III inactivates thrombin and other factors (Xa, IXa, XIa, XIIa, kallikrein) that are accelerated by heparin. AT III deficiency may be inherited or acquired. Inherited AT III deficiency is transmitted as autosomal dominant. Laboratory testing include functional assay (chromogenic) and immunologic assay.
- Hyperhomocysteinemia may be inherited or acquired. Inherited causes include deficiencies of 5,10-MTHFR or cystathionine beta synthase caused by gene mutations that encode these enzymes. Hyperhomocysteinemia is a risk factor for cardiovascular diseases and thrombophilia. Laboratory testing includes determination of homocysteine concentration in plasma using an automated analyzer and PCR testing (currently not fully accepted as part of the routine battery).
- Multiple epidemiologic studies have demonstrated that elevated factor VIII activity is a risk factor for thromboembolism.
- Lupus anticoagulant and ACA are immunoglobulins that prolong in vitro phospholipid-dependent clotting times. These antibodies are examples of antiphospholipid antibodies and may causes arterial or venous thrombosis or miscarriage.
- Laboratory testing for lupus anticoagulant include aPTT, mixing study, dRVVT, PNP, and hexagonal phase phospholipid neutralization test (Staclot LA). The dilute Russell viper venom test is a commonly used screening test for LA. Dilute means that a minimal amount of phospholipid is present, whose rate is limiting. If LA is present it binds to the phospholipid and prolongs the clotting time. The principle of DRVVT confirmation ratio, PNP, and Staclot LA is the same. If excess phospholipid is present in the test media, it should bind all LA, and the clotting should be reduced.
- Testing for ACA is performed using immunological methods (ELISA assays for IgG, IgM, and IgA types of antibodies).

References

[1] Hoppe C, Matsunaga A. Pediatric thrombosis. Pediatric Clin North Am 2002;49:1257–83.
[2] Darvall KA, Sam RC, Adam DJ, Silverman SH, et al. Higher prevalence of thrombophilia in patients with varicose veins and venous ulcers than controls. J Vasc Surg 2009;49:1235–41.
[3] Kenet G, Sadetzki S, Murad H, Martinowitz U, et al. Factor V Leiden and antiphospholipid antibodies are significant risk factors for ischemic stroke in children. Stroke 2000;31:1283–8.
[4] Khan S, Dickerman. Hereditary thrombophilia. Thromb J 2006;4:15.
[5] Stevens SM, Woller, Bauer, Kasthuri, et al. Guidance for the evaluation and treatment of hereditary and acquired thrombophilia. J Thromb Thrombolysis 2016;41:154–64.
[6] Dahlback B, Carlsson M, Svensson PJ. Familial thrombophilia due to a previously unrecognized mechanism characterized by poor anticoagulant response to activated protein C: prediction of a cofactor to activated protein C. Proc Natl Acad Sci USA 1993;90:1004–8.
[7] Kujovich JL. Factor V Leiden thrombophilia. Genet Med 2011;13:1–16.
[8] Jebeleanu G, Procopciuc L. GA 2010 A prothrombin gene mutation identified in patients with venous leg ulcers. J Cell Mol Med 2001;5:397–401.
[9] Chan YC, Valento D, Mansfield AO, Stansby G. Warfarin induced skin necrosis. Br J Surg 2000;87:266–72.
[10] Khor B, Van Cott EM. Laboratory test for protein C deficiency. Am J Hematol 2010;85:440–2.

[11] Maron BA, Loscalzo J. The treatment of hyperhomocystinemia. Annu Rev Med 2009;60:39—54.

[12] Jenkins PV, Rawley O, Smith OP, O'Donnell JS. Elevated factor VIII levels and risk of venous thrombosis. Br J Haematol 2012;157:653—63.

[13] Blickstein D, Blickstein I. Oral contraception and thrombophilia. Curr Opin Obstet Gynecol 2007;19:370—6.

[14] Gialeraki A, Valsami S, Pittaras T, Panayiotakopoulos G, Politou M. Oral contraceptives and HRT risk of thrombosis. Clin Appl Thromb Hemost 2018;24:217—25.

Sources of errors in hematology and coagulation

Introduction

Hematology tests such as complete blood count (CBC) and peripheral blood smear analysis are commonly ordered test in clinical laboratory, and sources of errors in such tests could be preanalytical, analytical, or postanalytical. Blood for CBC is typically collected in vacuum tubes that contain the anticoagulant dipotassium ethylenediaminetetraacetic acid (K_2-EDTA). The blood collected in the vacuum tube is analyzed on the automated hematology analyzer for CBC results. For coagulation, citrated plasma is the preferred specimen. Although minimization of analytical errors has been the main focus of developments in laboratory medicine, the other steps are more frequent sources of erroneous results. Carraro and Plebani M reported that in the laboratory, preanalytical errors accounted for 62% of all errors, with postanalytical representing 23% and analytical 15% of all laboratory errors [1]. Even using disodium EDTA versus dipotassium EDTA, tube may contribute to difference as values of hemoglobin, hematocrit (HCT), MCV (mean corpuscular volume), and lymphocyte count were higher when collected in disodium EDTA tube compared with dipotassium EDTA tube. In addition, tube must be filled with blood because HCT and MCV values may be falsely low [2]. Lima-Oliveira et al. investigated whether the use of different dry K_2 (dipotassium) EDTA vacuum tubes might represent a bias in hematological testing. Blood was collected in three dipotassium EDTA vacuum tubes from different manufacturers: Venosafe, Vacuette, and Vacutainer, followed by analysis of specimens. The authors observed significant differences in measurement of HCT, MCV, white blood cell (WBC) count, and platelet distribution width (PDW) when comparing Venosafe versus Vacuette. The authors also reported significant differences for MCV, WBC, and PDW when comparing Venosafe versus Vacutainer and for HCT and MCV when comparing Vacuette versus Vacutainer. Clinically significant variations were observed for HCT and PDW in Venosafe versus Vacuette; PDW in Venosafe versus Vacutainer; and HCT and MCV in Vacuette versus Vacutainer. The authors concluded that use of dipotassium EDTA vacuum tubes from different manufacturers represent a clinically relevant source of variation for HCT, MCV, and PDW [3]. Therefore, laboratory must validate blood collection tube before use for routine blood collection.

Errors related to sample collection, transport, and storage must be avoided to get accurate results in hematology testing. The amount and concentration of dipotassium EDTA in the collection tube require that blood should be collected up to a specific mark on the tube. If too little blood is collected, dilution of the sample can become an issue with alteration of parameters. Relative excess EDTA in such cases also affects the morphology of blood cells. Transport of specimen should ensure that high temperatures are avoided. Red cell fragmentation is a feature of excess heat. Prolonged storage may result in

Hematology and Coagulation. https://doi.org/10.1016/B978-0-12-814964-5.00018-8

degenerative changes in the WBCs. This is best illustrated in neutrophils where they have a round pyknotic nucleus. Abnormal lobulation of the lymphocyte nuclei is also another established phenomenon with prolonged storage of blood. These cells may be considered as atypical lymphocytes with an incorrect implication of an underlying lymphoproliferative disorder. Table 18.1 summarizes sources of laboratory errors in hematology testings.

Table 18.1 Sources of laboratory errors in hematology.

Falsely high hemoglobin
- Turbid sample (hyperlipidemia, parenteral nutrition, hypergammaglobulinemia, cryoglobulinemia, marked leukocytosis)
- Smokers (high carboxyhemoglobin)

Falsely low hemoglobin
- Rare

Falsely high red blood cell (RBC) count
- Large platelets
- Red cell fragments

Falsely low RBC count
- Cold agglutinin

Falsely high mean corpuscular volume
- Cold agglutinin
- Hyperosmolar state (uncontrolled diabetes mellitus)

Falsely low hemoglobin
- Large platelets
- Hypoosmolar state

Falsely high white blood cell (WBC) count
- Nucleated red cells
- Nonlysis of red cells (due to target cells in hemoglobinopathy)
- Giant platelets or platelet clumps (due to EDTA)
- Cryoglobulins
- Microorganisms

Falsely low WBC count
- Leukoagglutination (due to EDTA)
- Cold agglutinin

Falsely increased lymphocyte count
- Nucleated red cells
- Nonlysis of red cells (due to target cells in hemoglobinopathy)
- Giant platelets or platelet clumps (due to EDTA)
- Malarial parasites
- Dysplastic neutrophils (hypolobated neutrophils)
- Basophilia

Falsely decreased lymphocyte count
- Rare

Falsely high neutrophil count
- Rare

> **Table 18.1 Sources of laboratory errors in hematology.—cont'd**
>
> Falsely low neutrophil count
>
> - Neutrophil aggregation
> - Neutrophil with hemosiderin granules (counted as eosinophils)
>
> Falsely increased eosinophil count
>
> - Neutrophils with hemosiderin granules (counted as eosinophils)
> - Red cells with malarial pigments
>
> Falsely low eosinophil count
>
> - Hypogranular eosinophils
>
> Falsely increased monocyte count
>
> - Large reactive lymphocytes
> - Lymphoblasts
> - Lymphoma cells
> - Immature granulocytes
>
> Falsely low monocyte count
>
> - Rare
>
> Falsely high platelet count
>
> - Fragmented red cells (in microangiopathic hemolysis)
> - Fragmented white cells (in leukemia)
> - Microorganisms
> - Cryoglobulin
>
> Falsely low platelet counts
>
> - Partial clotting or platelet activation
> - Giant platelets
> - Platelet clumps and platelet satellitism (due to EDTA)
> - GP IIb/IIIa antagonists
> - Platelet agglutination (due to cold agglutinin)
>
> False-positive fragility test
>
> - Immunologically mediated hemolytic anemias

Errors in common hematology testing

Automated hematology analyzers can rapidly analyze whole blood specimens for the CBC and differential count, which include red blood cell (RBC) count, WBC count, platelet count, hemoglobin concentration, HCT, RBC indices, and leukocyte differential. In addition, modern hematology analyzers are capable of analyzing many leukocytes using flow cytometry–based method, some in combination with cytochemistry or fluorescences or conductivity to count different types of WBC including neutrophils, lymphocytes, monocytes, basophils, and eosinophils (five-part differential). Nucleated RBCs are also detected. Sources or errors in measurement of various hematological parameters are addressed in this section.

Errors in hemoglobin measurement and red blood cell count

Hemoglobin measurement is based on absorption of light at 540 nm. If the sample is turbid, this will produce higher hemoglobin levels. Examples of such state include hyperlipidemia [4], patients on parenteral nutrition, hypergammaglobulinemia, and cryoglobulinemia. Turbidity from very high WBC count can also falsely elevate hemoglobin levels. Smokers have high carboxyhemoglobin that may falsely elevate the measured hemoglobin level.

Fetal hemoglobin may also falsely elevate carboxyhemoglobin level. A previously healthy 3-month-old girl presented to the pediatric emergency department with smoke inhalation from a defective furnace. She was asymptomatic, but because of high carboxyhemoglobin level, she was treated for carbon monoxide poisoning using normobaric oxygen therapy. However, subsequent serum carboxyhemoglobin levels were persistently elevated, despite treatment and the infant appearing clinically well. As such, she had a prolonged stay in the emergency department. Further investigations showed that fetal hemoglobin interferes with the spectrophotometric method used to analyze serum carboxyhemoglobin levels [5]. Falsely low carboxyhemoglobin level after therapy with hydroxycobalamin has also been reported [6].

Large platelets may be counted by some instruments as RBCs. Also, red cell fragments greater than 40 fL will be counted as whole red cells. In both situations, the RBC count will be falsely high. Cold agglutinins will cause red cell agglutination in vitro and result in low RBC counts. If cold agglutinins are suspected, the sample should be warmed to obtain an accurate RBC count.

Errors in mean corpuscular volume and related measurements

If there is red cell agglutination, then red cell clumps will be counted as single red cells, but the volume of the estimated cell will be much higher. This may result in falsely high MCV values. If large platelets are counted as red cells, then these platelets typically have less volume than a normal red cell. This may also result in falsely low MCV values.

If the patient is in a state of high osmolarity, the cytoplasms of the red cells of such patients are also hyperosmolar. When diluents are added to the blood in the analyzer, water will move into the red cells causing them to swell in size. MCV values could be higher than that in the in vivo state. Examples of hyperosmolar states are uncontrolled diabetes mellitus, hypernatremia, and dehydration. The converse will occur in hypoosmolar states.

Values for HCT, MCH, and mean corpuscular hemoglobin concentration (MCHC) are obtained by calculation using hemoglobin levels, RBC counts, and MCV values. If there is an error in any of these values, the calculated values will also be inaccurate.

Errors in WBC counts and WBC differential counts

Falsely high WBC counts are more common than falsely low WBC counts. There are several situations where the WBC count may be falsely elevated. One of the most frequent situations is high WBC count in the presence of significant number of nucleated blood red cells (NRBC). If an accurate WBC count is required, then a corrected WBC count needs to be performed. This can be done by some hematology analyzers by running the sample again in the "NRBC mode" or performing a manual count. Platelet aggregates and nonlysis of red cells are other causes of spuriously high WBC counts. If the high WBC count is due to nonlysis of red cells, this may be a tip-off for hemoglobinopathies. Target cells seen in hemoglobinopathies are typically resistant to lysis. Platelet aggregates may be due to EDTA, and redrawing blood in citrate tube may be the solution in such cases. Erroneous WBC counts with spurious leukocytosis can be seen with the presence of cryoglobulins, and microorganisms. Spurious leukopenia can be seen in cold agglutinins and EDTA-dependent leukoagglutination. Moreover, when there are giant platelets or NRBCs or red cells resistant to lysis, these may be counted as lymphocytes in some instruments, giving rise to a falsely high lymphocyte count. Hemoglobinopathies and target cells are important causes of nonlysis of red cells.

Presence of malarial parasites in red cells has been also known to increase the lymphocyte count. During the clinically silent liver phase of malaria, an increase of peripheral total leukocyte count and differential lymphocytes and monocytes occurs. However, this increase is followed by the appearance of parasites in the peripheral blood after 2–3 days, accompanied by a marked decrease in total leukocyte count, lymphocyte count, and the neutrophil count and a rise of the neutrophil to lymphocyte count ratio [7].

In myelodysplastic syndrome, if the myeloid series is affected, then hypolobated and hypogranular neutrophils can be present. Automated analyzers may no longer count these dysplastic neutrophils as such; instead, they may be counted as lymphocytes.

Basophilia is typically seen in chronic myelogenous leukemia. Basophils are cells with coarse granules that may even obscure the nucleus. If the analyzer falsely recognizes all the dense granules of basophils as one single nucleus, then these cells could be counted as lymphocytes.

It is thus apparent that there can be multiple situations where the lymphocyte count is inappropriately elevated. Although falsely low lymphocyte count is rare, falsely low neutrophils counts are encountered more frequently. Neutrophil aggregation is a documented phenomenon and can result in low neutrophil count. Neutrophils have fine granules, whereas the granules of eosinophils are larger. Basophils have quite large granules. If neutrophils have hemosiderin granules, they may be counted as eosinophils. If eosinophils are hypogranular, then these eosinophils may be counted as neutrophils. Red cells infected by malarial parasites may contain malarial pigments. Malaria infected red cells are resistant to lysis. These red cells with malarial pigments may be counted as eosinophils.

A key difference between lymphocytes and monocytes is the difference in size, monocytes being significantly larger. Reactive (activated) lymphocytes typically have more abundant cytoplasm when compared with a nonreactive lymphocyte. Their size approaches that of a monocyte. These lymphocytes thus may be counted as monocytes.

Also, it has been reported that abnormal lymphocytes such as those seen in chronic lymphocytic leukemia, lymphoblasts, and leukemic or lymphoma cells can be miscounted as monocytes. When there is left shift in the WBC series, we tend to see slightly more immature cells such as bands and metamyelocytes. Cells that are more immature are naturally larger and may also be counted as monocytes. Storage of blood at room temperature and delay in running the sample on the analyzer may also contribute to inaccurate WBC differential values.

When differential counts obtained by automated analyzers are compared with differential counts performed manually, a difference in results is relatively frequent. Most often, they are clinically inconsequential. However, it is always important to correlate significantly abnormal results with morphological review of the peripheral smear.

Errors in platelet count

In certain situations, hematology analyzers are known to provide a falsely low platelet count when the true platelet count is adequate. This may give rise to suboptimal clinical management.

Partial clotting of specimen or platelet activation during venipuncture may cause platelet aggregation. Both mechanisms may lead to low platelet counts. Checking the specimen for any clot, analyzing the histograms, and reviewing the smear are all important steps to avoid misleading low platelet counts. There are various other mechanisms to explain falsely low platelet counts, otherwise referred to as pseudothrombocytopenia (anticoagulant-induced pseudothrombocytopenia), platelet satellitism, giant platelets, and cold agglutinin-induced platelet agglutination.

Falsely elevated platelet counts are much less common than falsely low counts. Fragmented red cells or white cell fragments may all be counted as platelets giving rise to high platelet counts. Fragmented red cells can be seen in states of microangiopathic hemolysis such as disseminated intravascular coagulation (DIC). White cell fragments can be seen in leukemic or lymphoma states. Patients with leukemia, especially acute leukemia, need supportive therapy in the form of blood component transfusions. If the platelet count is falsely elevated in a patient with acute leukemia, the decision to transfuse platelets may be delayed, with undesirable clinical consequences. Falsely high platelet counts may also be seen in presence of cryoglobulins and or microorganisms present in blood [8].

Errors in specific hematology testing

In the following section, specific selected test errors that are often seen in hematology laboratory are addressed. These errors may be related to the presence of cold agglutinins, cryoglobulins in the specimen, or related to conditions such as pseudothrombocytopenia.

Cold agglutinins

Cold agglutinins are polyclonal or monoclonal autoantibodies directed against RBC i or I antigens and preferentially binding erythrocytes at cold temperatures. These autoantibodies are typically immunoglobulin M subtype, which may be associated with malignant disorder (for example, B-cell neoplasm) or benign disorders (such as postinfection and collagen vascular disease) and can be manifested clinically as autoimmune hemolytic anemia. In the hematology laboratory with automated analyzers, cold agglutinins typically present as a discrepancy between the RBC indexes. The agglutinated erythrocytes may be recognized as single cells or may be too large to be counted as erythrocytes; subsequently measured MCV is falsely elevated and the RBC count is disproportionately low. Although the measured hemoglobin is correct due to its independence on cell count, the calculated indexes are incorrect as following: the HCT (red cell count × MCV) is low, whereas the mean corpuscular hemoglobin (MCH: hemoglobin/red cell count) and the mean corpuscular hemoglobin concentration (MCHC: hemoglobin/HCT) are elevated. Hemagglutination may be grossly visible to the unaided eye, and microscopic examination of the peripheral blood smear would show erythrocyte clumping. By rewarming the blood sample to 37°C, the erythrocyte agglutination is alleviated and correct values may be obtained. More severe cases of cold agglutinin may require saline replacement technique if rewarming the sample fails to resolve the RBC index discrepancy.

Spurious leukopenia due to cold agglutinin is also encountered occasionally with automated hematology analyzers. The mechanism is postulated to be an IgM autoantibody directed against components of the granulocyte membranes. Cold agglutinin-induced leukopenia should be recognized as a potential cause of pseudogranulocytopenia so that WBC counts can be accurately reported, and unnecessary evaluation of patients for leukopenia can be avoided.

Cryoglobulins

Cryoglobulins are typically IgM immunoglobulins that precipitate at temperatures below 37°C, producing aggregates of high molecular weight. The first clue to a diagnosis of cryoglobulinemia is laboratory artifacts detected in the automated blood cell counts. The precipitated cryoglobulins of various sizes may falsely be identified as leukocytes or platelets causing pseudoleukocytosis and pseudothrombocytosis. At the same time, the RBC indexes are generally unaffected. Correction of the artifacts for automated counts can be obtained by warming the blood to 37°C or by keeping the blood at 37°C from the time of collection to the time of testing. Peripheral blood smear typically shows slightly basophilic extracellular material, and leukocyte cytoplasmic inclusions are occasionally found.

Pseudothrombocytopenia

Pseudothrombocytopenia is caused by various etiologies, including giant platelets, anticoagulant-induced pseudothrombocytopenia, platelet satellitism, and cold agglutinin-induced platelet agglutination. Pseudothrombocytopenia may occur with giant platelets. Because of their large size, the giant platelets are excluded from electronic platelet counting causing pseudothrombocytopenia. This scenario is of particular clinical importance in patients with rapid consumption of platelets in the peripheral circulation such as observed in DIC, acute immune thrombocytopenic purpura, or thrombotic thrombocytopenic purpura. Effective platelet production by bone marrow in these cases should be present with many large platelets in peripheral blood, many of which may not be identified by automated analyzers. An accurate platelet count can be obtained with a manual count using phase-contrast microscopy.

Anticoagulant-induced pseudothrombocytopenia is an in vitro platelet agglutination phenomenon generally seen in specimens collected into EDTA tube. It has been reported both in healthy subjects and in patients with various diseases (including collagen vascular disease and neoplasm) and has an overall incidence of approximately 0.1%. Although the agglutination is most pronounced with EDTA, it may occasionally occur with other anticoagulants as well, such as heparin, citrate, or oxalate [9]. As the platelet aggregates are large, the automated hematology counters do not recognize them as platelets, leading to lower platelet counts. In some cases, the aggregates are large enough to be counted as leukocytes by automated instruments, causing a concomitant pseudoleukocytosis. The platelet aggregation in pseudothrombocytopenia is usually temperature-sensitive, with maximal activity at room temperature. The EDTA-induced pseudothrombocytopenia is mediated by autoantibodies of IgG, IgM, and IgA subclasses directed at an epitope on glycoprotein IIb. This epitope is normally hidden in the membrane GP IIb/IIIa due to ionized calcium maintaining the heterodimeric structure of the GP IIb/IIIa complex. EDTA, through its calcium chelating effect, dissociates the GP IIb/IIIa complex with GP IIb epitope exposure. It is noted that in Glanzmann's thrombasthenia, a disorder characterized by the quantitative and/or qualitative abnormality of glycoprotein IIb/IIIa, pseudo-thrombocytopenia does not occur. Interestingly, in recent years, Abciximab (a GP IIb/IIIa antagonist) has been found to be associated with pseudothrombocytopenia [10]. If anticoagulant-induced pseudothrombocytopenia is suspected, a peripheral blood smear should be examined for platelet clumping.

Platelet satellitism has features similar to anticoagulant-induced pseudothrombocytopenia. In the presence of EDTA, platelets bind to leukocytes and form rosettes. The binding is usually to neutrophils, but binding to other leukocytes has also been reported. The automated analyzers do not identify platelets that bind to leukocytes, resulting in pseudothrombocytopenia. Platelet satellitism is mediated by autoantibodies of IgG type directed at GP IIb/IIIa on the platelet membrane and to an Fc gamma receptor III on the neutrophil membrane.

Platelet agglutination due to cold agglutinins causing pseudothrombocytopenia is a rare condition. The platelet agglutination is anticoagulant-independent, usually occurs at 4°C, and is mediated by IgM autoantibodies directed against GP IIb/IIIa. As these autoantibodies have little activity at temperatures above 30°C, these are not associated with any clinical significance.

Spurious leukocytosis

The presence of microorganisms in the peripheral blood can result in spuriously high WBC counts or differentials by automated analyzers. Organisms that have been shown to be associated with this artifact include *Histoplasma capsulatum*, *Candida* sp., *Plasmodium* sp., and *Staphylococcus* sp. Spurious leukopenia due to EDTA is sometimes encountered [11]. Leukoagglutination has been reported as a transient phenomenon in neoplasia (especially lymphoma), infections (infectious mononucleosis, acute bacterial infection, etc.), alcoholic liver diseases, and autoimmune diseases (rheumatoid arthritis, etc.). It can also occur in the absence of any obvious underlying disease, even though an inflammatory condition is often found. Other EDTA-dependant counting errors are well known: platelet clumps and platelet-to-neutrophil satellitism. Association of neutrophil clumping and platelet satellitism has also been observed.

False-positive osmotic fragility test

The osmotic fragility test is useful for diagnosis of hereditary spherocytic hemolytic anemia. Spherocytes are osmotically fragile cells that rupture more easily in a hypotonic solution than do normal RBCs. As these cells have a low (surface area/volume) ratio, they lyse at a higher solution osmolarity than do normal RBCs with discoid morphology. After incubation in a hypotonic solution, a further increase in hemolysis is typically seen in hereditary spherocytosis. Cells that have a larger (surface area/volume) ratio, such as target cells or hypochromic cells are more resistant to lysing in a hypotonic solution.

Conditions associated with immunologically mediated hemolytic anemias may present with many microspherocytes in peripheral blood. Consequently, the fragility test can be positive in immunologically mediated hemolytic anemias besides hereditary spherocytosis, but the former would have a positive direct Coombs test and the latter would not.

Errors in coagulation testing

Patients with coagulation disorders may either bleed or form thromboses. Hemostasis involves activation of the clotting factors and platelets. Evaluation of platelet events may include a CBC, examination of the peripheral smear, bleeding time, and platelet aggregation test. Evaluation of the clotting factors is typically done by PT (partial prothrombin time) and aPTT (activated partial thromboplastin time; also abbreviated as APTT) measurements. Abnormal PT or aPTT will usually lead to mixing studies to determine whether the abnormal result is due to factor deficiency or inhibitors. If there is correction of the prolonged clotting time in the mixing study, then factor assays will be performed to identify the deficient factor(s). Inhibitors include specific clotting factor inhibitors and lupus anticoagulants. Inhibitor screen and inhibitor assays or confirmatory tests for lupus anti-coagulants will follow. Blood sample for coagulation tests are typically obtained in tubes with sodium citrate buffer as anticoagulant. There are various sources of erroneous test results in coagulation testing. These sources of errors are addressed in this section.

Errors in PT and aPTT measurements

The PT measures the time required for a fibrin clot to form after addition of tissue thromboplastin and calcium to platelet-poor plasma collected in a citrated tube. PT measures the activity of clotting factors VII, X, V, II, and fibrinogen. If the aPTT is normal, then a prolonged PT is due to factor VII deficiency. The PT is relatively insensitive to minor reductions in the clotting factors. The aPTT is prolonged with deficiencies of XII, XI, X, IX, VIII, V, II, and fibrinogen. Just like the PT, aPTT can be normal in minor deficiencies. In general, the deficient factor has to be around 20%–40% to cause a prolonged aPTT. Most laboratories use automated method for PT and aPTT measurements. Either optical or mechanical methods are employed to monitor clot formation. If measured with optical methods, shortened times may be seen with turbid plasma (e.g., hyperlipidemia and hyperbilirubinemia). The aPTT is a test conventionally used to monitor heparin therapy. It is important to properly separate plasma from platelets as soon as possible. Platelet factor 4 can neutralize heparin, thus spuriously reducing aPTT values. Factor VIII levels are reflected in aPTT measurements. Factor VIII is an acute-phase reactant. Again if the aPTT is being used to monitor heparin therapy in a patient who has an underlying cause for acute-phase reactants to be elevated, aPTT values may be falsely lower than expected.

Errors in thrombin time measurement

The thrombin time (TT) measures the time to convert fibrinogen to fibrin. Dysfibrinogenemia, elevated levels of fibrin degradation product, and paraproteins can interfere with fibrin polymerization, thus falsely prolonging the TT. Amyloidosis can inhibit the conversion of fibrinogen to fibrin, also pro-longing the TT. In certain malignancies, heparin-like anticoagulants have been known to be the cause of prolonged TT. In the following section, specific selected test errors that are often seen in coagulation laboratory will be discussed in more detail.

Incorrectly filled tubes

Citrate tubes for coagulation tests are designed for a 9:1 ratio of blood to citrate buffer. Underfilling or overfilling of the citrate tube results in imbalances in this blood to buffer ratio and produces artificially prolonged or shortened clotting times, respectively. Underfilling or overfilling results in too little or too much blood sample for fixed amount of anticoagulant in the tube. The amount of blood that fills the citrated tubes is controlled by vacuum, which maintains the proper 9:1 ratio of blood to anticoagulant. Underfilling may be caused by air bubbles in the tube, vacuum loss, or by not allowing the tube to completely fill during the blood collection process. If the tube stopper is removed, it also becomes difficult to obtain the correct amount of blood and attain the proper 9:1 ratio. If the patient's HCT is known in advance of blood collection to be greater than 55% (such as in patients with polycythemia) or below 21% (patients with severe anemia), the amount of sodium citrate must be adjusted using the following formula:

$$C = 0.00185 \times (100\,H) \times V$$

where C is the volume of 3.2% sodium citrate in milliliters, H is the HCT in percent, and V is the volume of blood in milliliters.

Errors in clotting tests due to HCT changes without adjustment in the citrate volume are most significant with elevated HCTs, as even severe anemia does not significantly change the PT or aPTT.

Dilution or contamination with anticoagulants

Blood collection from indwelling lines or catheters could be a potential source of testing error. Sample dilution from incomplete flushing or hemolysis caused by improper catheter insertion can alter coagulation test results. Heparinized lines should be avoided if blood must be drawn from an indwelling catheter. If using a heparinized line is absolutely necessary, adequate line flushing must be achieved before blood collection. The National Committee for Clinical Laboratory Standards recommends flushing lines with 5 mL of saline. At least 5 mL or six times the dead space volume of the catheter should be discarded before blood collection. The Intravenous Nursing Standards of Practice recommends that manufacturers' instructions should always be followed on the appropriate discard volume. These guidelines also state that blood should not be acquired from various types of indwelling cannula, venous administration sets, and indwelling cardiovascular or umbilical lines. Even when the initial volume drawn is discarded before blood collection according to these guidelines, specimens drawn from a heparinized line are still easily contaminated with heparin. Consequently, blood for coagulation tests should be best drawn directly from a peripheral vein, also avoiding the arm in which heparin, hirudin, or argatroban is being infused for therapy. Before coagulation testing, heparin may be removed or neutralized with polybrene in the coagulation laboratory; however, residual heparin may continue to cause testing interferences.

By enhancing antithrombin activity, heparin inhibits activated factors II (thrombin), X, IX, XI, XII, and kallikrein. In contrast, lepirudin, danaparoid, and argatroban inhibit only activated factor II. These anticoagulants (heparin, lepirudin, and argatroban) prolong the aPTT and interfere with coagulation tests such as factor assays and lupus anticoagulant assays. Factor assays may yield falsely low levels, whereas lupus anticoagulant may be falsely positive.

Traumatic phlebotomy

Traumatic phlebotomy can result in artificially shortened coagulation results such as PT and aPTT. This is due to excessive activation of coagulation factors and platelets by release of tissue thromboplastin from endothelial cells. A proper free-flowing puncture technique will avoid this release of tissue thromboplastin and avoid this artifact.

Fibrinolysis products and rheumatoid factor

Fibrinolysis is mediated by plasmin, which degrades fibrin clots into D-dimers and fibrin degradation products. Plasmin also degrades intact fibrinogen, generating fibrinogen degradation products. Fibrin degradation products and fibrinogen degradation products are collectively known as fibrin/fibrinogen degradation products (FDP) or fibrin/fibrinogen split products (FSP). Assays for D-dimer and FDP are semiquantitative or quantitative immunoassays.

Latex agglutination: Patient plasma is mixed with latex particles that are coated with monoclonal anti-FDP antibodies. If FDP is present in the patient plasma, the latex particles agglutinate as FDP binds to the antibodies on the latex particles. These agglutinated clumps are detected visually. Various dilutions of patient plasma can be tested to provide a semiquantitative result known as FDP titer. Latex agglutination assays are also available for D-dimers. Various automated and quantitative versions of this assay are commercially available for D-dimers in which the agglutination is detected turbidimetrically by a coagulation analyzer rather than visually by a technologist.

Enzyme-linked immunosorbent (ELISA) assays: Quantitative ELISA assays are also available for FDP and D-dimers. The traditional ELISA method is accurate but is not useful due to long analytical time. An automated, rapid ELISA assay for D-dimers is also available (VIDAS, bioMerieux Inc).

One of the most important limitations of D-dimer and FDP assays is interference by high rheumatoid factor (RF) levels. This may cause false-positive results with almost all available assays. The most useful clue to detect this interference is evaluation of the DIC panel. If all values in this panel (PT, aPTT, TT, and fibrinogen) are normal except for FDP or D-dimer, the presence of RF is most likely the cause.

Platelet aggregation testing with lipemic, hemolyzed, or thrombocytopenic samples

Platelet aggregation measures of the ability of platelets to adhere to one another and form the hemostatic plug, which is the key component of primary hemostasis. It can be performed using either platelet-rich plasma or whole blood. Substances such as collagen, ristocetin, arachidonic acid, ADP (adenosine $5'$-diphosphate), epinephrine, and thrombin can stimulate platelets and hence induce aggregation. Response to these aggregating agents (known as agonists) provides a diagnostic pattern for different disorders of platelet function. Measurement of aggregation response is typically based on changes in the optical density of the sample.

Platelet aggregation is affected by a number of confounding variables. Lipemic and hemolyzed samples complicate aggregation measurements as they obscure spectral changes due to platelet aggregation. Thrombocytopenia also makes platelet aggregation evaluations difficult to interpret as a low platelet count by itself may yield an abnormal aggregation pattern.

Challenges in anticoagulants and lupus anticoagulant tests

The International Society on Thrombosis and Haemostasis Scientific Subcommittee on Lupus Anticoagulant has recommended two sensitive screening tests for LAs that assess different components of the coagulation pathway [12,13]. These tests are clotting time—based assays, such as the dilute Russell viper venom time, aPTT-based assays, kaolin clotting time, or dilute prothrombin time (tissue thromboplastin inhibition test). Lupus anticoagulants prolong various phospholipid-dependent clotting times in the laboratory as they bind to phospholipid and thereby interfere with the ability of phospholipid to serve its essential cofactor function in the coagulation cascade. Lupus anticoagulant screening assays usually have a low concentration of phospholipid to enhance sensitivity. Any abnormal (prolonged) screening result typically requires a 1:1 mixing study in which the patient plasma is mixed with one equal volume of normal plasma to demonstrate that the clotting time remains prolonged on mixing. Confirmatory assays are performed if the screening assay remains abnormal after the 1:1 mixing. Confirmatory assays typically demonstrate that on addition of excess phospholipid, the clotting time shortens toward normal. The platelet neutralization procedure is a confirmatory assay in which the source of the excess phospholipid is freeze-thawed platelets. The hexagonal phospholipid neutralization procedure is also based on the same principle, i.e., the clotting time becomes corrected after addition of phospholipid in hexagonal phase. Note that the aPTT may or may not be prolonged, depending on the amount of phospholipid in the reagent.

In many lupus anticoagulant assays, heparin (including subcutaneous low-dose heparin) may cause false-positive lupus anticoagulant results. By enhancing antithrombin activity, heparin inhibits activated factors II (thrombin), X, IX, XI, XII, and kallikrein. Subsequently, clotting times such as PT and aPTT are prolonged and interfere with lupus anticoagulant assays. Lepirudin, danaparoid, and argatroban inhibit activated factor II and also can prolong clotting times. Before coagulation testing, heparin may be removed or neutralized with polybrene in the coagulation laboratory; however, residual heparin may continue to cause testing interferences. Results for lupus anticoagulant assays can be interpreted correctly in patients on Coumadin. Table 18.2 summarizes important laboratory errors in coagulation laboratory due to various entities.

Table 18.2 Sources of laboratory errors in coagulation.

Falsely prolonged clotting times
- Underfilling of citrate tube
- Polycythemia
- Sample from indwelling catheters (dilution or contamination with anticoagulant)

Falsely shortened clotting times
- Overfilling of citrate tube
- Traumatic phlebotomy
- Turbid plasma (e.g., hyperlipidemia, hyperbilirubinemia) in optical instrument

Falsely shortened activated partial thromboplastin time in patients on heparin
- Delay in separation of plasma from platelets
- Elevated factor VIII (acute phase reactant)

Falsely prolonged thrombin time
- Dysfibrinogenemia,
- Elevated levels of fibrinogen degradation product (FDP) and paraproteins
- Amyloidosis
- Heparin-like anticoagulants (in malignancy)

Falsely high FDP and D-dimer
- Rheumatoid factor

Falsely abnormal platelet function
- Lipidemia
- Hemolysis
- Thrombocytopenia

Falsely low factor levels
- Heparin
- Lepirudin
- Danaparoid
- Argatroban

False-positive results of lupus anticoagulant tests
- Heparin
- Lepirudin
- Danaparoid
- Argatroban

Key points

- Hemoglobin measurement is based on absorption of light at 540 nm. If the sample is turbid, this will produce higher hemoglobin levels. Examples of such state include hyperlipidemia, patients on parenteral nutrition, hypergammaglobulinemia, and cryoglobulinemia. Turbidity from very high WBC count can also falsely elevate hemoglobin levels.
- Smokers have high carboxyhemoglobin that may falsely elevate the measured hemoglobin level.
- Large platelets may be counted by some instruments as RBCs. Also, red cell fragments greater than 40 fL will be counted as whole red cells. In both situations, the RBC count will be falsely high.
- Cold agglutinins will cause red cell agglutination in vitro and result in low RBC counts. If cold agglutinins are suspected, the sample should be warmed to obtain an accurate RBC count.
- If there is red cell agglutination, then red cell clumps will be counted as single red cells, but the volume of the estimated cell will be much higher. This may result in falsely high MCV values. If large platelets are counted as red cells, then these platelets typically have less volume than a normal red cell. This may also result in falsely low MCV values.
- One of the most frequent situations is high NRBCs. If an accurate WBC count is required, then a corrected WBC count should be performed.
- Platelet aggregates and nonlysis of red cells are other causes of spuriously high WBC counts. Hemoglobinopathies and target cells are important causes of nonlysis of red cells.
- In myelodysplastic syndrome, if the myeloid series is affected, then hypolobated and hypogranular neutrophils can be present. Automated analyzers may no longer count these dysplastic neutrophils as such; instead, these cells may be counted as lymphocytes.
- A key difference between lymphocytes and monocytes is the difference in size, monocytes being significantly larger. Reactive (activated) lymphocytes typically have more abundant cytoplasm when compared with a nonreactive lymphocyte, and their size approaches that of a monocyte. These lymphocytes thus may be counted as monocytes.
- Partial clotting of specimen or platelet activation during venipuncture may cause platelet aggregation. Both mechanisms may lead to low platelet counts. Checking the specimen for any clot, analyzing the histograms, and reviewing the smear are all important steps to avoid misleading low platelet counts. There are various other mechanisms to explain falsely low platelet counts, otherwise referred to as pseudothrombocytopenia (anticoagulant-induced pseudothrombocytopenia), platelet satellitism, giant platelets, and cold agglutinin-induced platelet agglutination.
- Falsely elevated platelet counts are much less common than falsely low counts. Fragmented red cells or white cell fragments may all be counted as platelets giving rise to high platelet counts.
- Cryoglobulins are typically IgM immunoglobulins that precipitate at temperatures below 37°C, producing aggregates of high molecular weight. The first clue to a diagnosis of cryoglobulinemia is laboratory artifacts detected in the automated blood cell counts. The precipitated cryoglobulins of various sizes may falsely be identified as leukocytes or platelets causing pseudoleukocytosis and pseudothrombocytosis.
- Conditions associated with immunologically mediated hemolytic anemias may present with many microspherocytes in peripheral blood. Consequently, the fragility test can be positive in immunologically mediated hemolytic anemias besides hereditary spherocytosis, but the former would have a positive direct Coombs test and the latter would not.

- Citrate tubes for coagulation tests are designed for a 9:1 ratio of blood to citrate buffer. Underfilling or overfilling of the citrate tube results in imbalances in this blood to buffer ratio and produces artificially prolonged or shortened clotting times, respectively.
- The aPTT (also abbreviated as APTT) is a test conventionally used to monitor heparin therapy. It is important to properly separate plasma from platelets as soon as possible. Platelet factor 4 can neutralize heparin, thus spuriously reducing aPTT values. Factor VIII levels are reflected in aPTT measurements. Factor VIII is an acute-phase reactant. Again if the aPTT is being used to monitor heparin therapy in a patient who has an underlying cause for acute-phase reactants to be elevated, aPTT values may be falsely lower than expected.
- The TT measures the time to convert fibrinogen to fibrin. Dysfibrinogenemia, elevated levels of fibrin degradation product, and paraproteins can interfere with fibrin polymerization thus falsely prolonging the TT.
- Traumatic phlebotomy can result in artificially shortened coagulation results such as PT and aPTT. This is due to excessive activation of coagulation factors and platelets by release of tissue thromboplastin from endothelial cells.
- One of the most important limitations of D-dimer and FDP assays is interference by high RF levels. This may cause false-positive results with almost all available assays. The most useful clue to detect this interference is evaluation of the DIC panel. If all values in this panel (PT, aPTT, TT, and fibrinogen) are normal except for FDP or D-dimer, the presence of RF is most likely the cause.
- Platelet aggregation is affected by a number of confounding variables. Lipemic and hemolyzed samples complicate aggregation measurements as they obscure spectral changes due to platelet aggregation. Thrombocytopenia also makes platelet aggregation evaluations difficult to interpret as a low platelet count by itself may yield an abnormal aggregation pattern.

References

[1] Carraro P, Plebani M. Errors in a stat laboratory: types and frequencies 10 years later. Clin Chem 2007;53: 1338–42.

[2] Chen BH, Fong JF, Chiang CH. Effect of different anticoagulant, underfilling of blood sample and storage stability on selected hemogram. Kaohsiung J Med Sci 1999;15:87–93.

[3] Lima-Oliveira G, Lippi G, Salvagno GL, Montagnana M, et al. Brand of dipotassium EDTA vacuum tube as a new source of pre-analytical variability in routine hematology testing. Br J Biomed Sci 2013;70:6–9.

[4] Nosanchuk JS, Roark MF, Wanser C. Anemia masked by triglyceridemia. Am J Clin Pathol 1977;62:838–9.

[5] Mehrotra S, Edmonds M, Lim RK. False elevation of carboxyhemoglobin: case report. Pediatr Emerg Care 2011;27:138–40.

[6] Livshits Z, Lugassy DM, Shawn LK, Hoffman RS. Falsely low carboxyhemoglobin level after hydrox-ocobalamin therapy. N Engl J Med 2012;367:1270–1.

[7] van Wolfswinkel ME, Langenberg MCC, Wammes LJ, Sauerwein RW, et al. Changes in total and differential leukocyte counts during the clinically silent liver phase in a controlled human malaria infection in malaria-naïve Dutch volunteers. Malar J 2017;16(1):457.

[8] Kakkar N. Spurious rise in the automated platelet count because of bacteria. J Clin Pathol 2004;57:1096–7.

[9] Schrezenmeier H, Muller H, Gunsilius E, Heimpel H, Seifried E. Anticoagulant-induced pseudo-thrombocytopenia and pseudoleucocytosis. Thromb Haemostasis 1995;73:506–13.

[10] Stiegler H, Fischer Y, Steiner S, Strauer BE, Reinauer H. Sudden onset of EDTA-dependent pseudo-thrombocytopenia after therapy with the glycoprotein IIb/IIIa antagonist c7E3 Fab. Ann Hematol 2000;79:161−4.

[11] Sultan S, Irfan SM. In Vitro Leukoagglutination: a rare hematological cause of spurious leukopenia. Acta Med Iran 2017;55:408−10.

[12] Brandt JT, Triplett DA, Alving B, et al. Criteria for the diagnosis of lupus anticoagulants: an update. Thromb Haemostasis 1995;74(4):1185−90.

[13] Chighizola CB, Raschi E, Banzato A, Borghi MO, et al. The challenges of lupus anticoagulants. Expert Rev Hematol 2016;9:389−400.

Index